To Meredith

In appreciation of
her contribution
to this effort

Biodiversity

Springer

New York
Berlin
Heidelberg
Barcelona
Budapest
Hong Kong
London
Milan
Paris
Santa Clara
Singapore
Tokyo

Takuya Abe   Simon A. Levin   Masahiko Higashi
Editors

Biodiversity
An Ecological Perspective

With 75 figures

Springer

Takuya Abe
Center for Ecological Research
Kyoto University
4-1-23 Shimosakamoto
Otsu 520-01
Japan

Simon A. Levin
Department of Ecology and
   Evolutionary Biology
Princeton University
Princeton, NJ 08544-1003
USA

Masahiko Higashi
Center for Ecological Research
Kyoto University
Kitashirakawa, Sakyo-ku
Kyoto 606
Japan

Cover: Biomass size spectra for oceans and lakes. See page 205. From Verh. int Ver. Limnol. 23:234–240, used by permission of E. Schweizerbart'sche Verlagsbuchhandlung.

Library of Congress Cataloging-in-Publication Data
Biodiversity: an ecological perspective / [edited by] Takuya Abe,
   Simon A. Levin, Masahiko Higashi.
      p. cm.
   Papers from a conference held in Kyoto in December 1993.
   Includes bibliographical references and index.
   ISBN 0-387-94702-7 (hardcover: alk. paper)
      1. Biological diversity—Congresses. 2. Biotic communities—
   Congresses. I. Abe, Takuya, 1945–   II. Levin, Simon A.
   III. Higashi, M. (Masahiko)
   QH541.15.B56B56   1996
   574.5—dc20   96-13589

Printed on acid-free paper.

Production coordinated by Carlson Co. and managed by Francine McNeill; manufacturing supervised by Jeffrey Taub.
Typeset by Carlson Co., Yellow Springs, OH, from the authors' electronic files.
Printed and bound by Braun-Brumfield, Inc., Ann Arbor, MI.
Printed in the United States of America.

9 8 7 6 5 4 3 2 1

ISBN 0-387-94702-7 Springer-Verlag New York Berlin Heidelberg   SPIN 10523660

Preface

Biodiversity is and will remain an issue of pressing global concern. The more we know about how changes in biodiversity might affect ecosystem processes, community structure, and population dynamics, the more clearly we understand why and what kind of biodiversity we should conserve. Reciprocally, the more we know about how ecological processes and mechanisms affect the creation and maintenance of biodiversity, the more clearly we understand how we can conserve and even recover biodiversity. Thus, basic understanding of the ecological causes and consequences of changes in biodiversity provides a critical perspective on the applied challenges.

The main purpose of the volume is to present an ecological perspective on biodiversity by examining its two facets, the ecological causes and consequences of biodiversity changes. The book extends its scope to cover the linkage between science and management issues, again from an ecological perspective.

This volume commemorates the awarding of the 1993 International Prize for Biology to Professor Edward O. Wilson. He provides the introductory chapter for the volume and the focus for the other presentations. We hope this volume leads to a new stage of biodiversity research, which Ed Wilson pioneered, and will serve as a suitable tribute to him.

T. ABE, S.A. LEVIN, AND M. HIGASHI

Acknowledgments

This book grew from an international symposium entitled "Ecological Perspective of Biodiversity," held in December 1993 at Kyoto. It was sponsored by the Japan Society for the Promotion of Science and the Japan Ministry of Education, Science, and Culture, whose generous support made possible the symposium on which this volume is based. We would like to thank the organizing committee (chaired by Professor Hiroya Kawanabe) for providing a wonderful opportunity to discuss this issue of compelling current importance. All chapters in this volume have been peer reviewed and reflect a sampling of what was presented at the meeting. We would like to extend our gratitude to the anonymous reviewers for their help and cooperation. Last, but not least, is the debt we owe to our wives and families for their understanding and support.

Contents

Contributors

TAKUYA ABE   Center for Ecological Research, Kyoto University, Otsu 520-01, Japan

MARCEL DICKE   Department of Entomology, Wageningen Agricultural University, P.O. Box 8031, 6700EH, Wageningen, The Netherlands

NILES ELDREDGE   American Museum of Natural History, Central Park West at 79th St., New York, NY 10024-5192, USA

MASAHIKO HIGASHI   Center for Ecological Research, Kyoto University, Kyoto 606, Japan

KEIICHI KAWABATA   Faculty of Education, Kanazawa University, Kanazawa 920-11, Japan

JIRO KIKKAWA   Cooperative Research Centre for Tropical Rainforest Ecology and Management (CRE-TREM) and Department of Zoology, The University of Queensland, Brisbane, Queensland 4072, Australia

TAKASHI KOHYAMA   Graduate School of Environmental Earth Science, Hokkaido University, Sapporo 060, Japan

JOHN H. LAWTON   NERC Centre for Population Biology, Imperial College, Silwood Park, Ascot, Berkshire SL5 7PY, UK

SIMON A. LEVIN   Department of Ecology and Evolutionary Biology, Princeton University, Princeton, NJ 08544-1003, USA

MASAMI NAKANISHI   Center for Ecological Research, Kyoto University, Otsu 520-01, Japan

TAKAYUKI OHGUSHI   Institute of Low Temperature Science, Hokkaido University, Sapporo 060, Japan

DANIEL R. PAPAJ   Department of Ecology and Evolutionary Biology and Center for
                          Insect Science, University of Arizona, Tucson, AZ 85721, USA

COLIN S. REYNOLDS   Freshwater Biological Association, NERC Institute of Freshwater
                          Ecology, Ambleside, Cumbrea LA 220LP, UK

JUNJI TAKABAYASHI   Faculty of Agriculture, Kyoto University, Kyoto 606-01, Japan

YASUHIRO TAKEMON   Department of Life Sciences, University of Osaka Prefecture,
                          Sakai 593, Japan

MUTSUNORI TOKESHI   School of Biological Sciences, Queen Mary and Westfield Col-
                          lege, University of London, Mile End Rd., London E1 4NS, UK

IAN M. TURNER   Department of Botany, National University of Singapore, Singa-
                          pore 119260

EDWARD O. WILSON   The Museum of Comparative Zoology, Harvard University,
                          Cambridge, MA 02138, USA

NORIO YAMAMURA   Center for Ecological Research, Kyoto University, Sakyo-ku,
                          Kyoto 606-01, Japan

YURI N. ZHURAVLEV   Institute of Biology and Soil Science, Far East Branch of Russian
                          Academy of Sciences, Vladivostok 690022, Russia

Introduction

EDWARD O. WILSON

From the time of Aristotle, biodiversity has been regarded as a central quality of life. Yet only recently has it been made the subject of scientific specialization in its own right, to be measured and evaluated with abstract theory and experimentation. In 1959, G. E. Hutchinson contributed importantly to this end with his now-famous article, "Homage to Santa Rosalia, or Why are there so many kinds of animals?" In the 1960s, the development of the theory of island biogeography offered partial answers to the question posed in the title of Hutchinson's article through the use of models to characterize equilibria of immigration, evolution, and extinction (MacArthur and Wilson 1967). With this work, it became clear that the fundamental properties of biological diversity can be clarified only by judicious and exacting studies that combine systematics and ecology.

After a relatively leisurely beginning, the subject was propelled to widespread scientific and public attention by events occurring during the 1980s. One of the most important was the estimate made by Norman Myers of the rate of tropical deforestation. Adding data country by country, he calculated the global loss of cover during the 1970s to be about 1% per year (Myers 1980). This bit of bad news was especially alarming to conservationists, because the rain forests were (and remain) the premier reservoirs of diversity. They teem with more species of plants and animals than all the other biomes of the world combined, although occupying only a very small part of the world's land surface—7% at the time of Myers' report and 6% now (Wilson 1992). Their area was thus about the same as that of the contiguous 48 states of the United States, and the average amount of cover removed each year in the 1970s was roughly equal to half the area of Florida. The reduction in area translated, in terms of the general area–species relations worked out in oceanic islands and other circumscribed ecosystems, to approximately 0.25% of species extinguished immediately or committed to relatively early extinction each year. The cutting and burning appeared to be accelerating, primarily because of the incursions by land-hungry rural populations and the increasing global demand for timber products.

About this time a "new environmentalism" was emerging in major environmentalist organizations, which included especially the World Wildlife Fund, the International Union for Conservation of Nature and Natural Resources, and the United Nations Environmental Programme. Their staffs set out to shift the institu-

1

tional focus to place more emphasis on entire ecosystems and proportionately less on "star" species such as the panda and tiger. They also took a more pragmatic approach by combining conservation projects with economic advice and assistance to local human populations most affected by the salvage of biological diversity. By the mid-1980s, conservationists everywhere had come to accept two key principles of the new environmentalism: first, that reserves cannot be protected indefinitely from impoverished people who see no advantage in them, and second, and conversely, the long-term economic prospects of these same people will be imperiled to the extent that the surrounding natural environment is destroyed.

The novelty of the new environmentalism as a credo is indicated by the recency of its talisman, "biodiversity." The word, shorthand for biological diversity, did not exist until the National Forum on BioDiversity was held in Washington, D.C., September 21–24, 1986, under the auspices of the National Academy of Sciences and the Smithsonian Institution. The Forum proceedings, published as the book *BioDiversity* (Wilson and Peter 1988), comprised contributions from more than 60 scholars and officials whose expertise ranged from ecology to economics and government. It was distributed widely around the world, becoming an early vade mecum and textbook of biodiversity management.

In June 1992, when more than a hundred heads of state met at the Earth Summit in Rio de Janeiro to debate and ratify protocols on the global environment, biodiversity had approached the status of a household word. It also became well established as a favorite subject of museum exhibitions and college seminars. President Bush's refusal to sign the Convention on Biodiversity at the Earth Summit brought the subject into the mainstream of politics; controversies surrounding the Endangered Species Act and the northern spotted owl then confirmed it as a part of Americana.

Science, like art, follows patronage. Biologists, alerted to biodiversity as a subject important in both science and public policy, turned to its study in growing numbers. Subscriptions to *Conservation Biology* soared, and a new journal, *Biodiversity and Conservation,* was created. By the early 1990s articles on extinction and other aspects of biodiversity had become commonplace in *Nature, Science,* and other journals that reach a large scientific audience. Much of the writing moved back and forth across the boundaries of biology, the physical sciences, economics, and forestry. The research on which it was based formed the new discipline of biodiversity studies, defined as the systematic analysis of hereditary variation at all levels of biological organization, from genes within populations to species to ecosystems, together with the development of technologies to conserve and manage the diversity for the benefit of humanity (Ehrlich and Wilson 1991).

Sound reasons existed for the rising interest in the practical side of biodiversity studies. The loss of species is wholly distinct from toxic pollution, ozone depletion, climatic warming, and other changes in the physical environment. Unlike these secular trends, *extinction cannot be reversed,* nor can lost species be replaced by new ones created through evolution in any amount of time that has meaning for the human mind. The average life span of a species and its immediate descendants ranges, according to group (such as Tertiary African mammals or Mesozoic ammonites), from half a million to 10 million years (Raup 1984; reviewed by Wilson 1992). Hence the average extinction rate outside major extinction spasms is 10^{-5} to 10^{-7} species per year, orders of magnitude less than that now imposed by human action. For a species to diverge from its sister species sufficiently for the two taxa to be placed by systematists in different genera or families has usually required hundreds of thousands to millions of years. It will continue to do so in the future—providing, in the first place, that the new forms have access to environments large enough and stable enough for sustained existence.

During the 1970s and 1980s, while paleontologists were placing natural extinction within a reliable time scale, biologists and others expanded our understanding of the multiple practical benefits of biodiversity. It was already widely appreciated that agriculture and medicine draw heavily on domesticated organisms that have been derived ultimately from wild species. What became clear from the new research was the enormous potential remaining in the millions of still-unexamined species. The scientists argued that new pharmaceuticals, crops, fiber sources, petroleum substitutes, and other products await discovery and development. And in case it is not practicable to use entire particular species, genes might be taken from the organisms and inserted into already existing domesticated species to enhance favorable traits. In Thomas Eisner's metaphor, we need not employ the whole book, but can instead pull pages from one loose-leaf folder to add to another (Eisner 1985).

Biodiversity studies also contain problems that are of the first intellectual rank in science. We do not know, to mention one important topic, the number of species on Earth even to the nearest order of magnitude. That number is certainly greater than 1 million and less than a billion, but whether it is closer to 10 million or to 100 million cannot be calculated from existing data. The former figure may seem the more reasonable, at least the more prudent, even if we take into account the millions of undescribed insect species inhabiting tropical forests. But what, then, of the legions of symbionts living on the insects? Many of these minute organisms, including nematodes, yeasts, and bacteria, are associated with only one or a very few host species. And what of the myriad of bacterial species, mostly unstudied, that saturate the living world? In a gram of typical soil or aqueous silt live 10 billion individuals, representing as many as 5,000 species, almost none of which is known to science. In general, we know more about the stars in the sky than we do about the organisms at our feet.

Still, it would be a mistake to regard a mere censusing of Earth's species to be the central goal of biodiversity studies. What is needed far more urgently in the immediate future is the thorough biogeographic study of focal groups, such as birds and flowering plants, that are already well enough studied to allow rapid identification of specimens. They are our best entrée to the general patterns of diversity and abundance. Precisely what fraction of species in these proxy groups live, for example, in tropical African rain forests? Or in the dry grasslands of temperate South America? A heartening recent advance has been made by the International Council of Bird Preservation, which assembled the massive data on land birds to pinpoint localities around the world with the highest levels of diversity and endemicity (Bibby et al. 1992). The authors were able to show that the biotically richest 2% of the land surface, if set aside in reserves, would hold 20% of the known land bird species.

Their assessment of such "hot spots" is an important step forward in global conservation planning, but it is only a beginning. As shown by the recent analysis of British data, different groups such as birds, dragonflies, and flowering plants may show discordant patterns (Prendergast et al. 1993). It is therefore important to promote a spreading of systematics research to increase the list of focal groups so as to cultivate comparative biogeography.

Systematics and biogeographic mapping can be enormously enhanced by computer-based information technology. Data on the distribution of species are stored on Geographic Information System (GIS) software, then collated with maps of topography, climate, soil types, vegetation, and other environmental features. Repeated over months and years, the GIS analyses can identify the causes of success and failure of individual species.

Taxonomists will be able to update systematics monographs more easily with entries of new species and changing digitized maps and drawings in CD-ROMs. Integrated into Geographic Information Systems, such data can provide the growing empirical base for the more fundamental biological studies of diversity. The preferred point of entry in each analysis depends on evidence from field or laboratory studies of factors that determine the abundance of the species under study. If, for example, the correlative analysis of GIS data indicates that species in a moss genus fluctuate in densities in close concert with summer rainfall, then humidity and the timing of reproductive cycles are the appropriate focus for further study. If a genus of epiphytes varies in diversity in a pattern close to that of halictid bees, then the specialist can profitably search for oligolectic species among the bees and study their behavior. If certain gesneriads seem to flourish and diversify around ant nests, he may wish to consider symbiosis. And so on through hundreds or thousands of the species, genera, and higher taxa that compose ecosystems.

Most of what today is called ecology is actually a refined version of natural history. That is all to the good, because an intellectually solid ecology can never be built solely from theoretical models, even if faithfully congruent to the biology of a few well-selected species. Even less likely to succeed is the theoretician's abstract view of the wild environment. Ecology as a science will work only if constructed from the bottom up, by piecing together the idiosyncrasies of thousands of species, each lovingly examined for its own sake. Through the contrivance of population models to fit particular species, then guilds and biocoenoses, the true larger patterns and grander trends will emerge. If general laws of ecology exist, they are most likely to be written in the equations of demography shaped to address the issues of biodiversity.

In the interim, large-scale experiments are being effectively performed on entire communities, or sectors of them, to answer some of the most important questions of community ecology. For example, it has been possible to create new habitats or sterilize old ones to test the theory of diversity equilibrium. Watersheds have been deforested to determine the effects of the perturbation on nutrient cycles. And most recently, as reported by John Lawton and his associates in this volume (Chapter 12), diversity can be varied in climate-controlled microcosms to measure the effect of the numbers and composition of species on productivity. But even these successful endeavors still require identification of the elements of the fauna and flora. They can be most effectively extended by close studies of the physiology and demography of the constituent species, pieced together to depict the dynamism of ecosystems. Ultimately, all ecological research must be from the bottom up.

The contributions of the authors of the 1993 Kyoto symposium illustrate very well the scientific "mainstreaming" of natural history under the influence of the new emphasis on biodiversity. The convergence of the specialists was enhanced by a new sense of common purpose, to assemble concepts that unite the disciplines of systematics and ecology and, in so doing, to create a sound scientific basis for the future management of biodiversity.

Literature Cited

Bibby, C.J., N.J. Collar, M.J. Crosby, M.F. Heath, Ch. Imboden, T.H. Johnson, A.J. Long, A.J. Stattersfield, and S.J. Thirgood. 1992. Putting Biodiversity on the Map: Priority Areas for Global Conservation. International Council for Bird Preservation, Cambridge, England.

Ehrlich, P.R. and E.O. Wilson. 1991. Biodiversity studies: science and policy. Science 253:758–762.

Eisner, T. 1985. Chemical ecology and genetic engineering: the prospects for plant protection and the need for plant habitat conservation. In: Symposium on Tropical Biology and Agriculture. Monsanto Company, St. Louis.

Hutchinson, G.E. 1959. Homage to Santa Rosalia or Why are there so many kinds of animals? American Naturalist 93:145–159.

MacArthur, R.H. and E.O. Wilson. 1967. The Theory of Island Biogeography. Princeton University Press, Princeton.

Myers, N. 1980. Conversion of tropical moist forests. Report to the National Academy of Sciences. National Research Council, National Academy of Sciences, Washington, DC.

Prendergast, J.R., R.M. Quinn, J.H. Lawton, B.C. Eversham, and D.W. Gibbons. 1993. Rare species, the coincidence of diversity hotspots and conservation strategies. Nature (London) 365:335–337.

Raup, H. 1984. Evolutionary radiations and extinctions. In: H.D. Holland and A.F. Trandall, eds. Patterns of Change in Evolution, pp. 5–14. Dahlem Konferenzen, Abakon Verlagsgesellschaft, Berlin.

Wilson, E.O. 1992. The Diversity of Life. Belknap Press of Harvard University Press, Cambridge.

Wilson, E.O. and F.M. Peter, eds. 1988. BioDiversity. National Academy Press, Washington, DC.

I

Ecological Causes of Biodiversity

1

Biogeographic Patterns of Avian Diversity in Australia

Jiro Kikkawa

Introduction

The number of species found per unit area of land (species density) has a geographic pattern including a latitudinal gradient of decreasing density from the tropics to the polar regions (Stevens 1989). The great heterogeneity of habitat in the tropics led Mayr (1969) to conclude that different vegetation zones are the most important isolating barrier for tropical mainland birds. Vegetational refuges during the Pleistocene served as islands in which isolated populations could speciate on the mainland.

Despite the profusion of evolutionary and ecological hypotheses put forward to explain patterns of species density, the biogeographic significance of changes in species density noted in earlier work has not until recently received much attention from ecologists. Cailleux (1953), for example, gave a global distribution of bird species densities ranging from 10 to 30 species (oceanic islands and polar regions) to 680 species (the upper Amazon) per 10,000 km^2, and evoked the great power of dispersal of some groups (e.g., migratory species) and the highly sedentary habits of others to explain the pattern. On the continental scale, Simpson (1964) demonstrated longitudinal as well as latitudinal trends in species density of mammals in North America that were later shown to be associated with average range size and habitat specificity of species (Pagel et al. 1991). In Fjeldså's (1994) work on species density of birds in Africa and South America, areas of active speciation were identified from separate treatments of three species groups of different putative ages and phylogenetic origin as determined by DNA × DNA hybridization.

This present study examines species density of Australian birds on a continental scale and attempts to identify regions of biogeographical importance for biodiversity conservation. Such regions must include not only faunal concentrations but also significant faunal changes that reflect historical as well as environmental factors. In Australia, bird species density shows a concentric zonation over the continent (Pianka and Schall 1981). When the areas of similar faunal compositions were grouped by a numerical analysis (Kikkawa and Pearse 1969), the resultant continental pattern of species richness remained concentric (Fig. 1.1). In these analyses only the presence/absence data were used to construct species density, but the

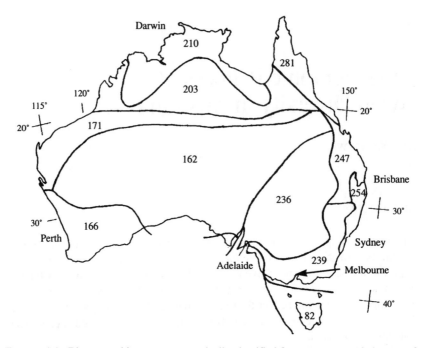

FIGURE 1.1. Biogeographic areas as numerically classified from presence and absence of 464 species of land birds at 121 sites, with the number of species (species richness) recorded for each classifed area. (From Kikkawa and Pearse 1969.)

species density of breeding birds or the turnover of species between zones might reveal further biogeographic patterns useful for the conservation of biodiversity.

Many species inhabiting inland Australia are known to fluctuate in numbers and distributional range in response to local climatic conditions, and some engage in nomadic movements (Serventy and Whittell 1962). However, even the most opportunistic breeders do not attempt to breed in all parts of their geographic range. For them to be able to breed, some specific requirements in their habitat selection would have to be met (Keast 1961). In fact, Schoener (1990) found that most Australian birds were more abundant in their breeding ranges than in non-breeding ranges, suggesting more favorable conditions within the breeding ground than outside. Species density of breeding birds would therefore reflect the conditions of the environment more precisely than the mere occurrence of species that might be unpredictably sporadic or ubiquitous.

A recent analysis of quantitative changes in bird species density also identified numerical clines and escarpments of species density over the Australian continent that are associated with rainfall and the height of vegetation (Gentilli 1992). However, biogeographic interpretation of species density distribution is still limited because changes in species composition, such as species turnover rates, are not considered.

In this chapter, the biogeographic significance of species density is analyzed by constructing a breeding bird density distribution and by calculating species turnover rates between grid squares over the Australian continent. An earlier work is also reexamined to assess the biogeographic significance of species turnover rates along an environmental gradient.

Methods

The distributional data of 463 species of strictly land birds in Australia as given in 100 × 100 km (1°) grid squares over the continent (Blakers et al. 1984) were supplied by the Royal Australasian Ornithologists Union (RAOU; the 1993 database). Excluding the introduced species and the distribution resulting from the known introduction of native species within the country, the sighting data consisted of 74,625 records over 825 grid squares, while the breeding data with the following additional entries consisted of 19,731 records over 702 squares. For the 55 species for which the sedentary population or isolated populations with subspecific status were not represented by a breeding square for lack of records, one nonbreeding square where sightings occurred most frequently was converted to a breeding square. If such an isolated population was known evenly over a number of squares, a center square of distribution was selected for conversion. The results are shown by a three-dimensional surface representation.

Because the data on breeding status were incomplete, the species turnover rate between grid squares was calculated only for the sighting data:

$$\text{Turnover rate} = \frac{\text{number of different species between two adjacent squares}}{\text{total number of species in two adjacent squares}}$$

Excluding the cases in which the adjacent square had no species (7 squares with no records on the continent and 139 squares on the periphery of the land mass), 1470 records of turnover rate between neighboring grid squares (north, south, east, west) were obtained. The results were smoothed by the weighted least-squares method for contouring.

For assessing local species diversity and turnover rates at a smaller scale, census data in different vegetation formations along a mesic–xeric transect in subtropical eastern Australia (Kikkawa 1968) were used to calculate the Shannon–Weaver index for local species diversity (α) and the dissimilarity index $(1 - C'_\lambda)$ of Morisita (1971) for between-habitat diversity (β).

Results and Discussion

Breeding Species Density

Distribution of breeding species density is shown in a three-dimensional map (Fig. 1.2) that plots relative numbers over the continent and Tasmania. Within the continent, forested areas have high species densities, reaching 90 species per 1° grid square in South West and 120 species in South East. The density decreases

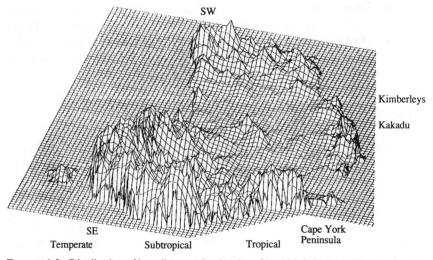

FIGURE 1.2. Distribution of breeding species density of land birds in Australia based on 1°
grid square data shown in three-dimensional surface representation.

northward on either side of the continent. This seemingly opposite trend to the
established pattern of latitudinal gradient is best explained by the theory of refugia
(Gentilli 1949; Keast 1961): that species concentrated in humid areas during the
periods of past aridity and relict-type distributions resulted in mountainous or
hilly terrains with diverse vegetation in South East and South West. Isolated peaks
of species density reaching 60 species in northern Australia and 40 species in the
center (Macdonnell Ranges) are also located in the well-known refuge areas for
savanna woodland species (Keast 1961). Dryness has generally been a marked
feature of Australia since the mid-Miocene, and modern communities are considered to be recent agglomerations of those that have survived in small, scattered
refuge areas throughout the late Cenozoic (Galloway and Kemp 1981).

Rain forest birds exhibit a different pattern of species density distribution on
the east coast of the continent, where rain forest vegetation remains in fragments
today (Webb and Tracey 1981). The species richness of rain forest birds is greatest in tropical Queensland, where 10.7% of 84 strictly rain forest birds are
endemic species (Kikkawa 1991). Subtropical and temperate rain forest fragments
contain fewer species with lesser degrees of endemism (Table 1.1). Thus, so far as
the rain forest birds are concerned, the latitudinal gradient of species diversity is
not reversed in Australia, and they account for relatively high species density in
North East (see Fig. 1.2). Monsoon and gallery forests of Cape York Peninsula
contain rain forest species, including 15 species (14.3%) of endemic taxa closely
related to New Guinea species. This fauna is attenuated in similar habitats in
North (Kakadu region) and North West (Kimberleys), as shown in Table 1.1. In
fact, these species produce a numerical cline from east to west along the increasingly fragmented closed-forest habitats.

The foregoing two trends of rain forest birds, the decline in species density

TABLE 1.1. Species richness of closed forest birds in Australia

	Tropical rain forest, northeastern Queensland	Subtropical rain forest, northeastern New South Wales	Temperate rain forest, Tasmania
Number of species	84	77	29
Species endemic to region	9	4	6
	Monsoon and gallery forest, Cape York Peninsula	Monsoon and gallery forest, Kakadu region	Monsoon and gallery forest, Kimberleys
Number of species	105	75	39
Taxa endemic to region	15 (New Guinea species)	3	0

Modified from Kikkawa (1991).

from north to south (increasing latitude) and from east to west (decreasing rainfall with increasing seasonality), parallel rain forest trees in their decline in both species density per hectare and species richness of the region from the tropical to the temperate and from Cape York Peninsula to Kimberleys (Whiffin and Kikkawa 1992). This fact, associated with a high degree of endemism in North East for both rain forest birds and trees, reflects historical biogeographic events and great habitat diversity within species-rich rain forest and strengthens the argument that North East had the largest refuge area of rain forest in Australia.

Paleoclimatic data support the notion that the widespread rain forest cover of the early Miocene retreated rapidly in the late Miocene, giving way to xerophytic forms in southeastern Australia (Kemp 1981). Consequently, subtropical parts of eastern Australia have been covered with sclerophyll vegetation throughout much of the Quaternary, and rain forest, except for occasional wetter periods, has been restricted to fragmented refugia near the coast. It is therefore not surprising that species density at this latitude reaches a maximum in semiarid formations.

The ecological significance of patterns found in the distribution of species density may be seen in the environmental factors associated with the gradient of species density, as found by Gentilli (1992) who called a steep change in species density (his "species frequency") an eco-biological escarpment. Gentilli (1992) found such escarpments along the species clines to form boundaries of biogeographic subregions over the continent, but the species density alone does not reveal biogeographic borders or the history of the fauna. One needs to consider the rate of replacement of fauna as well as the change in species density itself between the adjacent squares.

Species Turnover Rate

Because breeding data were incomplete in many squares, the species turnover rates were calculated using sighting data. The results are presented in Figure 1.3 as a smoothed contour map. Remote areas and small areas representing peripheral squares of the continent had records of only a few species and boosted the species turnover rate between the species-rich adjacent squares. The process of smoothing eliminated most of these anomalies, but very high turnover rates remained in North East and South West where such squares appeared in a group, being also influenced by true replacement of species.

On the whole, the species turnover rate between 1° grid squares of latitude and longitude (100 × 100 km) did not produce sharp boundaries between the known faunal subregions. However, replacement of the Torresian fauna with New Guinea affinities by the continental fauna occurred in Cape York Peninsula along the species-rich eastern region. A high species turnover rate in this region was also recorded by Cody (1993), who emphasized historical factors as determining the high local and regional diversity as well as between-site diversity in northeastern Australia. In the species-poor west, relatively high turnover rates were found where the coastal refugia species were replaced by arid-zone species along the steep gradients. In eastern Australia, the species turnover was gradual both to the south and to the west, and no clear boundaries were revealed. The Australian Alps and Snowy Mountains of South East did not show any increase in species turnover rate, confirming that vicarious species did not contribute significantly to the similar species density across this barrier (Gentilli 1992).

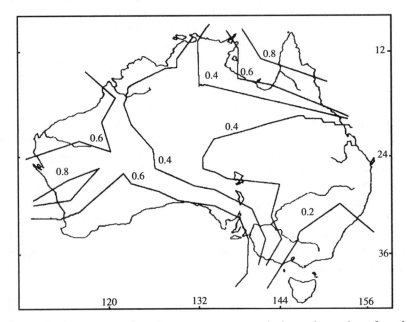

FIGURE 1.3. Contour map of species turnover rates smoothed over the continent from the 1° grid square sighting data of land birds.

Species Diversity Along a Vegetation Gradient

The question of high species density in semiarid formations of the subtropical region was examined by comparing bird census data along a transect from the coastal subtropical rain forest to savanna in northern New South Wales (Kikkawa 1968).

Figure 1.4 shows the mean number of species breeding in 8-ha plots in four different types of vegetation in northern New South Wales, together with mean α-diversity within each habitat and β-diversity between habitats (for calculations, see Kikkawa 1986). Both species density and species diversity were highest in tall woodland where as many as 55 species were breeding in 8 ha. In savanna, fewer than half this number bred in 8 ha, but the species composition was very different, overlapping only 17%–26% with tall woodland plots. This high species turnover rate is a feature of low species density areas, where two adjacent plots in the same environment may have different selections of the regional fauna.

In practice, this means that the number of new species being encountered increases slowly and that species association is unpredictable in the arid region (Kikkawa 1974). If a birdwatcher returns from a field trip and says he or she identified a yellow-throated scrubwren *(Sericornis lathami)*, then we could tell what habitat the birdwatcher visited and, with a very high probability, what other species he or she should have seen there. If the species identified was a black honeyeater *(Certhionyx niger)*, we could still discern the type of habitat in which it was seen but not other species that might have been seen at the same time; if the

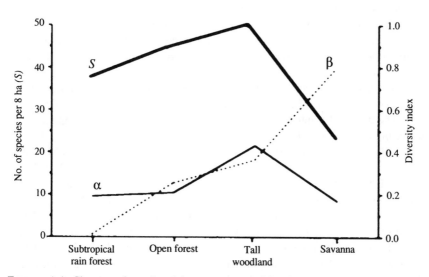

FIGURE 1.4. Changes of species richness represented by the mean number of species breeding in 8 ha of each habitat *(S)*, local species diversity represented by Shannon-Weaver index (α), and between-habitat diversity (β). (Data from Kikkawa 1968, 1986.)

species were a grey shrike-thrush *(Colluricincla harmonica)* we could determine neither the habitat nor other species. Subtropical rain forest contains mostly the first type of species (the yellow-throated scrubwren), and savanna contains the second type (the black honeyeater). Species are predictably recurrent in wet formations and most will be encountered fairly quickly, but in dry formations there does not appear to be a predictable association of species that may be encountered regularly. The second type of species will have a wide range of distribution if considered over many years, as they appear sporadically to different parts of the vast arid-to-semiarid areas of the continent. The third type of species (the grey shrike-thrush) is limited in number among the land birds.

The high species density and diversity in semiarid formations is explained in terms of past climate (Keast 1961; Brereton and Kikkawa 1963; Kikkawa and Pearse 1969). It appears that the isolation of refugia occurred several times in the past and that not all refugia occurred contemporaneously. The value of refugia obviously would have differed according to the requirements of species and whether the climatic zones shifted north–south or in a concentric manner (Kikkawa and Pearse 1969), leaving pockets of different relict species in different parts of the continent. Interestingly, a paucity of species occurs precisely in this region in subtropical Australia where moist-adapted forms would have been restricted to small refuge areas of the dividing range. In fact, if the Pleistocene climate caused concentric zones of aridity to move toward the periphery, many moist-adapted species would have become extinct as semiarid woodland formations reached the outermost zone. Semiarid vegetation then would have been fragmented as the wetter regions of the continental fringe are today. Later, when less arid conditions returned, these zones would again move toward the continental center, taking with them those forms that had been isolated in the periphery, so that where new species had been formed they might coexist as closely related forms partitioning the more varied semiarid habitats or fluctuating in numbers and distribution over the vast areas of the continent. For the arid center, no such mechanism of isolation and rejoining existed to the extent that new species, if formed allopatrically, could share the habitat when brought together. New Guinea, on the other hand, would produce moist-adapted forms because its rain forest areas would be fragmented into large isolates allowing rapid speciation within this habitat during the arid periods.

For the rain forest birds of Australia, recent work based on allozymes (Christidis et al. 1988) and phylogenetic analysis of the mitochondrial DNA (Joseph and Moritz 1993) showed that genetic differentiation of at least one group of birds *(Sericornis)* between isolated populations along the east coast of the continent greatly predates the Pleistocene. Within North East, however, the two endemic species *(S. keri* and *S. beccarii)* show low sequence divergences from the respective relatives, indicating recent speciation. On the other hand, Kershaw's (1994) evidence from pollen analysis reinforces the notion that even in North East, where the rain forest refugia were considered to be most extensive, allopatric speciation of rain forest flora did not occur during the Pleistocene. While this could be a result of the small size of the fragmented area compared to New Guinea or Amazonia, Kershaw (1994) considered that expansion of savanna woodland had progressively increased in this region in the late Pleistocene and isolated the Tertiary relicts of primitive angiosperms and other rain forest species.

Implications for Biodiversity Conservation

Following the International Union for the Conservation of Nature (IUCN) Convention on Biological Diversity (World Conservation Union 1992), many nations have been developing guidelines for the selection, establishment, and management of protected areas to conserve biodiversity. In Australia, protected areas, including national parks are mostly under state or territory jurisdictions and occupy about 5.2% of the land area (Hooy and Shaughnessy 1992). Because environmental and biophysical information is adequate for identifying major ecosystems in Australia, it is not difficult to organize protected areas of different categories to a system of ecosystem-based reserves. An ecosystem-based reserve system incorporating national parks, scientific and educational reserves, outstanding natural and cultural areas, and wildlife management areas would protect representative ecological assemblages. Gaps may be identified and reserves may be expanded to fill in the gaps.

However, ecological associations represented in ecosystems will not necessarily define groups important for biodiversity conservation. Current analysis shows species density and species turnover rates may also indicate the regions of importance for biodiversity conservation. The biogeographic significance of species diversity such as the origin and relict status of biota needs to be considered. For example, the "Workshop 90" held in Manaus, Brazil, to develop methodology for the conservation of the world's greatest biological diversity concluded that the Pleistocene refugia should be identified as high-priority conservation areas (Jorge Pádua and Alves Coutinho 1990). Data available indicate that such refugia are not identical for different groups of animals and plants, and thus a large number of conservation areas would have to be erected to cover biodiversity.

So far as the land birds of Australia are concerned, high species density areas of tropical and temperate regions, particularly in North, North East, South East, and South West are obviously important for biodiversity conservation, but in addition the areas of steep gradients and turnover of species in Cape York Peninsula and Western Australia are also important for providing protection of species from adverse consequences of climatic fluctuations or environmental catastrophes.

The unstable conditions of semiarid regions mean that the relatively large area of habitat or abundance of organisms is no guarantee that such ecosystems with all target species will survive in reserves when drought conditions persist or wildfire leaves only few patches of vegetation. Blondel and Vigne (1993) considered that large protected areas are necessary for the conservation of bird species diversity in the Mediterranean region where a mosaic of habitats is maintained by a regime of fire disturbance as in Australia. A high species diversity in the Mediterranean region, however, is not the result of active speciation or endemism developed in the Pleistocene; rather, the region is acting as a biogeographic sink with a variety of habitats attracting species from other regions.

There are still other patterns not revealed in the current analysis. One of these is shown by rare and endangered species, which often occur in isolation, creating hot spots of biodiversity. For example, the eight species of grasswren *(Amytornis)* show a relict pattern of distribution over the continent that does not match either an ecological or a biogeographic pattern. High species density or species turnover rates did not pick up particular areas of importance for grasswrens. Their conservation would require inclusion of these hot spots in the reserve system.

Summary

The biogeographic significance of species density was analyzed by constructing a breeding bird density distribution and by calculating species turnover rates between 1° grid squares over the Australian continent. The relationship between species diversity and turnover rates of birds was also examined along a mesic–xeric gradient in subtropical Australia.

It was found that species density of breeding birds is generally high in forested areas of continental fringes, particularly in South West and South East. Concentration of species in moist regions was consistent with the refugia theory, but species density in the subtropical region was greatest in semiarid formations, indicating the significance of severe past climate in Australia. A latitudinal gradient of rain forest species with effects of advancing aridity throughout the late Tertiary and the Quaternary was evident, and the species turnover rate was greatest where the wet-adapted New Guinea fauna was replaced by the Australian continental fauna. Association of species in arid-to-semiarid Australia was weak while it was strong among the moist-adapted species.

For conservation of biodiversity, the history of biota is considered important in selecting protected areas. Refugia, in particular, require close attention. Where environmental conditions are known to fluctuate and produce a mosaic of habitat, large areas need to be protected to maintain biodiversity. Hot spots of rare species distribution should also be included in a reserve system.

Acknowledgments. The bird atlas data used in the analysis were supplied by the Royal Australasian Ornithologists Union under copyright. I thank Peter Douglas for computing the species turnover rates and for data management; Bill Lyon and Mike Dale for their help with map representations; and Craig Moritz, John Wiens, and Steve Turton for their useful comments on the manuscript.

Literature Cited

Blakers, M., S.J.J.F. Davies, and P.N. Reilly. 1984. The Atlas of Australian Birds. Royal Australasian Ornithologists Union. Melbourne University Press, Carlton, Victoria.

Blondel, J. and J.-D. Vigne. 1993. Space, time, and man as determinants of diversity of birds and mammals in the Mediterranean region. In: R.E. Ricklefs and D. Schluter, eds.

Species Diversity in Ecological Communities. Historical and Geographical Perspectives, pp. 135–146. University of Chicago Press, Chicago.

Brereton, J. Le Gay and J. Kikkawa. 1963. Diversity of avian species. Australian Journal of Science 26:13–14.

Cailleux, A. 1953. Biogéographie Mondiale. Presses Universitaires de France, Paris.

Christidis, L., R. Schodde, and P.R. Baverstock. 1988. Genetic and morphological differentiation and phylogeny in the Australo-Papuan scrubwrens (Sericornis, Acanthizidae). Auk 105:616–629.

Cody, M.L. 1993. Bird diversity components within and between habitats in Australia. In: R.E. Ricklefs and D. Schluter, eds. Species Diversity in Ecological Communities. Historical and Geographical Perspectives, pp. 147–158. University of Chicago Press, Chicago.

Fjeldså, J. 1994. Geographical patterns for relict and young species of birds in Africa and South America and implications for conservation priorities. Biodiversity and Conservation 3:207–226.

Galloway, R.W. and E.M. Kemp. 1981. Late Cainozoic environment in Australia. In: A. Keast, ed. Ecological Biogeography of Australia, pp. 51–80. W. Junk, The Hague.

Gentilli, J. 1949. Foundations of Australian bird geography. Emu 49:85–129.

Gentilli, J. 1992. Numerical clines and escarpments in the geographical occurrence of avian species; and a search for relevant environmental factors. Emu 92:129–140.

Hooy, T. and G. Shaughnessy, eds. 1992. Terrestrial and Marine Protected Areas in Australia (1991). Australian National Parks and Wildlife Service, Canberra.

Jorge Pádua, M.T. and M.M. Alves Coutinho. 1990. Systematic Approaches to Conserving Biodiversity in Amazonia through Conservation Units. Report to International Union for Conservation of Nature and Natural Resources, Switzerland.

Joseph, L. and C. Moritz. 1993. Phylogeny and historical aspects of the ecology of eastern Australian scrubwrens Sericornis spp.: evidence from mitochondrial DNA. Molecular Ecology 2:161–170.

Keast, A. 1961. Bird speciation on the Australian continent. Bulletin of the Museum of Comparative Zoology, Harvard University 123:303–495.

Kemp, E.M. 1981. Tertiary palaeogeography and the evolution of Australian climate. In: A. Keast, ed. Ecological Biogeography of Australia, pp. 31–49. W. Junk, The Hague.

Kershaw, A.P. 1994. Pleistocene vegetation of the humid tropics of northeastern Queensland, Australia. Palaeogeography, Palaeoclimatology, Palaeoecology 109:399–412.

Kikkawa, J. 1968. Ecological association of bird species and habitats in eastern Australia; similarity analysis. Journal of Animal Ecology 37:143–165.

Kikkawa, J. 1974. Comparison of avian communities between wet and semiarid habitats of eastern Australia. Australian Wildlife Research 1:107–116.

Kikkawa, J. 1986. Complexity, diversity and stability. In: J. Kikkawa and D.J. Anderson, eds. Community Ecology. Pattern and Process, pp. 41–62. Blackwell Scientific, Melbourne.

Kikkawa, J. 1991. Avifauna of Australian rainforests. In: G. Werren and P. Kershaw, eds. The Rainforest Legacy, Vol. 2. Flora and Fauna of the Rainforests, pp. 187–196. Australian Heritage Commission, Commonwealth of Australia, Canberra.

Kikkawa, J. and K. Pearse. 1969. Geographical distribution of land birds in Australia—a numerical analysis. Australian Journal of Zoology 17:821–840.

Mayr, E. 1969. Bird speciation in the tropics. In: R.H. Lowe-McConnell, ed. Speciation in Tropical Environments. Biological Journal of the Linnean Society 1:1–17.

Morisita, M. 1971. Composition of Iδ-index. Researches on Population Ecology (Kyoto) 13:1–27.

Pagel, M.D., R.M. May, and A.R. Collie. 1991. Ecological aspects of the geographical distribution and diversity of mammalian species. American Naturalist 137:791–815.

Pianka, E.R. and J.J. Schall. 1981. Species densities of Australian vertebrates. In: A. Keast, ed. Ecological Biogeography of Australia, pp. 1675–1694. W. Junk, The Hague.

Royal Australasian Ornithologists Union. 1993. RAOU Atlas Database, the 1993 version.

Schoener, T.W. 1990. The geographical distribution of rarity: misinterpretation of atlas methods affects some empirical conclusions. Oecologia 82:567–568.

Serventy, D.L. and H.M. Whittell. 1962. Birds of Western Australia, 3rd Ed. Paterson Brokensha, Perth.

Simpson, G.G. 1964. Species density of North American recent mammals. Systematic Zoology 13:57–73.

Stevens, G.C. 1989. The latitudinal gradient in geographical range: how so many species coexist in the tropics. American Naturalist 133:240–256.

Webb, L.J. and J.G. Tracey. 1981. Australian rainforests: patterns and change. In: A. Keast, ed. Ecological Biogeography of Australia, pp. 605–694. W. Junk, The Hague.

Whiffin, T. and J. Kikkawa. 1992. The status of forest biodiversity in Oceania. Journal of Tropical Forest Science 5:155–172.

World Conservation Union. 1992. Global Biodiversity Strategy: Guidelines for Action to Save, Study and Use Earth's Biotic Wealth, Sustainably and Equitably. UNESCO, Paris.

2

The Role of Architecture in Enhancing Plant Species Diversity

TAKASHI KOHYAMA

Introduction

Terrestrial plants requiring light and soil nutrients have evolved their sessile and cumulative growth forms. The resultant architecture of vegetation creates spatial heterogeneity of microenvironments and resources. Thus, terrestrial plants are good examples of "autogenic engineers" (*sensu* John Lawton, Chapter 12, this volume) that modulate resource availability through a changing physical structure. Such niche dimensions as architectural differentiation among species (Cody 1986) and chance establishment in a temporally changing resource structure (Grubb 1977) play predominant roles in plant communities. Spatial heterogeneity is a critical aspect in describing the pattern and dynamics of natural populations and communities (Levin 1976). The importance of temporal fluctuation in conditions for plant coexistence (Chesson and Warner 1981; Shmida and Ellner 1984) can be also understood in the context of the sessile feature of plant systems.

However, the significance of the autogenic formation of structure by plants for their community maintenance has not yet been fully examined. In this chapter, I describe the role of vegetation architecture in the creation and maintenance of species diversity.

Patterns of Plant Species Diversity

Forests are good model systems when we think about the significance for biodiversity of a resource structure created by the organisms themselves. Trees create a tall and persistent three-dimensional architecture as a result of their cumulative growth habit. The extreme diversity of tree species in tropical rain forests provides us with a good lesson. There is a difference of a factor of 10 between tropical and extratropical regions in the maximum densities of tree species coexisting at a range of scales. Figure 2.1a shows an example: the number of tree species in South East Asian tropical forests exceeds by one order those in Japanese cool-temperate deciduous forests throughout the spatial scale. The difference in size among coexisting trees is remarkable in tropical rain forests, where the peak of diversity occurs among tree species of lower stature, less than 20 cm maximum dbh (trunk diameter at breast height), or about 20–30 m maximum height in forests 60–70 m tall (Fig. 2.1b). By reviewing many hypotheses proposed to explain the latitudinal gradient of species diversity, Currie (1991) and Rohde (1992) have revealed that only change in available energy along the gradient is sufficiently correlated to species diversity.

We must be aware that the energy correlation is different for different life-forms. The clear correlation for trees has been well documented by Currie (1991), among others. Lianas can also be in this category (Grubb 1987). However, this is not the case for epiphytes (Gentry 1988), shrubs, or herbs (Grubb 1987). A possible explanation for differences among life-forms is that trees, compared with other life-forms, create a persistent aboveground architecture and resource heterogeneity in a cumulative manner which effectively promotes the packing of tree species. Lianas and epiphytes are also dependent on the canopy architecture of forests. Further, epiphytes require moist microsites on high architectural features for establishment and survival, so that it is not surprising that their peak of diversity occurs at higher altitudes in the tropics where cloud forests develop (Gentry

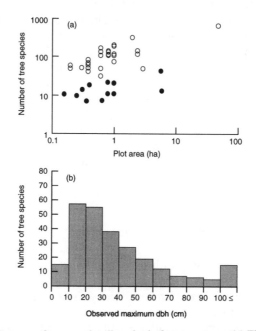

FIGURE 2.1a,b. Patterns of tree species diversity in forest systems. (a) The number of tree species in southeast Asian tropical rain forests (open circles) and Japanese cool-temperate deciduous forests (solid circles) for trees ≥10 cm breast-height-diameter (dbh) in various sized plots. Right-hand data point from 50-ha plot of Pasoh. (From data compiled by Tohru Nakashizuka from various sources, with permission.) (b) Distribution of the maximum trunk diameter for tree species with ≥300 individuals in the Pasoh 50-ha plot. (From data of Manokaran et al. 1992, with permission.)

1988). Herbs and shrubs create less persistent aboveground architecture; the heterogeneity in horizontal resource distribution, and its temporal fluctuations, will be a more important determinant of species diversity than the vertical resource gradient.

The prevailing correlation between available energy and tree species diversity can be explained by the vertical architecture of forests. The wider the vertical gradient of light resources along a forest profile, the larger the number of species that can coexist (Begon et al. 1986). This kind of explanation has, however, a substantial drawback. Even for emergent trees of tropical rain forests, their offspring must either start from a dark forest floor or depend on canopy gaps for regeneration. Analyses combining dynamic models with vegetation architecture are thus necessary to test this resource-gradient hypothesis.

Modeling Architecture and Species Dynamics

To model regeneration habits of tree populations in relation to the spatial distribution of light resources, we must take into account the architecture-dependent effect that suppresses three demographic processes, namely recruitment, growth, and survival. Foliage density for all trees irrespective of species taller than the focal tree will define the intensity of suppression.

Based on observed parameters of demographic processes, we can simulate the dynamics of multiple-species systems. A model of multispecies systems with size structure has shown that stable coexistence among species is possible (Kohyama 1992, 1993a). The background and the framework of the model have been described in these articles and relevant reviews (Chesson and Pantastico-Caldas 1994; Kohyama 1994). The following account summarizes results from the model in relation to the coexistence of species (Kohyama 1993a). The model simulation for a closed system, i.e., without influx of propagules from outside the system, has shown that the plant size structure and the size-structure-dependent regulation of demographic parameters contribute to the stable coexistence between species. In other words, the gradient of light resources from canopy top to stand floor, created by size-structured populations of plants, makes it possible for species to be stratified along the gradient. When there is no horizontal heterogeneity in plant size structure, the condition necessary for coexistence is a trade-off between maximum size and per capita reproduction rate. The larger the difference in size between species, the wider the range in reproduction rate required for coexistence becomes, and the time required for the system to attain the equilibrium state is shortened.

Another aspect of forest architecture is the shifting gap mosaic initiated by tree falls. Kohyama (1993a) incorporated his original size structure model with vertical structure into the patch demography model (Levin and Paine 1974; Levin 1976). When we take into account the horizontal heterogeneity of stand structure reflecting the shifting gap mosaic, various other trade-offs between demographic parameters also contribute to coexistence of species. For example, species can coexist when one has a faster potential size growth and the other is more tolerant to shading. Again, the more differences between species, the more opportunities exist for coexistence. The widened possibilities for coexistence in a gap-dynamic system can be also explained in terms of the heterogeneity of resource availability resulting from the three-dimensional vegetation structure.

Contributions of soil nutrients to community organization are different from those of light resources. Competition among plants for soil nutrients occurs in a two-sided manner because large plants as well as smaller plants distribute their fine roots in the same relatively fertile layer in the topsoil. By contrast, competition for light resources occurs in a one-sided manner whereby taller plants receive greater amounts of light than shorter ones. Reflecting the difference in the mode of competition between the essential resources of nutrients and light, there are differential adaptations between above- and belowground components of architecture among species (Kohyama and Grubb 1994). The differential pattern of abundance between resources is another attribute that facilitates coexistence among species, as demonstrated by Tilman (1988).

Systems with More Than Two Species

In the classic Lotka–Volterra equations of competition, coexistence between two species is explained in terms of asymmetrical species-to-species competition factors. The extension of this analysis for multispecies systems characterized by a species-by-species matrix of competition factors (i.e., community matrix) is complex and not easy. By contrast, a model that incorporates size structure implies that size dimensions, related to resource gradients, alone can provide opportunities for many species to coexist (Kohyama 1992, 1993a).

Applying the same model and procedure in Kohyama (1993a) for analysis of the conditions for coexistence between two species, I examine here the significance of vegetation structure in the maintenance of systems with more than two species. Results of the invasion of the third species and so forth are drawn in Figure 2.2. Let us assume that all species share identical demographic parameters other than potential recruitment rate and potential maximum size. Each species with its unique traits of recruitment and maximum size has a specific domain within which other species can coexist with it (Fig. 2.2a–c). Any species of which the parameters of recruitment rate and maximum size are large enough to be located beyond the coexistence domain in the Figure 2.2a–c coordinates takes the place of the first species. The system with two fixed species generates a new boundary connecting these species on the coordinates; the new domain for a third species to coexist with them is defined as consisting of the portions above the new boundary where respective potential domains of coexistence overlap (Fig. 2.2d). The same is true for a fourth species (Fig. 2.2e), and as a result the domains of coexistence are further subdivided and become smaller with an increasing number of coexisting species. A consequence is that a wider domain of coexistence between two species can be subdivided by more species packed into the same system.

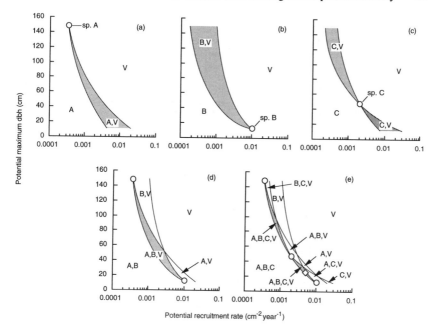

FIGURE 2.2a–e. Equilibrium states of a gap-averaged multispecies system in relation to two demographic parameters, potential maximum breast-height diameter (dbh) and potential per capita recruitment rate, of the visitor species (V), retaining all other parameters constant for all species. In the shaded domains, fixed species (open circles) and visitors coexist stably. (a–c) Two-species systems with fixed species A–C and visitor V. (d) A three-species system with A, B, and visitor V. (e) A four-species system composed of A, B, C, and visitor V; the open square corresponds to species D in Figure 2.3. (Details of the model and other parameters from Table 2, Kohyama 1993a, with permission.)

Figure 2.3 shows the time-course and stationary size distribution of a four-species system, with the parameter sets of the species shown in Figure 2.2e. It takes a long time to attain a stationary state in the present gap-averaged model system (Fig. 2.3a), but remarkably less time is required in gap-dynamic model systems (Kohyama 1993a). The simulated stationary size distributions of the four species show considerable size overlap among the four species (Fig. 2.3b), while stratification among the species plays an essential role in their coexistence. A dozen abundant tree species co-occurring in the warm-temperate rain forest show a clear trade-off relationship between per capita recruitment rate and maximum size (Fig. 2.4), suggesting that they satisfy this necessary condition.

The pattern of coexistence resulting from a trade-off between size and reproduction rate explains why maximum species diversity occurs in the lower strata of species-rich forests (Fig. 2.1b). As Figure 2.2 shows, the width of the domain of coexistence with other species in terms of reproduction rate (horizontal axis) increases with size difference (vertical axis) in an almost logarithmic manner. The

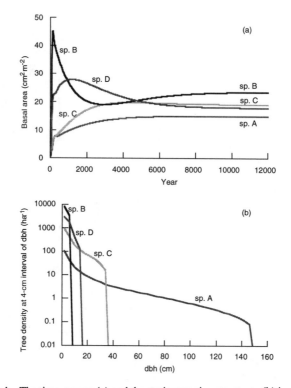

FIGURE 2.3a,b. The time-course (a) and the stationary size structures (b) in a gap-averaged system with four species. The differences between the four coexisting species are shown in Figure 2.2e; other demographic parameters are the same irrespective of species. (From Kohyama 1993b, with permission.)

wider domain reduces the time required to attain stationary states and provides further opportunities for other species to coexist. Therefore, it is likely that the separation of coexisting species along a vertical forest profile occurs at logarithmic intervals of maximum sizes. Trees of lower stature will enjoy a greater opportunity to coexist with others than those with taller stature. Thus, a log-normal distribution of the species' maximum size in Figure 2.1b can be explained as arising from the mechanism of coexistence along the vertical vegetation profile.

Another possible explanation of the size-diversity pattern in tropical rain forests (see Fig. 2.1b) is the contribution of horizontal heterogeneity of stand stratification. Lateral heterogeneity in light resources is particularly large below the uneven emergent/canopy layer (Koike and Syahbuddin 1993). The extended opportunities of coexistence in horizontally heterogeneous gap-dynamic systems (Kohyama 1993a) are likely to be greatest for mostly middle-layer species.

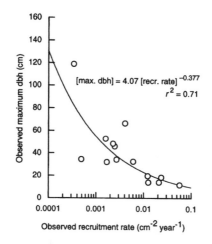

FIGURE 2.4. Relationship between the per capita recruitment rate (the observed recruitment rate divided by the basal area of the species) to dbh ≥2 cm and the observed maximum dbh, for the 14 most abundant species in a warm-temperate rain forest on Yakushima Island, southern Japan. (Based on censuses 1983–1993 in 0.45-ha plots in the Segire basin.)

Does High Mortality Enhance Diversity?

The model results imply that taller systems as well as systems with faster tree growth rates support more species (Kohyama 1993b, 1994). Figure 2.5 shows the sensitivity of a two-species system when potential tree growth rate (i.e., tree growth rate at small size when no shading by taller trees occurs) and mortality (constant irrespective of size and shading in this case) change simultaneously for both species. Increasing size growth rate and decreasing mortality widen the range of coexistence in a similar manner.

The depression of the domain of coexistence with increasing mortality may not be observed if the system is gap dynamic (as real forests) and thus allows the new dimension of coexistence between shade-tolerants and pioneers. This possibility can be tested by the gap-dynamic model of Kohyama (1993a). Here I examine a system composed of two species with identical demographic parameters other than the potential size growth rate and the relative susceptibility to shading. Simultaneous changes in tree mortality of both species and a stand turnover rate that is identical to tree mortality show that the domain of coexistence in terms of a trade-off between the two parameters of size growth again decreases with increasing turnover rate/mortality (Fig. 2.6). Therefore, the model does not support the possibility that a higher turnover rate, related to a higher frequency gap cycle and a larger gap-fraction in the total area, enhances species coexistence because increasing the gap-formation rate reduces the domain of coexistence.

Phillips et al. (1994) examined data sets from tropical rain forests throughout the world and found that tree species diversity was most correlated with the flux rate of tree populations (mortality and recruitment). They argued that this result

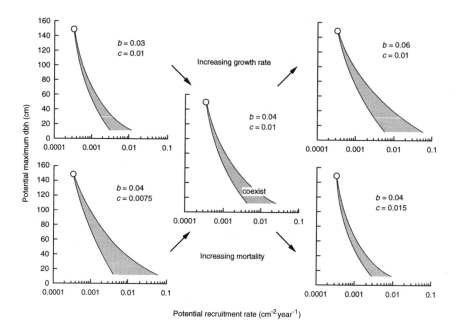

FIGURE 2.5. Dependence of the coexistence domain of gap-averaged two-species systems on potential maximum relative growth rate of dbh (b, cm cm^{-1} year^{-1}; defined for trees with dbh = 1 cm without shading) and mortality (c, year^{-1}; independent of tree size and shading) for both species. Other settings are the same as in Figure 2.2a. In the shaded domains, two species coexist. (Details of the model and other parameters are from Table 2, Kohyama 1993a, with permission.)

supported the importance of small-scale disturbance in maintaining tropical species diversity (cf. Connell 1978; Huston 1979; Denslow 1987). They also found there is a correlation between diversity and flux in terms of basal area, which is related to the primary production rate. Worldwide, there is no clear tendency for tropical rain forests to have a higher tree mortality than extratropical forests; for example, an average of mortality records for tropical rain forests compiled by Swaine et al. (1987) is 0.014 year^{-1} and that for temperate forests in the northern hemisphere by Nakashizuka (1991) is 0.017 year^{-1}. Taking into account the difference in available energy for assimilation, a prevailing difference among forest types is likely to be size growth rate (Kohyama 1991, 1993b). As Figure 2.5 shows, increasing mortality and leaving other parameters constant reduces the domain of coexistence, while increasing the size growth rate widens the coexistence domain.

Tropical forests are taller and faster growing systems because of the higher energy state of tropical regions. Thus, these two coupling characteristics will multiplicatively contribute to the observed 10-fold tree species diversity.

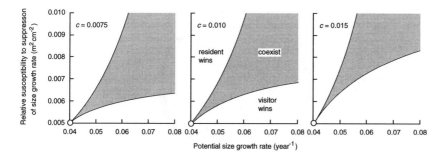

FIGURE 2.6. Equilibrium states of gap-dynamic two-species systems with changing tree mortality (c, year^{-1}; independent of tree size and shading) for both species. The gap formation rate is identical to tree mortality. The first resident species is fixed on the origin of the growth rate–susceptibility coordinates; the state of the system depends on the position of visitor species on the coordinates. In the shaded domains, two species coexist. (Details of the model and other parameters are from Table 2, Kohyama 1993a, with permission.)

The Physiological Basis of the Condition of Coexistence

The autonomic behavior of shoot modules is a feature of plants (Sprugel et al. 1991). It is worth mentioning that the feedback process of generating structural heterogeneity of the local environment and resources and regulating growth and survival as a result of this heterogeneity can be examined most precisely at the level of a shoot "population" (Fig. 2.7). Therefore, individual-based approaches are sometimes too crude to construct functional models of plant systems.

The development of a canopy is effectively simulated by the assumption that each shoot unit behaves independently in response to the capture of light resources (Takenaka 1994). It is possible to incorporate Takenaka's idea of modeling shoot populations within a three-dimensional light resource into individual-based models of plant populations and communities with size structure. Such an attempt does not necessarily make the theory complex: the interrelationships between demographic parameters observed in field censuses can be explained functionally by a smaller number of physiological parameters. For instance, improvements of production efficiency will increase size growth rates and reproduction rates and, at the same time, decrease mortality through abortion at a whole-individual level. The susceptibility of size growth rate to shading is likely to be correlated to that of recruitment rate. The trade-off between the maximum size and the per capita recruitment rate, that essentially contributes to species coexistence, reflects the alternative in allocation of assimilates between vegetative production and reproduction. The trade-off between potential growth rate and susceptibility to shading can be explained functionally by alternatives in the chloroplast arrangements within leaves (Horn 1971) or those in the allocation of resources within chloroplasts (Evans 1989). Thus, demographic parameters in size-structure-based models are mutually interdependent, and we can formulate this interdependence through physiological modeling.

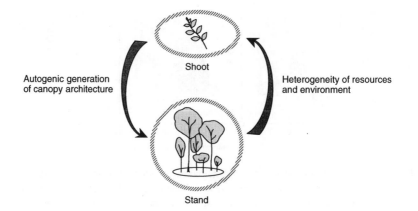

FIGURE 2.7. The aboveground feedback processes of plant populations. Dynamics of shoot units generate the vegetation architecture, and the architecture defines the three-dimensional distribution of light resources that regulate the behavior of shoot units.

Global and Evolutionary Aspects

The persistent architecture of vegetation, particularly of forests, introduces inertia or a considerable time lag in responses to environmental change at the ecosystem level. A geographically extended model of the dynamics of tree size distribution (Kohyama and Shigesada 1995) simulated the change in vegetation types along the latitudinal gradient with global warming over a century. The model predicted that biomass responded immediately in each resident forest type, but that no latitudinal movement of forest types followed. Even two millennia after a century-long period of warming, the realized latitudinal movement was only 30% of a final steady state. This rate of latitudinal movement was one order slower than the potential dispersal rate of seeds assumed in the simulation, 1–10 km per year.

This result does not mean that forest systems are insensitive to environmental change. Resident forest systems have experienced significant changes in dynamics. Phillips and Gentry (1994) reported the turnover rate of tropical rain forests has doubled in the past 10 years or so. The increasing turnover rate in terms of mortality and recruitment rate they found is undoubtedly related to increasing primary production rate, and thus the framework of the multispecies systems is changing rapidly. Limitations in the latitudinal movement of species should influence the reformation of plant communities. How species diversity and ecosystem efficiency respond in the all-changing global environment is a key question providing a challenge to us (Shugart et al. 1992) .

We should also refer to evolutionary processes. Taking into account a steady-state multispecies system with a trade-off between size and fertility (e.g., see Fig. 2.2e), invasion of a "super" species with improved physiological functions at any position in the top right-hand domain will exclude all other species. This super species will then define a new domain of coexistence with other (also improved) species on the size–fertility coordinates, and a reformation of the tree community will follow. At the same time, the whole ecosystem efficiency in terms of productivity and biomass will be expected to increase. Consequently, the evolution of terrestrial plant systems brings about the development of three-dimensional architecture. This process inevitably facilitates the increase in the number of regeneration niches.

In conclusion, the study of biological diversity will focus on the feedback between the regulation of physiological functions of individuals by resource heterogeneity and the autogenic creation of this heterogeneity by organisms (Fig. 2.7). In the short term, it is negative feedback that regulates the self-organization of ecosystems in given environments, and for the long term it is positive feedback that enhances the evolution of ecosystem efficiency and biodiversity. Two simultaneous approaches are needed: the description of processes and mechanisms of the organization of heterogeneous resource/environment structure on a physiological basis, and the examination of the role of heterogeneous structure at both community and ecosystem levels.

Summary

Recent studies on terrestrial plant systems show that the canopy architecture of vegetation can promote a state of stable coexistence among plant species. The vertical gradient of light resources and its horizontal variation, reflecting a shifting gap mosaic, play key roles. Trade-offs between demographic parameters required among species for coexistence can be explained on a physiological basis. Aboveground architecture is more persistent in forest systems than other plant systems; thus, its role in community organization is different among systems. Forest architecture effectively explains the prevailing global-scale correlation between available energy and tree species diversity. In architectural systems, biomass and productivity contribute multiplicatively to increase the number of species that can coexist. A high gap formation rate alone has an effect of reducing both architectural development and species diversity. For plants other than trees, less persistent aboveground architecture results in less clear global patterns of species diversity. The time scale of responses of plant systems to current global climatic change and to far longer evolution processes is dependent on their architectural properties.

Acknowledgments. I thank Peter Bellingham, Peter Grubb, Masahiko Higashi, and Tohru Nakashizuka for valuable comments on this chapter. I also acknowledge Tohru Nakashizuka and the Iwanami Shoten Publishers for permission to reproduce the material in Figure 2.1a and Figure 2.3.

Literature Cited

Begon, M., J.L. Harper and C.R. Townsend. 1986. Ecology: Individuals, Populations and Communities. Blackwell Scientific, Oxford.

Chesson, P. and M. Pantastico-Caldas. 1994. The forest architecture hypothesis for diversity maintenance. Trends in Ecology & Evolution 9:79–80.

Chesson, P.L. and R.R. Warner. 1981. Environmental variability promotes coexistence in lottery competitive systems. American Naturalist 117:923–943.

Cody, M.L. 1986. Structural niches in plant communities. In: J. Diamond and T.J. Case, eds. Community Ecology, pp. 381–405. Harper & Row, New York.

Connell, J.H. 1978. Diversity in tropical rain forests and coral reefs. Science 199: 1302–1310.

Currie, D.J. 1991. Energy and large-scale patterns of animal- and plant-species richness. American Naturalist 137:27–49.

Denslow, J.S. 1987. Tropical rainforest gaps and tree species diversity. Annual Review of Ecology and Systematics 18:431–451.

Evans, J.R. 1989. Photosynthesis and nitrogen relationships in leaves of C3 plants. Oecologia 78:9–19.

Gentry, A.H. 1988. Changes in plant community diversity and floristic composition on environmental and geographical gradients. Annals of the Missouri Botanical Garden 75:1–34.

Grubb, P.J. 1977. The maintenance of species-richness in plant communities: the importance of the regeneration niche. Biological Review 52:107–145.

Grubb, P.J. 1987. Global trends in species-richness in terrestrial vegetation: a view from the northern hemisphere. In: J.H.R. Gee and P.S. Giller, eds. The Organization of Communities, Past and Present, pp. 99–118. Blackwell Scientific, Oxford.

Horn, H.S. 1971. The Adaptive Geometry of Trees. Princeton University Press, Princeton.

Huston, M. 1979. A general hypothesis of species diversity. American Naturalist 113: 81–101.

Kohyama, T. 1991. Simulation of stationary size distribution of trees in rain forests. Annals of Botany (London) 68:173–180.

Kohyama, T. 1992. Size-structured multi-species model of rain forest trees. Functional Ecology 6:206–212.

Kohyama, T. 1993a. Size-structured tree populations in gap-dynamic forest—the forest architecture hypothesis for the stable coexistence of species. Journal of Ecology 81:131–143.

Kohyama, T. 1993b. Why can so many tree species coexist in tropical rain forests? (in Japanese). Kagaku 63:768–776.

Kohyama, T. 1994. Size-structure-based models of forest dynamics to interpret population- and community-level mechanisms. Journal of Plant Research 107:107–116.

Kohyama, T. and P.J. Grubb. 1994. Below- and above-ground allometries of shade-tolerant seedlings in a Japanese warm-temperate rain forest. Functional Ecology 8:229–236.

Kohyama, T. and N. Shigesada. 1995. A size-distribution-based model of forest dynamics along a latitudinal environmental gradient. Vegetatio 121:117–126.

Koike, F. and Syahbuddin. 1993. Canopy structure of a tropical rain forest and the nature of an unstratified upper layer. Functional Ecology 7:230–235.

Levin, S.A. 1976. Population dynamic models in heterogeneous environments. Annual Review of Ecology and Systematics 7:287–310.

Levin, S.A. and R.T. Paine. 1974. Disturbance, patch formation, and community structure. Proceedings of the National Academy of Sciences of the United States of America 71:2744–2747.

Manokaran, N., J.V. LaFrankie, K.M. Kochummen, E.S. Quah, J.E. Klahn, P.S. Ashton, and S.P. Hubbell. 1992. Stand table and distribution of species in the fifty hectare research plot at Pasoh Forest Reserve. FRIM Research Data 1, Forest Research Institute Malaysia, Kepong, Malaysia.

Nakashizuka, T. 1991. The importance of long-term studies of forest dynamics in large plots (in Japanese). Japanese Journal of Ecology (Sendai) 41:45–53.

Phillips, O.L. and A.H. Gentry. 1994. Increasing turnover through time in tropical forests. Science 263:954–958.

Phillips, O.L., P. Hall, A.H. Gentry, S.A. Sawyer, and R. Vásquez. 1994. Dynamics and species richness of tropical rain forests. Proceedings of the National Academy of Sciences of the United States of America 91:2805–2809.

Rohde, K. 1992. Latitudinal gradients in species diversity: the search for the primary case. Oikos 65:514–527.

Shmida, A. and S. Ellner. 1984. Coexistence of plant species with similar niches. Vegetatio 58:29–55.

Shugart, H.H., T.M. Smith, and W.M. Post. 1992. The potential for application of individual-based simulation models for assessing the effects of global change. Annual Review of Ecology and Systematics 23: 15-38.

Sprugel, D.G., T.M. Hinckley, and W. Schaap. 1991. The theory and practice of branch autonomy. Annual Review of Ecology and Systematics 22:309–334.

Swaine, M.D., S. Lieberman, and F.E. Putz. 1987. The dynamics of tree populations in tropical forests: a review. Journal of Tropical Ecology 3:359–366.

Takenaka, A. 1994. A simulation model of tree architecture development based on growth response to local light environment. Journal of Plant Research 106:321–330.

Tilman, D. 1988. Plant Strategies and the Dynamics and Structure of Plant Communities. Princeton University Press, Princeton.

3

Species Coexistence and Abundance: Patterns and Processes

MUTSUNORI TOKESHI

Introduction

One of the fundamental questions in community ecology is in what way different species coexist in a given area and how such an assemblage is maintained through time. This question obviously requires an explanation based on contemporary ecological processes, whether biotic or abiotic, which control the organization of ecological assemblages. Contemporary processes are in general amenable to manipulative experiments, and a large body of experimental studies carried out to date on various ecological communities testifies to the power and potential of such approaches. On the other hand, it is also important to recognize the fact that evolutionary history, which invariably lies behind and casts a shadow over contemporary ecological patterns, largely defies the application of manipulative experiments. Thus, to a larger or smaller extent contemporary patterns need to be investigated through nonexperimental approaches alongside experimental ones. In a broad perspective, irrespective of the difficulty of testing largely unreproducible phenomena such as historical contingency (Gould 1989), consideration of both evolutionary and contemporary processes is crucial to a better understanding of the mechanisms of species coexistence and diversity.

In this chapter, I consider some aspects of species coexistence and abundance as they relate to broader issues of biodiversity. The objective here is to draw a general picture that may stimulate further research, rather than a retrospective review of firmly established scientific facts. This is considered an essential step toward tackling as vast and fundamental a question as biodiversity; speculation over fundamental issues is no less important than hard answers to relatively trivial questions.

Basis of Biodiversity

The recent debate on global-scale environmental degradation and its consequences on the biota (e.g., Wilson 1988) has focused our attention to the importance of species richness as a basic measure of biodiversity. This trend notwithstanding, the term biodiversity in its broadest sense can be interpreted to encompass not only species-level diversity (i.e., species richness) but molecular-, gene-, phylogenetic-, population-, community-, and ecosystem-level diversity as well as functional diversity of the Earth's organismal entity. As such, research on biodiversity in its ideal form requires a multifaceted approach involving expertise from a variety of fields. This being said, it is worthwhile to emphasize that a species represents a biologically concrete, well-defined unit and that species richness constitutes a convenient measure of biodiversity with which comparisons can be made across a range of temporal and spatial scales. Biodiversity at other levels of organization may be handled, at least in a general technical sense, by extension or extrapolation from investigations into species richness.

Discussion of species richness and coexistence is often inexorably bound up with the *origin* and *maintenance* of diversity, two operationally distinct matters that may nevertheless be closely related within an evolutionary domain. The long-term maintenance of diversity can be taken as a balance between originations and extinctions of species, and an observed pattern of species assembly at a particular point in time is largely dependent on the cycles of originations and extinctions. Furthermore, the rates of species origination and extinction may be related to one another and to the current level of diversity, leading under certain circumstances to a fairly constant level of diversity as observed for the families of marine invertebrates from the Ordovician through the Permian (Sepkoski 1978, 1979, 1984). Species origination in this context is considered to be associated with three major aspects: (i) genetic constitutions of organisms (e.g., polyploidy among plants); (ii) geographic separation (on a large or small spatial scale, depending on the size and dispersal ability of organisms); and (iii) coevolution, in particular of angiosperms and insects since the Mesozoic, the latter taxa accounting for a large proportion of total species richness on Earth (Wilson 1992; see Labandeira and Sepkoski 1993 for a heterodox view on the issue). On a much shorter time scale, however, the maintenance of diversity is more of a function of contemporary biotic and abiotic processes, quite independent of origination and extinction of species (although the latter, but not the former, is fast acquiring an acutely contemporary status under human influences).

In connection with the origin and maintenance of diversity, one issue that constitutes a major problem in community ecology is how an observed level of diversity in a particular assemblage has been reached with respect to resource utilization. An increase in diversity may occur through one of two, not necessarily mutually exclusive, processes. First, organisms may acquire abilities to exploit new resources (or "niches") that have not previously been utilized by existing species, i.e., invasion of a new "adaptive zone" (*sensu* Simpson 1944, 1953) (Fig. 3.1a). Second, an existing resource pool may be more finely divided or shared by more species, or in other words, invasion or division of already exploited niches (Fig. 3.1b). If sympatric speciation has been the dominant mode in the evolutionary origination of species, this second process is perhaps an appropriate representation in many cases. An ancestral species that spreads over a wide resource spectrum, in terms of habitat space or food resources, may split into two or more species, each specializing in a narrower range of resources. These two processes,

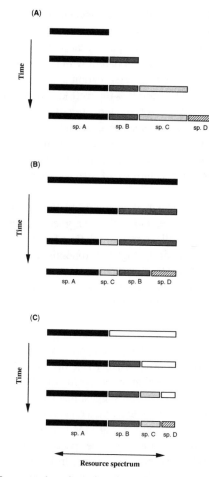

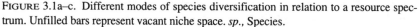

FIGURE 3.1a–c. Different modes of species diversification in relation to a resource spectrum. Unfilled bars represent vacant niche space. *sp.*, Species.

i.e., adaptive invasion and niche fragmentation, may operate simultaneously or merge together in and across various organismal groups. Broadly speaking, however, the former is more closely associated with diversification of higher taxa such as phyla, orders, and families, while the latter tends to apply to phylogenetically closer groupings such as genera, species, and subspecies or sibling species. For example, an increase in diversity of marine benthos (the number of families, in particular) from the early to mid-Palaeozoic (Sepkoski 1979) and during the Mesozoic (Bambach 1977, 1983) seems to have been accompanied by an increase in the range of morphology and life style exhibited by those organisms, which presumably indicate a widening resource spectrum.

On the other hand, an assemblage of species that broadly share the same resource spectrum (or form a trophic guild) is more often than not taxonomically closely related, frequently constituting a convenient entity to be subjected to a detailed ecological study on the present-day time scale. This exercise is rather difficult with extinct communities of the past because of the incompleteness of fossil records, which difficulty is more apparent with the lower taxa. It is likely that in many taxa these two modes of diversification and resource use grade into one another (Sepkoski 1988). An example is the well-known trophic diversification in the cichlid fishes in Lake Malawi (i.e., widening of resource spectrum), with the concurrent existence of groups of species apparently sharing the same resource in the same manner (Fryer and Iles 1972). Another salient example relates to distinct insect herbivore assemblages associated with different tropical plants which, at the same time, demonstrate a number of species sharing the same host plant (Strong 1982).

While the spectrum of resources available to the Earth's biota has steadily widened over evolutionary time, it may also be argued that for a period of less than 1 million years (which was still a long enough time for speciation in many groups, but too short for diversification of higher taxonomic categories), the size of the total resource pool available to a particular assembly of taxa has remained at roughly the same level. In this case an increase in local diversity may be envisaged as a more or less sequential filling-up of the potential total niche space, which may correspond to the formation of a guild (Fig. 3.1c). This situation can in fact be considered as intermediate between diversification based on the conquering of completely new niche space (Fig. 3.1a; potential total niche space completely unbound) and that based on a progressively fine division of an already exploited niche space (Fig. 3.1b; full exploitation of the total niche space since the beginning). There is some fossil evidence to suggest that such processes of community assembly might have occurred repeatedly on a relatively short time scale of 20,000 to 100,000 years following climatic changes (Milankovitch cycles, Bennet 1990; see also Coope 1994). Thus, invasion of niche space can take different forms over different time scales and taxonomic resolutions. In terms of guild formation, the situations depicted in Figures 3.1b and 3.1c perhaps represent two extreme cases, with different assemblages falling onto various points in between these.

In addition to the possible qualitative differences in the processes of diversification as just considered, it should be noted that diversity as it has traditionally been treated by ecologists encompasses not only species richness but relative abundances of different species in an assemblage. In a discussion of biodiversity, it is therefore useful to consider together species richness, niches and resources, and abundance, because these represent different sides of the same ecological coin. Indeed, the foregoing discussion of the resource-based mechanisms of diversification allows consideration of relative species abundance through the approach of niche-oriented species abundance models (see following).

Species Abundance Patterns As Evolutionary Processes

The need to integrate the concepts of species richness, niche, and abundance in the discussion of biodiversity points to one area of community ecology that has experienced relatively slow development until recently, being largely independent of the advances witnessed in other related fields. Investigation into so-called species abundance patterns traces its history back to the 1920s and 1930s, when it was treated with its allied subject, species–area relationships (Arrhenius 1921; Motomura 1932; MacArthur 1960; see review in Tokeshi 1993). Despite the number of models proposed and review articles written since then, the analysis of species abundance patterns has long remained a rather dry exercise with community data, little more than a superficial description of community patterns based on comparisons with a limited selection of models. This may largely be attributable to a lack of conceptual basis and deficiencies in the theoretical framework, making it difficult to relate observed patterns to various aspects of ecological processes (Tokeshi 1990, 1993). Development of niche-oriented species abundance models (Tokeshi 1990) raised, among other things, issues of resource/niche division and links between resource use and abundance, which have a close bearing on the debate on biodiversity. Furthermore, if the evolution of biodiversity is considered with reference to resource use, niche apportionment models offer a convenient framework in which to consider the aspects of community and guild formation through evolutionary time. For discussions on contemporary versus evolutionary processes of niche apportionment, see Tokeshi (1993).

Niche Apportionment Models: Conceptual Basis

Models of species abundance patterns can be classified into three major categories: statistically oriented, niche oriented, and other models (Table 3.1). Despite frequent reference in the ecological literature, statistically oriented models have inherent difficulty in relating the patterns described to ecological processes, and indeed they often represent large communities whose boundaries are loosely defined, i.e., a collection rather than a community (Tokeshi 1993). It is often very difficult to apply a resource-centered approach to such assemblages, because different (i.e., taxonomically distant) organisms may possess totally different resource bases and consequently total niche space is undefinable in practice (see Fig. 3.1a). This problem is of course one of degree, and no clear-cut demarcation exists between closely knit and disparate assemblages in terms of resource base. Nevertheless, it is worth noting that a resource-centered approach is more appropriate for assemblages with taxonomically closely related species. In the present context, such assemblages are ideal for analyses with niche-oriented models, as they allow a resource-based consideration of diversification and guild formation. Where the totality of a resource to be divided by an assemblage of species is assumed to be more or less fixed in quantity, the term niche apportionment models is used in preference to the more general niche-oriented models.

Conceptually, niche apportionment models envisage species abundance to be defined through a two-stage process (Fig. 3.2). In the first stage, the total niche is divided into species niches by some division rule. This is then followed by the second or translation stage in which species niches are converted to the abun-

TABLE 3.1. Models of space abundance patterns

Model	Reference
Statistically oriented	
Log series	Fisher et al. (1943)
Log-normal	Preston (1948, 1962)
Negative binomial	Anscombe (1950); Bliss and Fisher (1953)
Zipf–Mandelbrot	Zipf (1949, 1965); Mandelbrot (1977, 1982)
Niche oriented	
Geometric series	Motomura (1932)
Particulate niche	MacArthur (1957)
Overlapping niche	MacArthur (1957)
Broken stick	MacArthur (1957)
MacArthur fraction	Tokeshi (1990)
Dominance preemption	Tokeshi (1990)
Random fraction	Tokeshi (1990)
Sugihara's sequential breakage	Sugihara (1980)
Dominance decay	Tokeshi (1990)
Random assortment	Tokeshi (1990)
Composite	Tokeshi (1990)
Other	
Dynamic model (biological)	Hughes (1984, 1986)
Neutral model (nonbiological)	Caswell (1976)

After Tokeshi (1993).

dances of different species. In the latter process, the most straightforward and reasonable supposition is that abundance is directly related to niche space: a larger niche results in greater abundance. In reality, if niche is equated with resources, interspecific differences in the efficiency of converting resource into individuals may distort a linear relationship between niche space and abundance among a set of species. Note, however, that the potential of such risk is reduced if a taxonomically close assemblage is chosen for analysis. Further, depending on the organism, there may be some technical difficulty in choosing an appropriate measure of species abundance. Nevertheless, biomass and number of individuals are perhaps most appropriate in many cases; it is of course possible to use measures such as degree of spatial coverage so long as they can adequately reflect the magnitude of resource utilization (Tokeshi 1993). The crucial aspect, then, is the first stage—what possibilities are there for dividing the total niche into species niches?

Before the formal and integrative formulation of niche apportionment models (Tokeshi 1990, 1993), models such as Motomura's (1932) geometric series model and McArthur's (1957) broken stick model were sometimes referred to as the sequential and the simultaneous breakage model, respectively (Pielou 1975). However, I have shown that this distinction is unnecessary, and it is now possible to view these and other schemes of niche apportionment as sequential processes, thus relating more closely to evolutionary sequences of events (Tokeshi 1993). This allows a uniform treatment of different possibilities of niche apportionment and diversification.

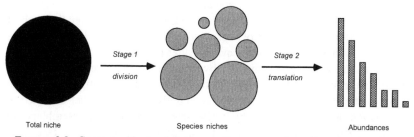

FIGURE 3.2. Conceptual basis of niche apportionment models of species abundance.

In an earlier discussion of guild formation within a fixed total niche space, two separate cases were represented, niche fragmentation (Fig. 3.1b) and niche filling (Fig. 3.1c). It should be noted here that, despite an apparent difference in the processes, these two schemes can in theory be considered within a single framework of niche apportionment. To see this point, we can ignore the identity of species associated with particular portions of niche space at a particular point in time. It then becomes clear that invasion of niche space may occur irrespective of the presence or absence of species in the portion of a niche to be invaded. In other words, the new species may carve out its own niche space from either vacant or already occupied niche space in exactly the same manner (i.e., ignore the shadings in Fig. 3.1c and the process converges to that of Fig. 3.1b). Thus, niche fragmentation and the sequential filling of a niche may be understood to represent two different facets of the same niche apportionment processes.

Demarcation between these two modes becomes progressively blurred the more species there are in an assemblage. On the other hand, niche fragmentation, whereby the total niche space occupied by all the species combined is taken to be unity at any one time, perhaps allows a simpler and more uniform treatment of different processes of apportionment than does sequential filling under most circumstances, and thus helps with the comparison of patterns across systems. Indeed, the original niche apportionment models (Tokeshi 1990) were all framed as niche fragmentation rather than niche-filling processes. The subtlety of interpretation indicated here represents one of the areas of the niche-oriented model approach that requires further investigations. Notwithstanding such a caveat, the argument presented serves as a conceptual basis on which niche apportionment models can be developed with reference to a supposed total niche space and its division. Therefore, emphasis is placed on broadly distinct ways in which the total niche is divided and apportioned among species.

Dominance Preemption and Dominance Decay Models

In considering niche apportionment, it is useful to envisage some extreme cases that would define the outer range of possibilities. One such case is represented by the dominance preemption model (Fig. 3.3), in which new species successively invade the niche space of the least abundant species in an existing assemblage (Fig. 3.3a) or, alternatively, new species successively carve out more than half of what is left vacant (i.e., invasion of vacant niche; Fig. 3.3b). The former interpretation refers to niche fragmentation and the latter to niche filling. In either situation, this model defines a pattern whereby successively dominant species (larger niche space) retain their dominant status by hierarchically preempting the niche. The smallest niche space in an assemblage at a given time, whether occupied or vacant, is the one to be invaded by a new species entering the assemblage. In Tokeshi's (1990) original formulation, a new species was assumed to take more than half the invaded niche space. This will make the dominance hierarchy of species exactly corresponding to the temporal sequence of invading species.

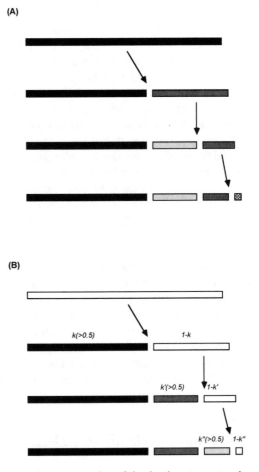

FIGURE 3.3a,b. Schematic representation of the dominance preemption model as derived from (a) a niche fragmentation process and (b) a niche-filling process. k (70.5), arbitrary value.

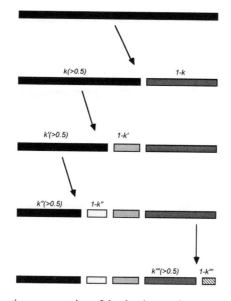

FIGURE 3.4. Schematic representation of the dominance decay model. k, arbitrary value.

However, if this process is considered in the context of niche fragmentation where the total niche is always fully occupied, the same pattern of dominance preemption can result from an invading species simply attacking the least abundant species and taking an arbitrary proportion of its niche space (i.e., Fig 3.3a). In this case, the sequence of invasion does not necessarily correspond with the dominance/abundance hierarchy.

Another extreme case, which represents the reverse of dominance preemption, is the dominance decay model (Fig. 3.4). In this model, instead of the smallest niche in an assemblage being invaded, the largest one is chosen by an invading species on successive occasions. Thus, in the niche fragmentation model, an invading species always attacks the most abundant species and takes an arbitrary proportion of its niche space. In contrast to the dominance preemption model, the sequential filling mode does not strictly apply to this model, as there is no guarantee that a remaining or vacant niche space at a particular point in time is always larger than those niches already occupied to allow invasion by new species. As the decaying process proceeds, the average niche size gradually decreases while the number of niches increases. At the same time, variation in niche size decreases and the whole assemblage asymptotically approaches quasi-equitability. Thus, abundances come to assume equitable values more strongly than in the broken stick model.

MacArthur Fraction and Random Fraction Models

In terms of niche fragmentation, dominance preemption and dominance decay assume that either the least or the most abundant species (i.e., with the smallest or

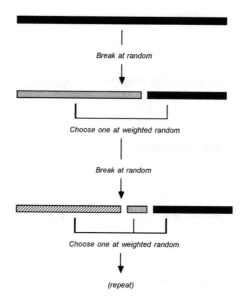

FIGURE 3.5. Schematic representation of the MacArthur fraction model.

the largest niche space, respectively) at a particular point in time is subject to invasion by a new species. Technically, this would necessitate mechanisms by which an invading species is able to assess niche sizes of different species, or a new speciation or invasion event is somehow restricted to occur in the most or least abundant species in an assemblage. That this is perhaps an unlikely phenomenon in most situations should not diminish the point that these models serve a useful purpose by defining two extremes within which the majority of cases will lie. Recognition of this point further helps us understand the relevance of other models.

The MacArthur fraction model (Fig. 3.5; Tokeshi 1990) is analogous to the broken stick model but stipulates a sequential rather than simultaneous process of niche division. All species in an assemblage are subject to invasion by a new species, with the probability of invasion being dependent on the abundance or niche size of each species. Thus, a species with a higher abundance is more likely to experience niche fragmentation than a less abundant species. In terms of the origination of new species, this may be expressed as the rate of speciation [splitting of an ancestral species to produce an additional (new) species] being directly proportional to the population size or abundance of a species. In contrast to the conceptual and technical difficulty of testing the simultaneous breakage (broken stick) model (Tokeshi 1993), the sequential MacArthur fraction model as described here does not appear overtly unrealistic within an evolutionary domain (Jablonski 1987). As a hypothesis, it certainly merits further analysis, in particular with assemblages of taxonomically close species that are ecologically tightly knit.

Closely allied with the MacArthur fraction model but slightly different from it with respect to the selection probabilities for successive niche division is the random fraction model (Fig. 3.6). In this model, all existing species in an assemblage

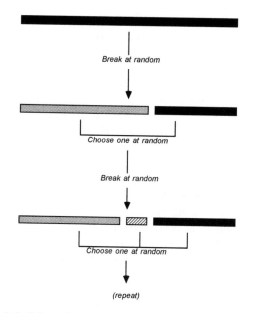

FIGURE 3.6. Schematic representation of the random fraction model.

are supposed to have the same probability of being selected for a subsequent niche division. In other words, the current abundance or niche size does not affect the chance that the species in question is challenged by an invading species. Further, if couched in the terms of speciation events, the model stipulates that splitting of one species into two occurs without recourse to the size of its niche. This suggestion may not be so strange as it might first appear. In the first place, there is no evidence to suggest that speciation probability always scales with niche size. Indeed, it is possible that those species with a relatively wide niche remain as they are because of their superior mobility or sufficiently versatile generalist feeding habits, thus effectively resisting speciation, while those with narrow niches are perhaps inherently more likely to "pursue" the tendency of specialization, resulting in further speciation. If this is coupled with the opposite trend of species with a wider niche to experience a higher chance of habitat fragmentation and isolation leading inevitably to speciation, speciation probability may effectively become independent of niche size. There seems to exist increasing evidence that some assemblages such as freshwater Chironomidae and Hydracari (P.E. Schmid and J. M. Schmid-Araya, personal communications) demonstrate patterns indistinguishable from the random fraction model, and more research is warranted on this matter.

Random Assortment and Composite Models

In an earlier discussion on the basis of biodiversity, I pointed out that one mode of increasing diversity in relation to niche/resource use is represented by an open-ended system (see Fig. 3.1a), in which invading species come to exploit new resources that were previously unused by existing species. In this case, different niches are essentially unrelated, especially with respect to the quantity/space of each niche available for exploitation; this makes the abundance of different species effectively independent of one another. An analogous situation may be seen in some contemporary assemblages in which local abundance experiences quasi-random variations on certain temporal scales because of the stochastically dynamic nature of some environmental factors (Tokeshi 1994). Under such circumstances, species niches cannot fill up and loosely trail behind the temporally variable total niche space. This was formulated as the random assortment model (Tokeshi 1990), which seemed to fit the numerical and biomass data of a chironomid assemblage. Interestingly, the same expected pattern of relative abundance may be produced by a niche-filling process whereby successive species carve out an arbitrary proportion of the remaining niche space from a total niche (Fig. 3.7; Tokeshi 1993). Indeed, in essence this suggests the existence of resource limits (which is reasonable if we consider the Earth as an ultimate resource boundary), while at the same time allowing the maximum level of independence of niches. Thus, although Tokeshi (1990) initially presented this model as being distinct from other niche apportionment models, it is most reasonable to consider this as a variation of niche apportionment (Tokeshi 1993). The model may acquire more relevance as further investigation into biodiversity is conducted.

A slight conceptual departure from all previous models is represented by the composite model (Tokeshi 1990), which incorporates different ways of niche apportionment. This departure comes from a consideration that an assemblage of species may be governed not only by one but two (or more) different processes. In

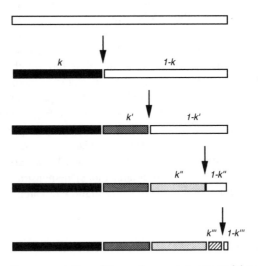

FIGURE 3.7. Schematic representation of the random assortment model as derived from a niche-filling process. k, k', k'', and k''' are arbitrary values.

the original composite model of Tokeshi (1990), it was assumed that the more abundant species in an assemblage were governed by one of the niche apportionment processes while less abundant species conformed to random assortment. The idea is that competitive interactions are more likely to be prevalent among numerically dominant species than among less numerous species. The composite model can of course be modified in various ways, even incorporating three different processes. This represents another area where further investigation is awaited.

Some Related Aspects

Relation to Statistically Oriented Models, Particularly the Log-Normal

As has been demonstrated, niche apportionment models are versatile in that they can be framed in terms of both evolutionary and contemporary processes. This makes a sharp contrast to statistically oriented models of species abundance patterns (see Table 3.1), which simply serve as descriptive measures of observed patterns without considering specific mechanisms and processes. Thus, so long as community ecology takes it as a grand objective to discern and elucidate recognizable ecological patterns and their underlying mechanisms, statistically oriented models remain a less satisfactory proposition than niche-oriented models. Nevertheless, statistically oriented models have a relatively long history of application, being employed as a convenient descriptor of patterns. This leaves an important question of the relationship between statistically oriented and niche-oriented models.

Among statistically oriented models, the log-normal model has received particular attention (May 1975; Pielou 1975; Gray 1987; Magurran 1988). This in large part resulted from the thorough, pioneering work by Preston (1948, 1962) and the perception that the model often fits assemblages with relatively large numbers of species. This idea has in turn encouraged attempts to formulate biological mechanisms that produce patterns conforming to the log-normal (e.g., Sugihara 1980; Gray 1987). In particular, Sugihara (1980) proposed a mechanism of niche division that was claimed to lead to the canonical log-normal and which he called a minimum community structure. However, Tokeshi (1990, 1993) pointed out that Sugihara's proposition loosely encompasses the two extreme cases of niche apportionment, making it almost inevitable to "approximate" the log-normal model, which lies in the middle range (in terms of the overall slope of the rank–abundance curve) of species abundance patterns. Furthermore, evidence to support a static breakage ratio of 0.75:0.25 across communities as assumed in the Sugihara model is extremely tenuous to say the least. Therefore, Sugihara's model cannot be considered to define a minimum structure, if that minimum is taken to mean something more tangible than "nonstructure."

In connection with this discussion, it is worthwhile to point out that Sugihara's model is not synonymous with the random fraction model, as has sometimes been assumed (Nee et al. 1991). Indeed, Sugihara (1980) himself explicitly rejected the idea of breakage points being distributed uniformly randomly in (0, 1). Interestingly, however, data of high quality on bird assemblages in general and British birds in particular (Nee et al. 1991; Gregory 1994) conform well to the random fraction model (a slightly better fit than Sugihara's model; M. Tokeshi, unpublished). Further progress is anticipated with regard to the application of niche apportionment models to such species-rich assemblages.

In a broader perspective, the major difficulty in giving biological explanations to the log-normal relates to the fact that, for such explanations to be valid, it should first be shown explicitly that mathematical artifacts of large numbers do not play a significant role in producing what are recognized as log-normal patterns (Tokeshi 1993). That the log-normal may result from the central limit theorem of large numbers was first suggested by May (1975), but this insight has not received sufficient attention or sparked further lines of investigation, despite the fact that May's work has been widely cited in ecological literature for the past two decades. The question of large numbers in a statistical sense also has a close bearing on the truly biological issue of how we define the taxonomic boundary of an assemblage under investigation (Tokeshi 1993).

The log-normal model (and also the log series) is in its principal form demonstrated as the frequency distribution of number of species against geometric abundance classes called octaves. When expressed as a rank–abundance curve, the abundance values inherently contain a zone of variation, the magnitude of which is not easy to assess. The net result is that the log-normal tends to encompass a wide range of patterns and is in consequence relatively insensitive to differences in abundance patterns, which niche apportionment models can possibly distinguish. In this respect, it is necessary to carry out a detailed examination of the relationship between the log-normal and various niche apportionment models. Apart from the log-normal, little is known about the biological relevance, if any, of other statistically oriented models and their relationship to niche apportionment models. Indeed, it can be said that the failure of the approach of species abundance patterns to go beyond a superficial description of community patterns is inherent in the nature of statistically oriented models. Whether or not we can break this mold depends largely on further, insightful research.

Relation to Stochasticity and Fractal Geometry

It has been increasingly recognized that stochasticity often plays a significant role in producing or modifying community patterns (Tokeshi 1994). Although the importance of stochasticity has been acknowledged in evolutionary biology, in particular in some areas of population genetics (e.g., Kimura 1983), contemporary ecologists have been rather slow to turn their attention to this direction. In a phenomenological sense, stochasticity in ecology was often interpreted to represent no more than an admission that we did not entirely know causal mechanisms underlying a pattern (e.g., stochasticity as incorporated in earlier fisheries models). Recently, however, attention has been drawn to stochasticity as a pervasive element of some ecological processes, worthy of being given a distinct identity of its own. In the case of species abundance patterns, all niche apportionment models except the geometric series incorporate stochastic processes. It may even be argued that failure of the geometric series to engage the serious interests of practicing ecologists stems largely from its rigid, deterministic division rule that is almost anathema to ecological common sense.

Incorporation of stochasticity in niche-oriented models makes the testing of these models more problematical, although this is not insurmountable for analyzing contemporary communities (Tokeshi 1993). On the other hand, if models are treated in an evolutionary context only and observed patterns are interpreted strictly as an evolutionary outcome, model testing becomes fairly difficult. This is especially so when each and every community is considered a unique entity embodying a unique process of formation, rather than representing a replicate of some sort. Thus, the application of niche apportionment models to purely evolutionary contexts has a practical problem of testing to be given a careful consideration. In a broader perspective, this represents a general problem of analyzing evolutionary phenomena, not being unique to the evolutionary patterns of species abundances.

In the case of statistically oriented models, the issue of stochasticity is even less clearly understood. In particular, no attempt has been made to link the inherent variability of abundance values associated with these models to an independent aspect of stochasticity. It is considered doubtful that a clear and straightforward answer exists. This represents another gray area in this discipline.

In parallel with stochasticity, recent attention has also been drawn to the relevance of fractal concepts in dealing with ecological patterns (Frontier 1987). This has particularly been useful in assessing landscape/vegetation patterns or the habitat structures for phytophagous/arboreal arthropods of various body sizes (e.g., Hastings et al. 1982; Morse et al. 1985; Krummel et al. 1987; Sugihara and May 1990). It is worth noting that, although application to species abundance patterns is yet to be explored (Williamson and Lawton 1991), the concept of fractal geometry is remarkably akin to niche apportionment models in that simple, recurrent rules that cross spatial scales lead to fairly complex patterns. This is apparent if one looks at classical fractal processes such as the Cantor sets (1872) or fractal tilings (Grünbaum and Shephard 1987). Fractal geometry also allows incorporation of stochasticity, a direction that may result in a fruitful avenue of research in conjunction with stochastic niche apportionment models.

Rare Species: Problems of Extinction

Notwithstanding the foregoing discussion indicating the conceptual similarity between fractal geometry and niche apportionment, there is one salient point that distinguishes between these two: the former is in essence an infinite process in which a pattern produces a smaller replicate of itself in an unending spiral of descending scales, while the latter is inevitably truncated at a lower end by the basic characteristics of life, i.e., the existence of a minimum physical entity needed to define a self-replicating organism. Thus, species abundance patterns possess a minimum value, corresponding to a single individual, beyond which further niche or resource division becomes practically meaningless.

Somewhat related to this point, there also exists a certain minimum number of individuals to guarantee the survival of a species population. So long as a community is made up of species populations, the importance of such "minimum viable population size" as an ingredient of community patterns cannot totally be neglected (Lawton 1990). Nevertheless, it should also be recognized that there is no guarantee that an assemblage to be studied contains a minimum viable number for each and every species encountered. Indeed, it is more likely that some species are below such population size if an ecological assemblage needs to be defined, to a larger or smaller extent, within somewhat arbitrary spatial boundary for the purpose of research. Under such circumstances, immigration from outlying areas is likely to be the major mechanism sustaining species populations within a defined area. This has a close bearing on one of the basic issues in community ecology, i.e., the definition of community boundaries (Tokeshi 1993).

Besides the issue of community boundaries, there is another, perhaps more fundamental, question about species survival and variability in population size. From an evolutionary point of view, each species may have a more or less fixed period of existence before going extinct. If this is the case, in any contemporary assemblage some species are likely to be on their declining course toward extinction, while others may be in their prime with relatively steady or even increasing population size. Thus, we may be observing those species that have already passed the point of no return, i.e., the minimum viable population size. Note that this represents a natural process of extinction, rather than a process precipitated by human activities. It is notable that, on an evolutionary time scale, extinction is not an uncommon phenomenon.

As has been discussed, biodiversity on an evolutionary time scale entails balance between species originations and extinctions. Thus, the issues of species with low abundance (i.e., rare species) and of extinction (as an extreme case of rareness) are of fundamental importance to the consideration of biodiversity. Total species richness in a particular assemblage often depends heavily on a relatively large number of rare species. In this respect, although niche-oriented models are in principle applicable to a community with any level of species richness, some technical aspects tend to make them less versatile in accounting for the patterns of rare species. One of the difficulties with stochastic niche apportionment models relates to the fact that the abundance patterns of rare species cannot easily be tested in as rigorous a manner as those of more abundant species. This may encourage exploration of other approaches to the analysis of rare species in a community. One such approach is that of metapopulation modeling, which considers a population as occupying a habitat consisting of a mosaic of smaller habitat units (Hanski and Gilpin 1991).

Rare Species and Metapopulation Models

Although the theoretical study of metapopulation dynamics has not specifically focused on the ecology of rare species in a community, a class of metapopulation models (Levins 1969, 1970; Hanski 1982; Tokeshi 1992) has come to have a close bearing on the issue. These models are simple in structure (although with some complex additional assumptions; see Tokeshi 1992) and no doubt ignore many (perhaps important) biological details of different species; nevertheless, they serve as a useful starting point for considering community-wide patterns. In particular, Tokeshi's (1992) model generally predicts a predominance of rare species in an assemblage as a result of interactions between local immigration/colonization and emigration/extinction in a stochastic environment. Rare species in this context refer to those with small population size and limited spatial occupancy. The model describes changes in the proportion of habitat units occupied (p) as a balance between the rates of immigration and of local extinction:

$$dp/dt = ip(1-p) - e(1-p)$$

where i and e are constants relating to immigration and extinction, respectively. An assemblage of species following the stochastic version of this model would typically lead to a single-modal frequency distribution of the number of species occupying different proportions of habitat units (p), with the mode skewed toward $p \approx 0$ (i.e., preponderance of rare species; Fig. 3.8). The abundance of rare species in a community in terms of species richness appears to be fairly common in both animal and plant assemblages (Tokeshi 1992 and unpublished data; S. Scheiner, personal communication).

One unresolved issue in this modeling approach concerns the question as to how species avoid going extinct, having come close to $p = 0$. The model itself does not provide an answer to this question. A number of possibilities seem to exist in this respect. First, these rare species may indeed be on their way to extinction, whatever period of time each species may eventually take to reach an ultimate equilibrium state. In this point of view, the majority of species in an assemblage are considered to be on their slow but steady decline toward extinction. Second, there may be yet unknown factors that operate in a density-dependent manner to enhance the survival of a species population when its size is small. This goes somewhat against our current perception of species populations, but perhaps it is still premature to negate this possibility totally. Third, it is possible that the majority of rare species in an assemblage are maintained by constant immigration of individuals from outside the area under investigation. In a sense, this goes back to an earlier discussion about the definition of community boundaries for research purposes and the near-impossibility of assuming that most, if not all, species populations occur exclusively in a single defined area. This point cannot be clarified until we conduct an integrated analysis encompassing different spatial scales, incorporating microhabitat to continental/global patterns. Thus, the importance of spatial scales cannot be neglected in the consideration of rare species.

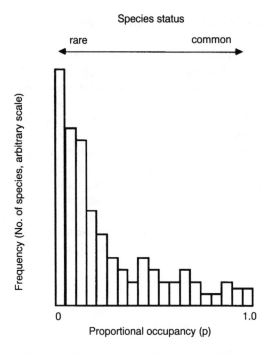

FIGURE 3.8. A typical example of the species frequency vs. habitat occupancy diagram of ecological communities, with an excess of rare species (i.e., of limited spatial occurrence) forming a conspicuous mode.

In contrast to the approach of species abundance patterns, metapopulation modeling cannot specify the (relative) abundances of different species in a community. Therefore, in this respect, rare species are again not sufficiently treated. In essence, the problem comes down to none other than the familiar difficulty of making a quantitative, statistical assessment with low numbers and frequencies. Some breakthrough is sorely needed in this discipline.

Conclusions

Species abundance patterns and allied aspects are considered to provide a useful framework with which to consider the development of biodiversity and community organization in general. Thus, the time has already come to regard these approaches not as separate, somewhat esoteric, lines of investigation but as potentially important avenues of research that can fruitfully be integrated into the mainstream of community ecology. In particular, novel attempts to combine theory and field-based research are awaited in this field; the insights thus gained may help us to tackle the ever-increasing problems of biodiversity on Earth.

Literature Cited

Anscombe, F.J. 1950. Sampling theory of the negative binomial and logarithmic series distributions. Biometrika 37:358–382.

Arrhenius, O. 1921. Species and area. Journal of Ecology 9:95–99.

Bambach, R.K. 1977. Species richness in marine benthic environments through the Phanerozoic. Paleobiology 3:152–167.

Bambach, R.K. 1983. Ecospace utilization and guilds in marine communities through the Phanerozoic. In: M.J.S. Tavesz and P.L. McCall, eds. Biotic Interactions in Recent and Fossil Benthic Communities, pp. 719–746. Plenum, New York.

Bennett, K.D. 1990. Milankovitch cycles and their effects on species in ecological and evolutionary time. Paleobiology 16:11–21.

Bliss, C.L. and R.A. Fisher. 1953. Fitting the binomial distribution to biological data and a note on the efficient fitting of the negative binomial. Biometrics 9:176–206.

Cantor, G. 1872. Über die Ausdehnung eines Satzes aus der Theorie der Trigonometrischen Reihen. Mathematische Annalen 5:123–132.

Caswell, H. 1976. Community structure: a neutral model analysis. Ecological Monographs 46:327–354.

Coope, G.R. 1994. The response of insect faunas to glacial-interglacial climatic fluctuations. Philosophical Transactions of the Royal Society of London B, Biological Sciences 344:19–26.

Fisher, R.A., A.S. Corbet, and C.B. Williams, 1943. The relation between the number of species and the number of individuals in a random sample from an animal population. Journal of Animal Ecology 12:42–58.

Frontier, S. 1987. Applications of fractal theory to ecology. In: P. Legendre and L. Legendre, eds. Developments in Numerical Ecology, pp. 335–378. Springer-Verlag, Berlin.

Fryer, G. and T.D. Iles. 1972. The Cichlid Fishes of the Great Lakes of Africa. Oliver & Boyd, Edinburgh.

Gould, S.J. 1989. Wonderful Life. Century Hutchinson, London.

Gray, J.S. 1987. Species-abundance patterns. In: J.H.R. Gee and P.S. Giller, eds. Organization of Communities—Past and Present, pp. 53–67. Blackwell Scientific, Oxford.

Gregory, R. 1994. Species abundance patterns of British birds. Proceedings of the Royal Society of London Series B, Biological Sciences 257:299–301.

Grünbaum, B. and G.C. Shephard. 1987. Tilings and Patterns. W.H. Freeman, New York.

Hanski, I. 1982. Dynamics of regional distribution: the core and satellite species hypothesis. Oikos 38:210–221.

Hanski, I. and M. Gilpin 1991. Metapopulation dynamics: brief history and conceptual domain. Biological Journal of the Linnean Society 42:3–16.

Hastings, H.M., R. Pekelney, R. Monticciolo, D. vun Kannon, and D. del Monte. 1982. Time scales, persistence and patchiness. BioSystems 15:281–289.

Hughes, R.G. 1984. A model of the structure and dynamics of benthic marine invertebrate communities. Marine Ecology Progress Series 15:1–11.

Hughes, R.G. 1986. Theories and models of species abundance. American Naturalist 128:879–899.

Jablonski, D. 1987. Heritability at the species level: analysis of geographical ranges of Cretaceous mollusks. Science 238:360–363.

Kimura, M. 1983. The Neutral Theory of Molecular Evolution. Cambridge University Press, Cambridge.

Krummel, J.R., R.H. Gardner, G. Sugihara, and R.V. O'Neill. 1987. Landscape patterns in a disturbed environment. Oikos 48:321–384.

Labandeira, C.C. and J.J. Sepkoski, Jr. 1993. Insect diversity in the fossil record. Science 261:310–315.

Lawton, J.H. 1990. Species richness and population dynamics of animal assemblages. Patterns in body size: abundance space. Philosophical Transactions of the Royal Society of London, B Biological Sciences 330:283–291.

Levins, R. 1969. Some demographic and genetic consequences of environmental heterogeneity for biological control. Bulletin of the Entomological Society of America 15: 237-240.

Levins, R. 1970. Extinction. Lectures on Mathematics in the Life Sciences 2:77–107.

MacArthur, R.H. 1957. On the relative abundance of bird species. Proceedings of the National Academy of Sciences of the United States of America 43:293–295.

Magurran, A.E. 1988. Ecological Diversity and Its Measurement. Croom Helm, London.

Mandelbrot, B.B. 1977. Fractals, Fun, Chance and Dimension. W.H. Freeman, San Francisco.

Mandelbrot, B.B. 1982. The Fractal Geometry of Nature. W.H. Freeman, San Francisco.

May, R.M. 1975. Patterns of species abundance and diversity. In: M.L. Cody and J.M. Diamond, eds. Ecology and Evolution of Communities, pp. 81–120. Belknap Press of Harvard University Press, Cambridge.

Morse, D.R., J.H. Lawton, M.M. Dodson, and M.H. Williamson. 1985. Fractal dimension of vegetation and the distribution of arthropod body lengths. Nature (London) 314:731–733.

Motomura, I. 1932. On the statistical treatment of communities (in Japanese). Zoological Magazine (Tokyo) 44:379–383.

Nee, S., P.H. Harvey, and R.M. May. 1991. Lifting the veil on abundance patterns. Proceedings of the Royal Society of London, B Biological Sciences 243:161–163.

Pielou, E.C. 1975. Ecological Diversity. Wiley, New York.

Preston, F.W. 1948. The commonness and rarity of species. Ecology 29:254–283.

Preston, F.W. 1962. The canonical distribution of commonness and rarity. Parts I and II. Ecology 43:185–215, 410–432.

Sepkoski, J.J. Jr. 1978. A kinetic model of Phanerozoic taxonomic diversity. I. Analysis of marine orders. Paleobiology 4:223–251.

Sepkoski, J.J. Jr. 1979. A kinetic model of Phanerozoic taxonomic diversity. II. Early Phanerozoic families and multiple equilibria. Paleobiology 5:222–251.

Sepkoski, J.J. Jr. 1984. A kinetic model of Phanerozoic taxonomic diversity. III. Post-Paleozoic families and mass extinctions. Paleobiology 10:246–267.

Sepkoski, J.J. Jr. 1988. Alpha, beta, or gamma: where does all the diversity go? Paleobiology 14:221–234.

Simpson, G.G. 1944. Tempo and Mode in Evolution. Columbia University Press, New York.

Simpson, G.G. 1953. The Major Features of Evolution. Columbia University Press, New York.

Strong, D.R. Jr. 1982. Harmonious coexistence of hispine beetles on *Heliconia* in experimental and natural communities. Ecology 63:1039–1049.

Sugihara, G. 1980. Minimal community structure: an explanation of species abundance patterns. American Naturalist 116:770–787.

Sugihara, G. and R.M. May. 1990. Application of fractals in ecology. Trends in Ecology & Evolution 5:79–86.

Tokeshi, M. 1990. Niche apportionment or random assortment: species abundance patterns revisited. Journal of Animal Ecology 59:1129–1146.

Tokeshi, M. 1992. Dynamics of distribution in animal communities: theory and analysis. Researches on Population Ecology (Kyoto) 34:249–273.

Tokeshi, M. 1993. Species abundance patterns and community structure. Advances in Ecological Research 24:111–186.

Tokeshi, M. 1994. Community ecology and patchy freshwater habitats. In: P.S. Giller, A.G. Hildrew, and D. G. Raffaelli, eds. Aquatic Ecology: Scale, Pattern and Process, pp. 63–91. Blackwell Scientific, Oxford.

Williamson, M.H. and J.H. Lawton. 1991. Fractal geometry of ecological habitats. In: S.S. Bell, E.D. McCoy, and H.R. Mushinsky, eds. Habitat Structure: Physical Arrangement of Objects in Space, pp. 69–86. Chapman & Hall, London.

Wilson, E.O., ed. 1988. Biodiversity. National Academy Press, Washington, DC.

Wilson, E.O. 1992. The Diversity of Life. Belknap Press of Harvard University Press, Cambridge.

Zipf, G.K. 1949. Human Behaviour and the Principle of Least Effort, 1st Ed. Hafner, New York.

Zipf, G.K. 1965. Human Behaviour and the Principle of Least Effort. 2nd Ed. Hafner, New York.

II

Evolutionary Causes of Biodiversity

4

Extinction and the Evolutionary Process

NILES ELDREDGE

Introduction

Extinction is currently receiving more intense scientific scrutiny than at any time since Cuvier (1812) established its empirical reality in his *Discours sur les Révolutions de la Surface du Globe*. The reasons for this heightened interest are fairly clear and appear to me to be twofold: environmental concerns over mounting species loss have increased, albeit sporadically, as the twentieth century has progressed. With the millennium now fast approaching, the next episode of mass extinction, this time human induced, is fast upon us. We simply can no longer afford *not* to focus on extinction.

The second reason for the renewed interest in extinctions comes from the fields of geology, paleontology, and, of all things, nuclear physics. I refer to the original paper by Alvarez et al. (1980) in which the *mass extinction* at the end of the Cretaceous Period some 65 million years ago—the event that defines the boundary between the Mesozoic and Cenozoic eras—was linked, via high concentrations of the rare earth element iridium, to a possible collision between the earth and one or more large extraterrestrial objects. In general, the geological time scale (Fig. 4.1) is the single best evidence that extinctions, some truly global in scale and massive in biotic impact, have occurred repeatedly during the past 530 million or so years of complex life on earth. Divisions of geological time are empirical, and reflect, for the most part, extinction events of varying degrees of severity.

These are fairly straightforward reasons why the topic of extinction has undergone something of a renaissance in scientific circles. But why was extinction all but ignored for the better part of two entire centuries until just recently? To be sure, there has been sporadic interest in the subject, but it remains true that little progress toward understanding causes of the mass extinctions of the past, their possible connections with present-day extinction events, and the relation between extinction and the evolutionary process was made throughout the late nineteenth and the majority of the twentieth century. For example, invertebrate paleontologist Norman D. Newell, of the American Museum of Natural History, was virtually alone among midtwentieth-century scholars in his attempts (e.g., Newell 1967) to bring the importance of extinction events in the history of life to wider attention.

Millions of years before the present	ERAS	PERIODS	EPOCHS	Duration of Eras (in millions of years)
.01	CENOZOIC	Quaternary	Recent	65
1.65			Pleistocene	
5		Tertiary	Pliocene	
23			Miocene	
35			Oligocene	
56.6			Eocene	
65			Paleocene	
145	MESOZOIC	Cretaceous		180
208		Jurassic		
245		Triassic		
290	PALEOZOIC	Permian	Carboniferous	325
323		Pennsylvanian		
362		Mississippian		
408		Devonian		
439		Silurian		
510		Ordovician		
570		Cambrian		
	PRE-CAMBRIAN			3,990
4.550				

FIGURE 4.1. A simplified chart of geological time. Arrows indicate episodes of global mass extinction. In general, divisions of geological time since the beginning of the Paleozoic are based on extinction and subsequent evolutionary diversifications of varying magnitude. Note that recent evidence suggests a date closer to 530 million years for the beginning of the Cambrian Period. (From Eldredge 1991, p. xvi.)

I believe there is a direct correlation, a cause-and-effect relation, between the rise of Darwinian evolutionary theory and the eclipse of extinction theory. Darwin (1859) was writing primarily to establish the simple validity of the proposition that all life has descended from a single common ancestor by a process he called "descent with modification" and which we all universally know as "evolution." Darwin, it is widely claimed, succeeded where others before him had failed in establishing the scientific respectability of the very notion of evolution because he was able to specify a credible mechanism—natural selection—that could account for the observed diversity of life. I do note that natural selection sustained withering attacks among Darwin's critics, many of whom, however, continued to accept the more general notion of evolution itself. But it is nonetheless true that Darwin *did* succeed and that natural selection itself has survived all its challenges. I regard natural selection as an intensely corroborated proposition, tantamount to a "law" of nature.

Thus, Darwin transformed world thinking from "evolution is impossible" to "evolution is inevitable," given the mere passage of time and modification of environments, itself an inevitability discovered by Darwin's contemporaries in the fledgling science of geology. Many years later, George Gaylord Simpson (1953) echoed the judgment of nearly a full century of post-1859 research when he exclaimed that natural selection is so powerful that it is impossible to imagine evolution *not* occurring.

The fossil record has proven to be something of an embarrassment to the original Darwinian account. Darwin (1859) himself devoted two entire chapters to geology and paleontology in his *On the Origin of Species*. "On the Imperfection of the Geological Record" (Chapter 9) presented Darwin's case for the enormity of geological time needed for natural selection to produce the stunning diversity of the modern biota. Darwin also established the modern science of taphonomy in this chapter. His discussion is a brilliant exegesis of all the pitfalls of preservation and discovery of fossils that Darwin developed to explain the apparent lack of many examples of the "insensibly graded series" which he believed had to be the necessary consequence of evolution through natural selection. Rashly (and wholly unnecessarily), Darwin predicted that his entire theory would stand or fall on the eventual recovery of many examples of gradual evolutionary change once the infant science of paleontology had time to become firmly established.

In the following chapter, "On the Geological Succession of Organic Beings," Darwin (1859) argued his case for the gradual evolution of species. It is in this chapter that he confronted the subject of extinction, effectively dismissing the writings of Cuvier and other early naturalists who argued for the existence of catastrophic, cross-genealogical extinctions. Instead, Darwin attributed extinction largely to the inability of primitive, ancestral taxa to compete effectively with more highly developed descendant counterparts.

And thus the stage was set: biologists ever since Darwin have seen extinction as a topic of at best minor importance, a side effect of the evolutionary process itself. Extinction became a minor topic in evolutionary discourse. Its effects are basically nil; its causes are simply the evolutionary success of other groups. New groups drive out the old, and extinction is caused by purely biological factors. Evolution is a *positive* process, while extinction is merely the inevitable disappearance of the no longer latest evolutionary model. Small wonder that extinction had attracted so little attention until comparatively recently!

But all this has changed radically in the past 20 years. Stasis—the strong tendency of species to remain relatively stable throughout their histories—is the most critical empirical generalization underlying the notion of "punctuated equilibria" (Eldredge and Gould 1972; see Eldredge 1985a, 1995 for general accounts, and Gould and Eldredge 1993 for a recent discussion). The paleontological and evolutionary biological communities now (for the most part) agree that species do not illustrate the "evolutionary change is inevitable given the mere passage of time" sort of evolutionary pattern that Darwin insisted must be the cornerstone of his theory. Long since out of its infancy, paleontology avers the great stability of species, a pattern that shines through despite the acknowledged pitfalls of fossilization. The fossil record is gappy, but the great stability of species, far from being an artifact of those gaps, is manifest in spite of all the problems inherent in preservation and recovery of fossils.

So too with extinction. I have already alluded to the "empirical" underpinnings of extinction phenomena. The old pre-Darwinian geologists and paleontologists (all amateur naturalists, in reality) were right when they emphasized the often-abrupt disappearance of most elements of entire faunas—followed by the equally abrupt appearance of entirely different faunal elements. They were wrong when they asserted (as Cuvier, for example, held) that such catastrophes completely wiped out all existing species and that consequent appearances of new forms represent the independent creative acts of some supernatural Divine Being.

We now know, of course, that so far at least some species have managed to survive even the most globally devastating mass extinction event. D. M. Raup (1979) has estimated that perhaps as many as 96% of all species extant 245 million years ago may have become extinct, therein suggesting an altogether different interpretation of the relation between extinction and evolution. Imagine restocking the world's ecosystems with new taxa *derived from just 4% of the genetic variation present in the immediately preceding set of ecosystems!* We now begin to see that extinction is far from a simple, unimportant consequence of evolution. Indeed, we begin to see something of the reverse: *evolution, in many ways, seems absolutely contingent on episodes of extinction.*

We do acknowledge that extinction may be caused by biological agencies. But cross-genealogical, ecosystem-wide extinctions, the sort that have been by far the most important in the history of life, are caused by loss of habitat through physical environmental change: bolide impacts (perhaps), massive volcanism (perhaps), physical changes engendered by plate tectonics, and—most commonly—episodes of climate change. Most extinction, in other words, is rooted in abiotic, physico-chemical degradations of habitat, and not in the sort of biotic interactions Darwin and his descendants have traditionally maintained.

Extinction and Evolution: Some Preliminary Considerations

In switching the Darwinian argument 180°—by suggesting that evolution is more a function of extinction than the other way around—we need first to establish a few caveats and disclaimers. Jablonski (1986) made a useful distinction between background extinction and true mass extinction. For that matter, a similar distinction might be made between "background" evolution (especially speciation) and the grand pulses of evolution affecting many different lineages simultaneously, the very pulses that follow so heavily on the heels of episodes of true mass extinction.

In other words, Darwin was by no means entirely wrong when he maintained that evolution and extinction are both going on at all times. Certainly natural selection is operating within populations on a generation-by-generation basis. So, too, is speciation occurring within isolated lineages from time to time, with no apparent correlation with prior extinction (of close relatives or of more general components of a biota). And extinction, the occasional disappearance of a species through any number of possible causes, is also occurring with regularity, albeit at relatively low levels of incidence. This is Jablonski's "background extinction." Coupled with "background speciation," these phenomena make it clear that *not all evolution is contingent on extinction.* I will continue to make this distinction throughout this chapter, especially in conjunction with some data that are presented later. Suffice to say at this juncture that to claim that evolution is heavily contingent on extinction is by no means to claim that all evolution is so contingent.

In a similar vein, historical data clearly reveal that background extinction and evolution both proceed at vastly lower rates than those observed in mass extinction episodes and during the ensuing "rebound" period when evolutionary rates are very high indeed. The causes of species stasis are currently hotly debated, and discussion of this phenomenon lies outside the purview of this chapter. However, the observed, comparatively very slow rates of background extinction and evolution during times of "ecological normalcy," related to the stasis phenomenon, bear some mention. For it seems clear, at least in general terms, that evolution (and extinction) rates are quite low in such times because component species and ecosystems are successful—and remain intact, largely unchanged for even millions of years, even in the face of environmental change.

Why does ecological stability—the virtually intact persistence of species associated in ecosystem context—inhibit evolutionary rate? The temptation is great simply to assert that there is no "room," in an ecological sense, for new species to arise, or certainly to become permanently established. And, in the main, this is the explanation I prefer. But here again we must exercise caution: nature is not so mechanistic that all successful speciation is contingent on prior extinction of another species in some rigidly constructed ecosystem. Indeed, Sepkoski et al. (1981) have presented convincing evidence that for the marine biosphere, at least, diversity has indeed *increased,* at least since the mid-Mesozoic. More recently, Benton (1995) reported that the fossil record indicates that diversity of microbes, algae, fungi, protists, plants, and animals has been increasing exponentially in both terrestrial and marine realms since the end of the Precambrian. Such data indicate that there are no absolute limits to "niche" numbers forever locked into the evolution–extinction equation. We must take care to avoid formulating rigid models that make absolute rules about the contingency of evolution on prior events of extinction.

Extinction and Evolution: Three Generalizations

With these provisos borne in mind, we can now explore some generalizations of the dependency of evolution on extinction that seem to hold true. My plan is simply to list these three generalizations and then to present examples that illustrate each of these general points.

Magnitude of Taxonomic Effect

Mass extinctions are cross-genealogical, ecosystem affairs. However, we measure extinction by counting up numbers of species that disappear. The more species affected, the greater the likelihood that taxa of higher rank—genera, families, orders, even classes—will become extinct. *In general, the more global and encompassing a mass extinction event, the greater the number and ranks of higher taxa that become extinct.* By the same token, in the phase of evolutionary diversification following a mass extinction event, *the number and rank of higher taxa that newly evolve is generally proportional to those eliminated in the extinction event.*

There are, of course, difficulties with this generalization. For one, in a very real sense it is only species that evolve, not higher taxa. Higher taxa are simply monophyletic skeins of species. Another difficulty arises from precladistic taxonomic practice, in which not only do taxa need not be monophyletic (thus biasing some of our data pertaining to mass extinctions), but rank designations tend to be arbitrary, thus potentially only artificially "proportional," as stated in this generalization.

These difficulties aside, I present two cases (dinosaurs vs. mammals; Paleozoic vs. modern corals) in which extinction and diversification are arguably at comparable levels and are reflected as such in the extinction and subsequent evolution of taxa of roughly equal categorical rank. Taxonomy here is just a rough estimator of the real principle: subsequent evolutionary diversification, however measured, is proportional to prior extinction.

Lag Effect

The greater the magnitude of an extinction event, the longer the period of time required for evolutionary diversification. Local extirpation of species (whether through true extinction or emigration; see following, Case History No. 4) is generally followed by rapid immigration and speciation. In geological terms, the time lag is often hardly measurable; apparently only a few thousands of years are required.

Time lags following more massive, global extinction events, however, frequently encompass millions of years before new ecosystems are established, staffed by newly evolved taxa. For example, the Lower Triassic Scythian stage, with its unusual fauna hardly typical of full-blown, normal marine benthic communities, is estimated to have lasted some 4 million years. Only after the Scythian did marine benthic communities assume the general configuration typical of much of the rest of the Mesozoic.

The Ecological Wheel

Mass extinctions result from the dismantling of ecosystems engendered through physical, usually climatic, causes. Correspondingly, subsequent periods of evolutionary diversification are largely realized in terms of developing replacement ecosystems.

Such periods seldom entail the origin of entirely novel adaptations. There are counterexamples to the generalization that evolution is largely contingent on prior extinction. Some higher taxa (such as teleost fishes, angiosperm plants) appear to have originated independently of prior extinction events. In accordance with Simpson's (1959) notion of "key innovations," some higher taxa are successful groups whose diversification appears to have hinged on the acquisition of a novel adaptation.

Indeed, postextinction evolutionary diversification rarely seems to have yielded wholly novel adaptations. Rather, the smaller in scale the extinction event, the more similar the subsequent ecosystems, composed largely of close relatives of recently extinct species. The greater the extinction, the more dissimilar the taxa that evolve, and the more dissimilar the resulting ecosystems. There are simply greater similarities between successive benthic marine faunas of the Paleozoic of eastern North America (see following example, Case History No. 3) than there are between benthic marine communities following episodes of truly global mass extinction. *Nonetheless, whatever the degree of taxonomic propinquity and similarity of community structure, the rebound evolutionary period following extinction amounts to "reinventing the ecological wheel"—setting up basic, essentially similarly structured ecological systems rather than the appearance of truly novel adaptations.*

Each of the following four case histories illustrates all three of these generalizations.

Case History No. 1: Vertebrates and the Cretaceous–Tertiary Boundary
Extinction Event

In the movie "Godzilla vs. King Kong," the future of the world is threatened by the unstoppable, destructive monster Godzilla, a fire-breathing version of *Tyrannosaurus rex*. The only creature capable of standing up to this fantastic being, which is able to withstand the most powerful of human weaponry, is King Kong—the equally imaginary gigantic ape of the world of motion pictures. And at first, things do not go at all well, because Godzilla is far stronger and far more dangerously equipped than the big ape. Until, of course, King Kong brings his superior intellect to bear on the problem. Coupled with his own prodigious strength, King Kong's brain saves the day, and the world is rid of that hulking reptilian menace, Godzilla.

Pure fantasy, of course, but this scenario draws heavily on traditional views of the evolutionary process in general, and on the nature of the transition from a dinosaur- to a mammal-dominated world in particular. The Darwinian tradition proclaims the inevitable transformation from the relatively less to the relatively more advanced—and "advanced" means "improved." In this tradition, it was inevitable that mammals, with their superior intelligence and reproductive capacities, would evolve from more primitive "reptiles." And it was further inevitable that mammals, once evolved, would eventually supplant the more primitive, less sophisticated reptiles. The Godzilla–King Kong battle of movie fiction was simply a fanciful enactment, at the individual gladiator level, of the general supposition of what happened when dinosaurs eventually gave way and mammals inherited the earth.

Charles Lyell, whose uniformitarian principles were so important to the formulation of Darwin's own evolutionary views, resisted accepting the very idea of evolution, much to Darwin's dismay and chagrin. It was Lyell who pointed out that mammals and dinosaurs both appeared at the same time in Upper Triassic sediments in Europe. Well over a century later, all the newer data have not altered the picture very much: the earliest known mammals and dinosaurs still date from the Upper Triassic.

For reasons that remain obscure, it was the dinosaurs, not the mammals, that radiated and became the dominant vertebrate components of Mesozoic terrestrial ecosystems. Collateral relatives—flying reptiles, plus plesiosaurs, ichthyosaurs and mosasaurs (the latter true lizards)—became important vertebrate constituents of aerial and aquatic ecosystems.

There was a major extinction spasm at the end of the Triassic, and many other significant extinction events at intervals throughout the remainder of the Mesozoic, a period covering some 170 million years. Each time, it was the dinosaurs and related reptilian groups, although severely cut back, that came back to reoccupy rebuilt ecosystems.

For all this time, mammals, although represented in several major distinct lineages, remained minor elements of most Mesozoic ecosystems. Mammals remained for the most part small, presumably nocturnal omnivores and carnivores. There are simply no large Mesozoic mammalian herbivores or carnivores—and, for that matter, no Mesozoic whales or bats. All those equivalent ecological roles were fulfilled by dinosaurian and other reptilian lineages.

Until, that is, the final days of the Cretaceous period. Whatever caused the global mass extinction at the end of the Mesozoic [the extraterrestrial bolide hypothesis of Alvarez et al. (1980) remains the leading contender], the last of the dinosaurs and congeners fell victim to extinction. After a 2- or 3-million-year lag, as terrestrial ecosystems began to reform, there was at last an entirely new cast of characters: mammals had, finally, inherited the earth.

Clearly, mammals "inherited" the earth, not because of inherent superiority, but through sheer luck. Smaller bodied taxa, for reasons not entirely clear, seem to have a higher probability of surviving episodes of mass extinction than relatively larger bodied taxa. Mammals survived (as did crocodiles, lizards, snakes, turtles, and birds among "Reptilia"). But this time, in the absence of competition from dinosaurs, which had heretofore always rebounded, evolving and restaffing Mesozoic ecosystems, it was the mammals that quickly radiated. By the Upper Paleocene, there were several groups of large-bodied mammals, all destined quickly to become extinct, and eventually replaced in Eocene times by the actual forerunners of modern mammalian lineages. There is no question about it: mammals radiated into their present diverse array only after dinosaurs had become extinct. The great Tertiary mammalian radiations could only have happened because the dinosaurs had become extinct.

This is Gould's (e.g., 1989) argument of *contingency:* the Godzilla–King Kong version of evolutionary events, including especially the relation between extinction and evolution, is simply incorrect.

Case History No. 2: Paleozoic Versus Modern Corals

Corals form an important component of the marine fossil record from Middle Ordovician times (approximately 470 million years ago) more or less continuously on up through to the present. The anthozoan order Rugosa—also known as "horn" or "tetra" corals—are the corals of the Paleozoic, on occasion forming reefs as massive as those of the Australian Great Barrier Reef of today. Another Paleozoic group, the order Tabulata, have also traditionally been considered as corals, but recent work strongly suggests that the tabulate "corals" more likely are allied to calcareous sponges.

Rugose corals became extinct at the end of the Paleozoic, victims of the great Permo-Triassic extinction that so far has been the most devastating of all mass extinction events in the history of life. After a lag of some 4 million years, scleractinian corals appear in the Lower Triassic and continue to the Recent as our modern corals.

The differences between rugosan and scleractinian corals are simple, yet profound. The fundamental symmetry of the septa (the radiating stony elements that support the soft tissue mesenteries of the coralline body) is fourfold in "tetracorals." In contrast, scleractinians display hexagonal symmetry, hence their informal name, "hexacorals." Moreover, the mineralogy of the calcium carbonate coralline skeletons differs in the two groups: rugose corals secreted the relatively more stable form calcite, while the scleractinian corallite is constructed of aragonite.

Paleontologists have debated the relations between these two groups of corals for years. Despite lack of intermediates, and the large morphological and temporal gaps between the two groups, most paleontologists have nonetheless assumed there to have been an ancestral–descendant relationship directly linking the Paleozoic Rugosa and the Mesozoic-Recent Scleractinia.

Some early paleontologists, however, noted the close similarity in body plan between Actiniaria (sea anemones) and Scleractinia, declaring them close relatives and the Rugosa to be only distantly related. New evidence, in the form of the sixfold symmetry of corallites discovered in Cambrian rocks of Australia, supports this latter interpretation.

In brief, it now seems most likely that Anthozoa of fourfold and sixfold symmetry existed in early Paleozoic times. Noncalcified sea anemones are virtually nonexistent in the Paleozoic fossil record. The conclusion is that, after a brief "experiment" in calcification by members of the sixfold symmetry lineage, the coral habit was exploited by fourfold symmetry Anthozoa: the Rugosa. Once again, as in the case of the Mesozoic dinosaurs, it was only after the final demise of the Rugosa 245 million years ago that corals were, in effect, "reinvented" in the evolutionary process. This time, a lineage within the sixfold symmetry group once again developed the ability to secrete a calcified exoskeleton. But this time, in the

Triassic, this lineage became successful, radiating into the diverse array of fossil and Recent Scleractinia. The Scleractinia are not derived from the Rugosa. But they did not, and presumably could not, have evolved until after the successful corals of the Paleozoic, the Rugosa, had been driven to extinction.

Of Species and Ecosystems: Dynamics of the Extinction–Evolution Interaction

Vivid as have been the gross effects of extinction on subsequent evolutionary episodes in the history of life, it nonetheless remains true that both ecological and evolutionary processes take place at the level of populations and species. I have examined the relation between ecological and evolutionary entities and processes in detail in recent years (e.g., Eldredge 1985b, 1989; Eldredge and Salthe 1984; Vrba and Eldredge 1984). To elucidate interactions between ecological and evolutionary (or "economic" versus "genealogical") realms, it is first necessary to understand the essential separateness of the ecological and genealogical (evolutionary) hierarchies. In particular, it is important to grasp the differences between entities that comprise these separate systems.

To summarize briefly, organisms engage in two classes of activity: (1) *economic* actions, which pertain to obtaining and utilizing energy resources to differentiate, grow, and maintain the soma; and (2) *genealogical* actions, i.e., reproduction. As a direct consequence, organisms find themselves parts of two larger sorts of systems. On the one hand, conspecific sexually reproducing organisms are parts of local populations (*avatars;* Damuth 1985), which themselves are integrated, functional parts of local ecosystems (or communities, depending on precise ecological perspective); local ecosystems are integrated into regional ecosystems and ultimately into the entire global economic system. In contrast, reproductive activities cause sexually reproducing organisms to be parts of local breeding populations *(demes),* which are parts of species, which themselves are parts of large-scale monophyletic groups, the so-called higher taxa.

One consequence of this general organization of biological nature is especially critical to our detailed analysis of the relation between extinction and evolution. *Species as entire entities are not parts of ecosystems.* Parts of species (avatars) are parts of ecosystems. Thus, in my following discussion, when I discuss the number of species present in Paleozoic faunas, what is meant is that each species is integrated at the population level within local ecosystems. The distinction between populations and entire species is essential to determining how evolution responds to extinction events.

Case History No. 3: "Coordinated Stasis" in Paleozoic Faunas

My goal in this section is simply to establish the very general pattern of extinction followed by evolution at the species level. To this end, I simply present some statistics generated by the work of Carlton Brett, of the University of Rochester, and his colleagues.

The original example of "punctuated equilibria" that came from my own work concerned the evolutionary history of the *Phacops rana* species complex (Eldredge 1972; see Eldredge 1971, 1985a). *Phacops rana* and its close relatives are Middle Devonian trilobites. The geological setting is the Hamilton Group of New York state and its correlatives in adjacent regions to the south along the Appalachian Mountains, and stretching west approximately 1000 miles to the central regions of the North American continent. The temporal interval spans some 5 or 6 million years, beginning roughly 380 million years ago. The species *Phacops rana* exhibits stasis throughout nearly this entire interval of time.

Preliminary results from a current exhaustive survey of the Hamilton fauna suggest that most of the nearly 300 invertebrate species of that fauna likewise span the entire interval; most show little or no sign of evolutionary change. Brett and Baird (1995) call this pattern "coordinated stasis," in which virtually all known species-level taxa are in apparent stasis. Ecosystems (or at least recurrent assemblages of species populations) are likewise markedly stable, with similar communities appearing over and over again in comparable habitats throughout the interval.

Of especial interest here is the fact that the Hamilton Group fauna is but one of a series of 10 such faunas that span a geological interval of approximately 65 million years, from the Silurian on up through the Middle Devonian. Each fauna contains at least 50 to as many as 335 documented species. Each fauna lasts from 5 to 7 million years. In each fauna, from 70% to as many as 85% of the species are present throughout the entire interval. Background extinction and evolution (speciation) both occur, but are minor aspects of the pattern when compared to the basic stable persistence of such a high percentage of the component species taxa.

And there is one final, compelling statistic to this story: *there is, on average, only 20% species overlap between adjacent faunas.* Only 20% of the species of one interval survive into the next interval, nor is there much of a detectable temporal lag between the disappearance of the species of an earlier fauna and the appearance of those of the next succeeding fauna. At the species level, temporal gaps become almost too small to be easily detected, at least between the temporally remote faunas of the Paleozoic.

Thus, we have stability interrupted, at fairly regular intervals, by the abrupt simultaneous disappearance of species that had remained collectively nearly unchanged for an average of 6 million years. We have the very abrupt appearance of a large number of new species immediately following this abrupt disappearance of the older species. And this happens no fewer than 10 times with great regularity throughout a long segment of Middle Paleozoic time. Indeed, the pattern is only interrupted by the eventual complete elimination of marine environmental conditions in the Appalachian regions.

This pattern of successive faunas at the species level is the rule, not the exception, for the entire Phanerozoic, the past 530 million years or so since the advent of multicellular life and the beginnings of a dense, rich fossil record. But it is dif-

ficult to decipher mechanisms in detail in a fossil record as far back as the Paleozoic, hundreds of millions of years ago. The actual dynamic mechanisms at work
in such faunal turnovers are more easily grasped with more recent examples,
involving more familiar species and ecosystems.

Case History No. 4: Miocene African Ecosystems and Vrba's Turnover
Pulse Hypothesis

The first of four major glacial advances that mark the Pleistocene epoch began
approximately 1.6 million years ago. Recently, however, evidence of an earlier,
abrupt drop in global temperatures about 2.5 million years ago has come to light.
Whether or not this earlier phase of global cooling was accompanied by growth of
continental ice sheets is not yet determined, but the ecosystems of many parts of
the world were in any case severely effected by this event.

Vrba (1985, 1993) has discussed the effects of this event on African ecosystems and on apparently correlative patterns of extinction and evolution. In sum,
she has found that about 2.5 million years ago there was a sudden, dramatic
switch-over from tropical forest to more open grasslands typical of a more arid
climate. The fossil record, especially sediments deposited in the newly opened
East African Rift System, documents this abrupt change in stark terms. Sudden
shifts in pollen content at 2.5 million years ago record the critical change in predominant vegetation. And Vrba (1985) records a concomitant revolution in the
vertebrate fauna that is just as striking and pronounced. Interestingly, even early
hominids appear to have been affected: *Australopithecus africanus* disappears
from the record at this time, while earliest members of the *Paranthropus robustus*
lineage (in the form of *Paranthropus aethiopicus*—especially the famous "black
skull") and the earliest *Homo habilis* (with stone tools) are now reliably dated as
appearing 2.5 million years ago.

The ecological requirements of modern African vertebrates are, of course, well
documented and shed great light on the needs and behaviors of their close
Pliocene relatives. Many woodland species abruptly drop from the record at
around 2.5 million years, and Vrba points out the two reasons why this is so.
Familiar habitat is lost, and some species undoubtedly became extinct right then.
Others, however, certainly persisted elsewhere (where the fossil record is poor or
altogether lacking) because they were able to locate familiar habitat.

Indeed, such habitat tracking, in my opinion, is a major cause of species stasis.
The old Darwinian notion that environmental change will automatically lead to
evolutionary change as natural selection "tracks" the environment has now
yielded to the realization that what is actually tracked is shifting habitats: populations of species will persist unchanged so long as recognizable habitat can be
found by at least some of their component populations. Thus, Vrba concludes, disappearance of species from the record need not imply total extinction in all cases;
local extinction is a part of the pattern as well.

The fossil record of 2.5 million years ago in the African Rift resembles the
boundaries between faunas in the Paleozoic more than 300 million years ago: not

only do most long-lasting species abruptly disappear, but "new" species suddenly appear to take their place. New ecosystems are constructed from new "players," populations of different species that had for the most part not been present just before 2.5 million years ago.

Where do those new players come from? Logically enough, Vrba sees the parallels between the earlier disappearances of the older species and their replacement by new species. Some must come in from elsewhere, simply tracking their own familiar habitats. This has been documented for a number of taxa, and is especially clear in an even earlier event 5 million years ago, when Eurasian antelopes first invaded the African continent (Vrba, 1993).

But, to complete the picture, Vrba (1985) makes the very important point that habitat disruption and degradation, which so easily lead to local and sometimes total species extinction, *may also lead directly to speciation.* And this is correct: allopatric speciation entails the fragmentation of existing species into two or more divisions that may (if reproductive isolation occurs) lead to the emergence of one or more "daughter" species. Thus many of the species that make their first appearance after the 2.5-million-year sudden drop in global temperature are in fact newly evolved, as a direct response not to the extinction of earlier close relatives, but as a direct reflection of isolation wrought by radical habitat disruption.

These, then, are the basic elements of Vrba's (1985) "turnover pulse hypothesis": like the Paleozoic marine fossil record, there are many instances of abrupt faunal turnover reflecting sudden, abrupt environmental change. Many species disappear, to be replaced by "new" ones. Some of these new species are simply migrants coming in from similar habitats elsewhere, but others are newly evolved in the environmental disruptions of the moment. Likewise, many of the older species do in fact become directly extinct, but others undoubtedly persist for some time in suitable habitats elsewhere. Such patterns of turnover in the fossil record have a complex ecological and evolutionary explanation, and environmental change is thus seen to be the main driving factor in evolution.

Dynamics of Extinction and Evolution: Some Concluding Thoughts

In an earlier era of biology, it was common to speak of "evolution into empty niches." Thus, with renewed emphasis on the importance of extinction to the evolutionary process, it might be tempting simply to say that evolution is somehow an automatic response to newly vacated niches.

But species do not have niches: as I have briefly outlined, species are not ecological entities. Vrba's turnover pulse hypothesis helps greatly in this regard, for we can now see that speciation is not a direct response to extinction, but rather an outcome of habitat fragmentation.

How could this, after all, be otherwise? Speciation is quintessentially the process of formation of new reproductive communities. That is what species *are:*

reproductive communities, genealogical entities. Niches are an economic concept: the economic role played by local populations within local communities or ecosystems. There is no conceivable way an "economic opportunity" could *cause* the origin of a new reproductive community. Thus Vrba's suggestion of the role of habitat fragmentation nicely removes a serious obstacle to understanding the actual mechanisms at work when extinction occurs and new species evolve apparently in their stead.

But there is one lingering difficulty: the temporal lags. The more localized a faunal turnover, the less apparent are the lags. But there is always some lag, and, as we have seen, it may entail millions of years in instances of truly global mass extinction. How do we reconcile such lags with Vrba's model of more-or-less immediate speciational response to environmental change and habitat fragmentation?

Consider the fate of most fledgling species: small populations, unless ecologically well differentiated from the parental species, stand little chance of becoming established, of surviving long enough, becoming successful enough, to be encountered in the fossil record. The fate of most fledgling species, I propose, is extinction, simply because most fledgling species are insufficiently differentiated in terms of basic economic requirements of component organisms to allow survival.

However, after major ecosystem disruption and extinction, circumstances are quite different: I imagine that the survival probability of fledgling species must increase enormously. [I concede that such a proposition is difficult in principle to test, but I (Eldredge, 1989) have cited the work of Lewis (1966) that seems to me to bear directly on this idea.] The rate of speciation remains the same, but the rate of accumulation of new species will tend to grow logarithmically, rather slowly at first and later at a more rapid pace, all a reflection of enhanced survival probability of new species following episodes of extinction.

Such is the state of existing theory. No doubt we will continue to improve our explanations of the evolutionary responses to episodes of extinction. But we have taken the very important first step: we have looked to nature—and especially to the fossil record—and have concluded that there are patterns there worthy of our most serious scientific attention. The interaction between extinction and evolution is complex, but we have finally conceded that the subject is vital to a full understanding of both the ecological and evolutionary workings of nature.

Literature Cited

Alvarez, L.W., W. Alvarez, F. Asaro, and H.V. Michel. 1980. Extraterrestrial cause for the Cretaceous–Tertiary extinction. Science 208:1095–1108.

Benton, M. J. 1995. Diversification and extinction in the history of life. Science 268:52–58.

Brett, C.E. and G.C. Baird. 1995. Coordinated stasis and evolutionary ecology of Silurian to Middle Devonian faunas in the Appalachian Basin. In: D.H. Erwin and R.L. Anstey, eds. New Approaches to Speciation in the Fossil Record, pp. 285–315. Columbia University Press, New York.

Cuvier, G. 1812. Discours sur les Révolutions de la Surface du Globe. (Originally published as Vol. 1, Discours Preliminaire, Recherches sur les Ossemens Fossiles.) Deterville, Paris.

Damuth, J. 1985. Selection among "species": a formulation in terms of natural functional units. Evolution 39:1132–1146.

Darwin, C. 1859. On the Origin of Species. John Murray, London.

Eldredge, N. 1971. The allopatric model and phylogeny in Paleozoic invertebrates. Evolution 25:156–167.

Eldredge, N. 1972. Systematics and evolution of *Phacops rana* (Green, 1832) and *Phacops iowensis* Delo, 1935 (Trilobita) from the Middle Devonian of North America. Bulletin of the American Museum of Natural History 147:45–114.

Eldredge, N. 1985a. Time Frames. Simon and Schuster, New York.

Eldredge, N. 1985b. Unfinished Synthesis. Biological Hierarchies and Modern Evolutionary Thought. Oxford University Press, New York.

Eldredge, N. 1989. Macroevolutionary Dynamics. Species, Niches and Adaptive Peaks. McGraw-Hill, New York.

Eldredge, N. 1991. Fossils. The Evolution and Extinction of Species. Abrams, New York.

Eldredge, N. 1995. Reinventing Darwin. The Great Debate at the High Table of Evolutionary Theory. Wiley, New York.

Eldredge, N. and S.J. Gould. 1972. Punctuated equilibria: an alternative to phyletic gradualism. In: T.J.M. Schopf, ed. Models in Paleobiology, pp. 82–115. Freeman, Cooper, San Francisco.

Eldredge, N. and S.N. Salthe. 1984. Hierarchy and evolution. Oxford Surveys in Evolutionary Biology 1:182–206.

Gould, S.J. 1989. Wonderful Life. The Burgess Shale and the Nature of History. W.W. Norton, New York.

Gould, S. J. and N. Eldredge. 1993. The majority of punctuated equilibrium. Nature 366(6452):223–227.

Jablonski, D. 1986. Background and mass extinctions: the alteration of macroevolutionary regimes. Science 231:129–133.

Lewis, H. 1966. Speciation in flowering plants. Science 152:167–172.

Newell, N.D. 1967. Revolutions in the history of life. Geolological Society of America, Special Paper 89:63–91.

Raup, D.M. 1979. Size of the Permo-Triassic bottleneck and its evolutionary implications. Science 206:217–218.

Sepkoski, J.J. Jr., R.K. Bambach, D.M. Raup, and J.W. Valentine. 1981. Phanerozoic marine diversity and the fossil record. Nature (London) 293:435–437.

Simpson, G.G. 1953. The Major Features of Evolution. Columbia University Press, New York.

Simpson, G.G. 1959. The nature and origin of supraspecific taxa. Cold Spring Harbor Symposia on Quantitative Biology 24:255–271.

Vrba, E.S. 1985. Environment and evolution: alternative causes of the temporal distribution of evolutionary events. South African Journal of Science 81:229–236.

Vrba, E.S. 1993. The pulse that produced us. Natural History 102(5):47–51.

Vrba, E.S. and N. Eldredge. 1984. Individuals, hierarchies and processes: towards a more complete evolutionary theory. Paleobiology 10:146–171.

5

Diversity and Evolution of Symbiotic Interactions

Norio Yamamura

Introduction

In ecological time scales and local spatial scales, biodiversity is the issue of coexistence of different species. However, in evolutionary time scales and global spatial scales, biodiversity is a result of the balance of speciation and extinction of species. In geological history, rapid speciation, called adaptive radiation, has occurred repeatedly. The origins of such newly developing taxa often evolved from the symbiosis of two different species (Price 1991). Thus, symbiosis can be considered an important source of the Earth's current biodiversity, and in this context symbiosis can be called evolutionary innovation (Margulis 1993).

In this chapter, I touch on several examples of symbiosis as a source of evolutionary innovation that was followed by adaptive radiation. Mutualistic symbioses are generally considered to evolve from parasitic relationships (Ewald 1987; Price 1991). I introduce here a mathematical model that identifies the conditions for the evolution of mutualism from parasitism, in which vertical transmission, defined as the direct transfer of infection from parent host to its progeny, is a key factor (Yamamura 1993). Finally, I discuss the evolutionary factors of mutualism without vertical transmission.

Symbiosis Followed by Adaptive Radiation

Most biologists now accept that the evolution of eukaryotes resulted from the symbiotic union of several independent ancestors (a host cell, an ancestor of mitochondria, an ancestor of the chloroplast), although the origins of other cellular organs such as the undulipodia, microtubules, centrioles, spindle, and microbody remain controversial (Maynard Smith 1989; Margulis 1993). There is no doubt that the evolution of eukaryotes provided the important source for the Earth's current biodiversity.

After the evolution of eukaryotes, there were probably several epoch-making instances in which the evolution of symbiosis was followed by adaptive radiation (Price 1991). Terrestrial plants might have evolved through the symbiosis of aquatic algae and fungi. Large terrestrial plants such as trees might also have evolved through the symbiosis of small terrestrial plants and fungi, the mycorrhizae. It is said that insects became able to use low-nutrition food through symbiosis with microorganisms that aid in digestion. The adaptive radiation of parasitic wasps is suggested to have occurred through symbiosis with a virus. Large herbivorous mammals would also have evolved through symbiosis with microorganisms that aid in digestion. According to Price (1991), the number of species originating from such symbioses constitutes 54% of all the current species on Earth.

Coral, through its acquisition of photosynthetic symbionts, brought about adaptive radiation in geological history several times (Stanley 1981). Some fish use light organs to assist vision and to give signals, and the light is generated by luminous bacteria. The adaptive radiation of fish having such light organs occurred within different taxa of euteleostei (McFall-Ngai 1991). Termite species have established a decomposer niche in tropical forests and tropical savanna, and the prosperity of termites probably results from symbiosis with microorganisms such as protozoa, bacteria, and fungi.

Evolution of Mutualistic Symbiosis

In symbiosis that results in adaptive radiation, both interacting species help each other in a surprisingly intimate manner, and this can be termed mutualistic symbiosis. How could such neat mutualism have evolved in the first place? Many authors have suggested the evolution of mutualism from parasitism. Roughgarden (1975) stated that parasites do not harm their hosts as much as they could because parasites owe their habitats and energy resources to their hosts. Ishikawa (1988) stated that parasitism naturally evolves into mutualism and described the process in a phrase: an enemy today is a friend tomorrow. Ewald (1987) also claimed that parasitism should eventually evolve toward commensalism, and commensal relationships are viewed as the raw material for evolution of mutualism.

The origin of some symbioses may be predation of a larger organism on a smaller organism. When the smaller organism evolves such that it can survive inside the larger organism, we can say that endosymbiosis has evolved. The first phase, when the smaller organism establishes a part of its life cycle in the larger organism, would result in some degree of deleterious effect on the host as the result of resource competition or simple occupation of some space inside the host. In my definition as used here, interaction with such deleterious effects on the fitness of the host is classified as parasitism. Evolution of mutualism from an origin of this type is thus included in the case of evolution from parasitism to mutualism.

The time-course of the evolution of reduced virulence has been documented in various parasites and diseases, among which the myxoma virus in Australian rabbits is well known (Fenner 1965). Jeon (1972) made a more exact observation: a virulent bacterium that infected an ameba changed into a mutualistic symbiont of the ameba by artificial selection of the surviving ameba. However, many parasites and diseases clearly continue to be virulent to their hosts (Ewald 1994). In conclusion, the reality would be that some parasites persist in antagonistic relationships with their hosts, while others have evolved into mutualistic relationships through reduction of their virulence.

Vertical Transmission

Many authors have suggested that vertical transmission, defined as the direct transfer of infection from a parent organism to its progeny, is a key factor for the evolution of mutualism that is highly beneficial to both organisms. For example, Ewald (1987) cited mutualism described by Buchner (1965), mainly between insect hosts and bacteria symbionts, and found that all of these accompany vertical transmission, such as transovarial transmission. Protozoa in termite guts are transferred among individuals through direct feces feeding (Breznak 1975). Colony-founding queens of some ants accompany their mutualistic partner, the mealy bug (Klein et al. 1992). Some parasitic wasps inject symbiotic virus into insect larvae when ovipositing (Stolts and Vinson 1979). This virus helps the parasitic wasp to defeat the defense mechanism of the host larvae. Surprisingly, the genome of the virus seems to be incorporated into the genome of the wasp (Fleming and Summers 1991).

Ewald (1987) pointed out that among other various modes of transmission, vertical transmission is the most remarkable characteristic of intimate mutualism between a symbiont and its host. Perhaps what occurred in the evolutionary history was a conflict in the vertical transmission rate between a host and a parasite. In the evolutionary sense, a host would not want its parasite to infect its progeny, while the parasite would want to infect the progeny. If the host continues to dominate this evolutionary race, the antagonistic relationship between the host and the parasite will also continue. If the parasite, however, wins the race and attains a higher vertical transmission rate, it could not continue to exploit the host so severely, because its progeny owe their lives mainly to the progeny of the host.

A Mathematical Model for the Evolution of Mutualism

I have made a simple mathematical model, based on Darwinian fitness, to identify the conditions under which evolution of mutualism from parasitism may actually occur (Yamamura 1993). The model also clarifies whether or not a long interrelationship between different species necessarily develops into evolution of mutualism, that is, whether or not an enemy today is a friend tomorrow, as Ishikawa (1988) stated.

The model is constructed as follows. An infecting parasite reproduces $a(x)$ offspring, where x is a parasite strategy that is the degree of exploitation on its host, for example, in energy units. Because the parasite can reproduce more offspring if exploiting the host more severely, $a(x)$ is an increasing function of x:

$$a(x) = a_0 + a_1 x \tag{1}$$

On the other hand, an infected host reproduces $b(x)$ offspring, where $b(x)$ is a decreasing function of x:

$$b(x) = b_0 - b_1 x \tag{2}$$

The values of a_1, b_0, and b_1 are naturally positive, but a_0 may be zero or even negative because parasites may be unable to reproduce without exploiting a host resource more than some critical amount. I extended x to negative values, which means that the parasite gives the host some positive effects in the sacrifice of its own fitness. For the negative values of x, the parasite is not parasitic but is actually beneficial to the host. Therefore, I will use "symbiont" hereafter to refer to both parasitic and beneficial cases. Because the numbers of offspring, $a(x)$ and $b(x)$, should be positive, the value of x is restricted in the interval, $-a_0/a_1 < x < b_0/b_1$. Therefore, negative values of x are admissible only when a_0 is positive.

An uninfected host reproduces b_n offspring. I discriminated b_n from b_0, the number of offspring of an infected host when $x = 0$, because I consider an evolutionary change in b_0 later. The vertical transmission rate r is defined as the proportion of offspring directly infected from their parent among $b(x)$ offspring. Offspring escaping from vertical transmission are exposed to infection by the parasites from other infected hosts along with offspring of uninfected hosts.

I have calculated, as the parasite strategy, the evolutionarily stable degree of exploitation, which is finally reached through evolution when the parameters involved remain constant. The solution x^* is a decreasing function of the vertical transmission rate r, as shown in Figure 5.1. It is easily realized that the parasite should reduce its virulence as r increases, and it should become a mutualist when r is larger than a critical value r_1 because x^* is negative in this range.

Next, I calculated the evolutionary direction of r from the point of view of both the symbiont and the host (see Fig. 5.1). When the vertical transmission rate r is controlled only by the symbiont, its value always increases. On the other hand, if it is controlled only by the host, its value decreases when r is smaller than a critical value r_2, and increases otherwise. The critical value of vertical transmission is represented as

$$r_2 = \frac{a_1 b_n}{a_1 b_0 + a_0 b_1} \tag{3}$$

This evolutionary pattern implies that there is a conflict between the symbiont and the host for smaller values of r and that there is no conflict for larger values of r. Suppose that r starts from a value in the conflict region. Once the symbiont wins the vertical transmission conflict and r increases to a value larger than r_2, one-way evolution begins toward a higher vertical transmission rate and highly mutualistic

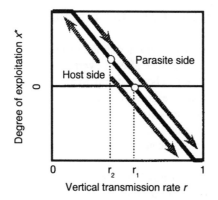

FIGURE 5.1. The evolutionary stable degree of exploitation x^* is a decreasing function of the vertical transmission rate r. There is a critical value r_1 above which the value of x^* is negative. There is another critical value, r_2, represented by Eq. 3, below which the evolutionary directions of r for the parasite and the host are conflicting, and above which they correspond to each other. When r exceeds r_2, one-way evolution begins toward a higher vertical transmission rate and highly mutualistic corporation.

cooperation because x^* is negative. Finally, the value of r reaches to 1, which indicates a perfect vertical transmission.

Even if r itself does not change, one-way evolution still begins when the critical value r_2 is reduced to a value smaller than r. The conditions whereby r_2 decreases can easily be identified from Eq. 3. The parameters a_0, a_1, b_0, and b_1 are coefficients of $a(x)$ and $b(x)$ in Eqs. 1 and 2, which may change through evolution during a long interaction of symbiosis. It is probable that a parasite increases a_0 by evolving toward use of some metabolic or digestive wastes excreted by the host. The parasite may increase a_1 by improving the conversion efficiency of exploiting resources to its fitness. A host may also evolve toward use of resources excreted by the parasite, leading to some increase in b_0, or to reduce the fitness effect of exploitation by the parasite, leading to some decrease in b_1. All the probable changes in parameters do not necessarily lead to the evolution of mutualism. It is certain that improvement of a_0 and b_0 (an effective use of the partner's disuse) favors the evolution of mutualism because r_2 is a decreasing function of a_0 and b_0. However, improvement of a_1 (the efficiency of resource usage) conversely reduces the possibility of mutualism because r_2 is an increasing function of a_1. Improvement of b_1 also has the same effect on r_2 as a_1.

The mathematical model here suggests that the vertical transmission rate is the most important factor for reduction of parasite virulence, and that initiation of the one-way evolutionary process from parasitism to mutualism may occur when the symbiont wins the conflict against the host on vertical transmission rate or when either side utilizes wastes of the partner. In the latter case, the attained mutualistic interaction can be regarded as a miniecosystem because matter circulation across different species is maintained.

Mutualism Without Vertical Transmission

In the model, vertical transmission is defined as the transfer of symbionts from generation to generation. The same effect as vertical transmission can be expected when symbionts reproduce in a host growing to a large size. For example, mycorrhizae reproduce following the growth of the roots of a tree. Because the model predicts that mutualism evolves under a high vertical transmission rate, it is likely to evolve when the host grows to a large size and lives a long life. Coral and a large colony of social insects can be categorized as such hosts with a long life span.

However, true vertical transmission is not common among this type of mutualism. Why are mycorrhizae not transmitted vertically through the seeds of the tree? More generally, why do not all mutualisms evolve into perfect vertical transmission? Mutualism without vertical transmission necessarily requires that symbionts have a free-living stage, independent of their hosts. Otherwise, newly born host individuals rarely could gain the symbionts. I discuss here three factors to explain the evolution of mutualism without vertical transmission.

The first possibility is that the apparent mutualism may actually be a parasitism or a manipulation. Leguminous bacteria might be parasitic on bean plants. The plants obtain nitrogen from bacteria, while giving sugar to the bacteria. The total effect on the plant might be negative. On the contrary, plants might manipulate bacteria and use them as tools for preparing nutrients. In this case, there is no benefit for the bacteria. When only one side gains, the other side should disturb the evolution of vertical transmission.

The second possibility is that the symbiotic association is beneficial for both partners under restricted conditions. In the phenomenon called bleaching, which is well known in coral, the photosynthetic symbionts depart from the host coral at high water temperatures (Brown and Ogden 1993). It is unclear in bleaching whether symbionts escape by themselves or the coral drives out the symbionts. In the case in which the benefits of symbiosis are temporarily changing, vertical transmission may not pay for its cost, and therefore it is less likely to evolve.

The last possibility, which I think is the most important, is that the symbiont adopts a bet-hedging strategy in which it uses the host as a refuge. Luminous bacteria can reproduce by themselves in seawater (McFall-Ngai 1991), but conditions for reproduction in seawater may fluctuate greatly. The inside of fish must be a stable environment for luminous bacteria, and it may serve like a dormant seed bank of weed grasses where a grass population can recover by germination of the dormant seeds even after a complete destruction. The host need not evolve vertical transmission in the case in which free-living symbionts are common in the habitat of the host, although the density of the free-living symbionts may fluctuate.

In conclusion, the full life cycles of most symbionts with a free-living stage have not been sufficiently investigated. In particular, studies on the mechanism of encounter and departure between the symbiont and its host are greatly lacking. Extensive ecological studies on these organisms will clarify how symbionts without vertical transmission are mutualistic to their hosts. At the same time, such studies, along with further studies of cases with vertical transmission, will give us basic biological information necessary for maintenance of biodiversity because the various symbioses involved in all ecosystems serve vital components of these ecosystems, and human activities in ignorance may disrupt parts of such important symbiotic relationships.

Summary

In geological history, rapid speciation, called adaptive radiation, has occurred repeatedly. The origins of such newly developing taxa often evolved from the symbiosis of different species. Mutualistic symbioses are generally considered to evolve from parasitic relationships. A mathematical model has suggested that the vertical transmission rate is the most important factor for reduction of parasite virulence, and that initiation of the one-way evolutionary process from parasitism to mutualism may occur when the symbiont wins the conflict against the host on vertical transmission rate, or when either side utilizes the wastes of the partner. In the latter case, the attained mutualistic interaction can be regarded as a miniecosystem because matter circulation across different species is maintained. Finally, a bet-hedging strategy of the symbiont was stressed as an evolutionary factor of mutualism without vertical transmission.

Literature Cited

Breznak, J.A. 1975. Symbiotic relationships between termites and their intestinal microbiota. Symposia of the Society for Experimental Biology 29:559–580.

Brown, B.E. and J.C. Ogden. 1993. Coral bleaching. Scientific American 268:44–50.

Buchner, P. 1965. Endosymbiosis of Animals with Plant Microorganisms. Wiley, New York.

Ewald, P.W. 1987. Transmission modes and evolution of the parasite-mutualism continuum. Annals of the New York Academy of Sciences 503(Endocytobiology III):295–305.

Ewald, P.W. 1994. Evolution of Infectious Disease. Oxford University Press, Oxford.

Fenner, F. 1965. Myxoma virus and *Oryctolagus cuniculus*. In: H. G. Baker and A. L. Stebbins, eds. The Genetics of Colonizing Species, pp. 485–501. Academic Press, New York.

Fleming, J.G.W. and M.D. Summers. 1991. Polydnavirus DNA is integrated in the DNA of its parasitoid wasp host. Proceedings of the National Academy of Sciences of the United States of America 88:9770–9774.

Ishikawa, H. 1988. Symbiosis and Evolution (in Japanese). Baihukan, Tokyo.

Jeon, K.W. 1972. Development of cellular dependence in infective organisms: microsurgical studies in amoebas. Science 176:1122–1123.

Klein, R.W., D. Kovac, A. Schellerich, and U. Maschwitz. 1992. Mealybug-carrying by swarming queens of a southeast Asian bamboo-inhabiting ant. Naturwissenschaften 79:422–423.

Margulis, L. 1993. Symbiosis in Cell Evolution, 2nd Ed. W.H. Freeman, New York.

Maynard Smith, J. 1989. Generating novelty by symbiosis. Nature (London) 341: 284–285.

McFall-Ngai, M. 1991. Luminous bacterial symbiosis in fish evolution: adaptive radiation among the leiognathid fishes. In: L. Margulis and R. Fester, eds. Symbiosis as a Source of Evolutionary Innovation, pp. 381-409. MIT Press, Cambridge.

Price, P.W. 1991. The web of life: development over 3.8 billion years of trophic relationships. In: L. Margulis and R. Fester, eds. Symbiosis As a Source of Evolutionary Innovation, pp. 262–272. MIT Press, Cambridge.

Roughgarden, J. 1975. Evolution of marine symbiosis—a simple cost-benefit model. Ecology 56:1201–1208.

Stanley, G.D. Jr. 1981. Early history of scleractinian corals and its geological consequences. Geology 9:507–511.

Stolts, D.B. and S.B. Vinson. Viruses and parasitism in insects. Advances in Virus Research 24:125–171.

Yamamura, N. 1993. Vertical transmission and evolution of mutualism from parasitism. Theoretical Population Biology 44:95–109.

6

Global Diversification of Termites Driven by the Evolution of Symbiosis and Sociality

Masahiko Higashi and Takuya Abe

Introduction

An effective way to search out the causal mechanisms that drive the macroevolution of a higher taxonomic group (e.g., an order) is to examine its past and present spatial distribution. On the evolutionary time scale, this may illustrate the dynamics of radiation and competition, pointing to those innovations that have been most crucial in achieving the contemporary diversity.

Synergetic traits, for example, mutualism and sociality, produce novel abilities that cannot be attained by the partners alone. These traits represent a contribution to biodiversity, but they may promote the diversification of a taxonomic group by causing its evolutionary (adaptive) radiation (Margulis and Fester 1991). Symbiosis between higher organisms and microorganisms, in particular, often creates capabilities for exploiting new food resources that are abundant but previously were not effectively utilized, thus opening a new niche. On the other hand, sociality enhances the efficiency of the exploitation (Wilson 1975). Either of these (or both together) may cause a rapid growth and spatial expansion of the species population of the higher organism, leading to a burst of diversification. Hymenoptera, including bees and ants, present an example of global diversification driven by the evolution of sociality and symbiosis with many animals and plants (Wilson 1971, 1975; Roubik 1989; Hölldobler and Wilson 1990). Marine corals are another example of global diversification driven by the evolution of symbiosis with zooxanthellar algae (Wilson 1975).

A third example is termites (Isoptera). Termites are superabundant in the tropic terrestrial environment and are widely spread from humid forests to the savanna and even arid areas (Lee and Wood 1971; Wood and Sands 1978; Swift et al. 1979; Brian 1983; Josens 1985; Wilson 1990; Eggleton et al. 1994). Their biomass approaches 10 g/m^2 (wet weight) where they are found (Wood and Sands 1978; Abe and Matsumoto 1979). Only a few animal groups may claim this order of biomass: humans, earthworms, herbivorous mammals in the African savanna, and ants (Fittkau and Klinge 1973; Sinclair and Norton-Griffiths 1979). The basis of this extraordinary abundance of termites lies in their highly developed social organization (Myles 1988; Noirot 1990; Wilson 1990; Nalepa 1994) and symbiosis with microorganisms (Grassé and Noirot 1959; O'Brien and Slaytor 1982;

Martin 1987; Wood and Thomas 1989; Breznak and Brune 1994). This symbiosis, together with their superabundance, brings termites to play a keystone role of "super-decomposer" and "carbon–nitrogen balancer" in the ecosystems of which they are a biotic constituent (Higashi et al. 1992). Termites consume a large portion of the dead plant material produced in terrestrial ecosystems of the tropics (Wood and Sands 1978; Matsumoto and Abe 1979; Collins 1981a, 1983a; Lepage 1981a; Martius 1994), and are in turn consumed by a great variety of animals ranging from ants and spiders to chimpanzee and humans, thus forming the basis for a large food web (Longhurst et al. 1978; Lepage 1981b; Deligne et al. 1981). Furthermore, they are ecosystem engineers (Jones et al. 1994; Lawton 1994; see also J. Lawton, Chapter 12, this volume) that modify the soil structure by constructing mounds and subterranean nests and galleries (Lee and Wood 1971; Wood and Sands 1978; Lal 1987; Wood 1988), providing many species of animals and plants with diverse habitats (Glover et al. 1964). Because of this keystone role, the global diversity of termites may be of great importance in understanding carbon fluxes between tropical land masses and the atmosphere with respect to both carbon dioxide and methane (Zimmerman et al. 1982; Collins and Wood 1984; Khalil et al. 1990).

The main objective of this chapter is to examine the theme of global diversification driven by the evolution of sociality and symbiosis in termites, which may serve as a model for generalization.

The Symbiosis and Sociality of Termites

Introduction

Termites, consisting of about 2200 species and 7 families (Fig. 6.1), are divided into two large groups: lower termites (Mastotermitidae, Kalotermitidae, Termopsidae, Hodotermitidae, Rhinotermitidae, and Seritermitidae) and higher termites (Termitidae). The lower termites are characterized by the presence of cellulolytic Protozoa in their hindgut, and mainly consume wood, while the higher termites are characterized by the absence of those Protozoa, and variously consume a range of dead and decaying plant material, including sound wood, standing and fallen plant shoots and leaves, decaying wood, and soil (Grassé and Noirot 1959; Wood 1978). The fossil records of certain termites (Termopsidae) can be traced back to the Cretaceous period of the Mesozoic era (Emerson 1967; Jarzembowski 1981), and the origin of termites might be traced back to the Triassic period (Emerson 1955). The higher termites, consisting of one family (the Termitidae), are the richest in species diversity (1639 species), occupying 75% of all species, while the Kalotermitidae (consisting of 350 species) are richest in species diversity among the lower termite families (Wood and Johnson 1986).

With the exception of species that show intermediate life styles, most termite species can be classified into two types (Abe 1984, 1987): those that forage away

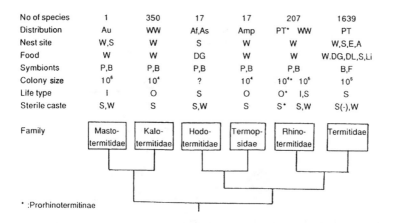

No of species	1	350	17	17	207	1639
Distribution	Au	WW	Af,As	Amp	PT* WW	PT
Nest site	W,S	W	S	W	W	W,S,E,A
Food	W	W	DG	W	W	W,DG,DL,S,Li
Symbionts	P,B	P,B	P,B	P,B	P,B	B,F
Colony size	10⁶	10⁴	?	10⁴	10⁴* 10⁶	10⁶
Life type	I	O	S	O	O* I,S	S
Sterile caste	S,W	S	S,W	S	S* S,W	S(-),W

Family	Masto-termitidae	Kalo-termitidae	Hodo-termitidae	Termop-sidae	Rhino-termitidae	Termitidae

* :Prorhinotermitinae

FIGURE 6.1. Six families of termites and a summary of their characterization in terms of distribution, nest site, food, symbionts, colony size, life type, and sterile caste. Distribution: *Au*, Australia; *WW*, worldwide; *Af*, Africa; *As*, Asia; *Amp*, amphitropical; *PT*, pantropical. Nest site: *W*, wood; *S*, subterranean; *E*, epigeal mound; *A*, arboreal. Food: *W*, wood; *DG*, dead grasses; *DL*, dead leaves; *S*, soil; *Li*, lichen. Symbionts: *P*, Protozoa; *B*, bacteria; *F*, fungi. Life type: *O*, one-piece; *I*, intermediate; *S*, separate. Sterile caste: *S*, soldier; *W*, worker. The number of species is after Wood and Johnson (1986).

from their nests, the so-called separate type, and those which nest in wood (e.g., a dead limb of a tree) and consume it as food, the so-called one-piece type. Intermediate-type termites nest in wood and consume it as food, but also construct galleries in the soil or on the ground for outside foraging. Figure 6.2 illustrates this classification into three life types. The maximum colony size of separate-type termites is some millions, while that of the one-piece type is at most 10,000 (Brian 1983; Maki and Abe 1986 ; Darlington 1990; Lenz 1994). One-piece-type termites are further divided into three groups in relation to the locality and nature of resource wood: (i) damp wood termites (all species of the Termopsidae except *Porotermes adamsoni,* which sometimes makes subterranean galleries), which nest in damp dead wood on the inland forest floor; (ii) ordinary wood termites (*Prorhinotermes* of the Rhinotermitidae), which nest in ordinary dead wood on the forest floor; and (iii) dry wood termites (most species of the Kalotermitidae), which mainly nest in the dry dead wood of standing trees.

The Symbiosis of Termites

Termites feed on wood and other dead plant material, and not usually on living plants (Wood 1978; Abe 1979; Collins 1983b). In dead plants, cellulose and other cell-wall substances are the dominant components, while cytoplasmic substances are mostly lost (Abe and Higashi 1991). Thus, termites have to decompose these

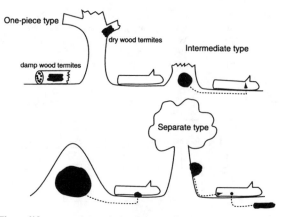

FIGURE 6.2. Three life types (one-piece, intermediate, and separate) exploiting wood pieces. *Closed areas,* nests; *broken lines,* foraging galleries.

hard cell-wall substances to utilize dead plant material, the superabundant potential food resources in their habitats. The means by which termites and other "cell-wall consumers" (Abe and Higashi 1991) decompose resistant cell-wall constituents are relatively well known (LaFage and Nutting 1978; Anderson and Ineson 1984; Breznak 1984; Martin 1987; Slaytor 1992; Breznak and Brune 1994).

However, the decomposition of hard substances is not the only problem that termites have to solve before making available food of dead plants (Collins 1983b; Waller and LaFage 1987). The carbon-to-nitrogen ratio (C:N) of dead plants, consisting of mostly cell-wall substances, is much higher than that of animal tissue. Fresh dead wood contains less than 0.5% nitrogen and has a C:N of 350–1000 (Cowling and Merrill 1966; LaFage and Nutting 1978), while termite tissues contain 8%–13% nitrogen (ash free) and have a C:N of 4–12 (Matsumoto 1976). To utilize dead plants as food resources, termites have to fill this gap in C:N with their potential food (i.e., reduce the C:N from the extremely high level of their food down to that of their own); that is, they have to solve the "carbon–nitrogen balance" (C–N balance) problem (Higashi et al. 1992).

There are logically two ways for solving this common problem of C–N balance: one is to add nitrogen (N), i.e., increase the nitrogen input to the consumer, and the other is to delete carbon (C), i.e., selectively output compounds with a higher C:N than that of the consumer's food. For example, kelp and other seaweeds excrete organic carbon produced in photosynthesis (Mann 1973). Aphids exude "honeydew" rich in carbon, and in the case of some species, symbiotic ants harvest the secretions, cleaning the aphids' bodies to prevent fungal infection (Buckley 1987). In contrast, terrestrial plants store carbon in the form of cellulose and other cell-wall substances. They can be thought of as dividing their biomass into two parts, cell wall and cytoplasm (Abe and Higashi 1991), with all the C–N imbalance in the former. This, in effect, transfers the C–N balance problem to cell-wall consumers. Many species conserve nitrogen by recycling it within their body, sometimes with the aid of microorganisms (Potrikus and Breznak 1981), which is one way to in effect add nitrogen to their food.

Termites have solved the problems of both cellulose digestion and C–N balance by the association with microorganisms. Thus, they have two kinds of symbioses with microorganisms: "cellulose digestion symbiosis" and "C–N balance symbiosis" (Higashi et al. 1992). The cellulose digestion problem of termites has been dominant in termite symbiosis research (O'Brien and Slaytor 1982; Breznak 1984; Slaytor 1992; Breznak and Brune 1994), but C–N balance symbiosis has received little attention until recently (Higashi et al. 1992), although the importance of nitrogen economy in termites has been recognized (LaFage and Nuttig 1978; Collins 1983b; Waller and LaFage 1987; Slaytor and Chappell 1994; Nalepa 1994). It has recently been argued, however, that no present-day termites need symbionts for cellulose digestion. This is because all modern termites, including lower termites, which contain cellulolytic Protozoa in the hindgut, have their own cellulase activity and, more importantly, they have no way to gain glucose other than digesting cellulose for themselves (Slaytor 1992). [In modern termites, phosphoenolpyruvate (PEP) carboxykinase, the controlling enzyme of gluconeogenesis, the synthesis of glucose from pyruvate, is inactive so that any pyruvate produced by cellulolytic Protozoa cannot be used by the termites (O'Brien and Breznak 1984; Slaytor 1992).]

On the other hand, all termites still need symbionts for C–N balance. The Isoptera, the order of insects comprising the termites, exhibit a great diversity of gastrointestinal symbionts (Grassé and Noirot 1959). This is expected to be closely related to variation in the mechanisms they possess for solving the C–N balance problem, because termites are not capable to solve this problem alone, and must have help from their symbionts. Thus, the C–N balance problem provides an effective point of view from which the diverse relationships that termites as a whole order (Isoptera) have with their symbionts can be elucidated. Figure 6.3 is a summary that superimposes the possible mechanisms for C–N balance, each of which is available to only a subset of termite species, with the particular symbionts responsible (Higashi et al. 1992). These symbiotic mechanisms for C–N balance can be grouped into two categories corresponding to the two logical choices for achieving a C–N balance: mechanisms for adding N to the food, referred to in Figure 6.3 as routes N-1 and N-2a, N-2b (Lee and Wood 1971; Matsumoto 1976; LaFage and Nutting 1978; Wood 1978; Breznak 1984), and mechanisms for selectively outputting C, referred to in Figure 6.3 as routes C-1 and C-2 (Wood and Sands 1978; Messer and Lee 1989).

Termites solve the C–N balance problem through association with their symbionts, but the C–N balance requirement imposes a constraint on the utilization of resources by termites. Any increase in the amount of assimilated food enlarges the amount of excess C in the assimilated food, which represents the C–N imbalance that the termite has to make up by adding N into its input or by selectively eliminating C. On the other hand, the amount of N that can be added to the input and the amount of C that can be selectively eliminated with the aid of a termite's sym-

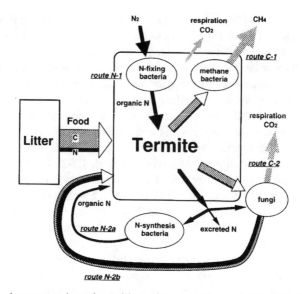

FIGURE 6.3. A summary (superimposition) of the possible mechanisms for C–N balance, with the responsible symbionts. *Black symbols,* nitrogen (N); *hatched symbols,* carbon (C).

bionts are finitely bounded. Thus, the maximum amount of food that the termite can utilize is confined to the level that corresponds to the upper bound of the amount of N that can be added and the amount of C that can be eliminated.

The mechanisms for C–N balance by selectively eliminating C require termites, which feed on food with an extremely high C:N ratio, to process (thus, consume) much more food to obtain a given amount of C–N balanced resource. One-piece-type termites, which stay in their wood nest and consume it, cannot afford to take the evolutionary direction toward enhancing these "wood-consuming" mechanisms for C–N balance by selectively eliminating C, which would ruin their nest. The obvious evolutionary course is to develop mechanisms to add N, such as symbiosis with nitrogen-fixing bacteria. Many studies using the acetylene reduction assay showed that termites are able to fix atmospheric nitrogen with the aid of facultative anaerobic bacteria in their hindguts (Benemann 1973; Brezanak et al. 1973; Prestwich et al. 1980; Prestwich and Bentley 1981). A more recent study with the ^{15}N natural abundance method has shown that at least 30%, and probably more than 50%, of the nitrogen of a one-piece-type (dry wood) termite, *Neotermes koshunensis,* was derived from the atmosphere (Tayasu et al. 1994).

On the other hand, those termites species that separate their nests from their feeding habitats are free from this limit in resource utilization and make a full choice among possible mechanisms for C–N balance, including those by selectively eliminating C. Among the separate-type termites has also appeared the habit of growing fungi (Fig. 6.4). This has brought additional mechanisms for C–N balance; by growing fungi, termites both add N through a local N-recycling (route N-2b in Fig. 6.3) and selectively eliminate C through fungal respiration

FIGURE 6.4. Nest *(left)* and fungus comb *(right)* of *Macrotermes michaelseni*, a fungus-growing termite, in the grassland of Kenya.

(route C-2 in Fig. 6.3). Recalling that the capability of N addition and C elimination determines the level of attainable productivity, the rank order of the degree of freedom and advantage inherent in the mechanisms of C–N balance is consistent with and may explain the observed rank order in productivity and colony size of the different termite types; one-piece-type termites are less abundant than separate-type termites, and where they exist, i.e., in tropical Africa and Asia, fungus-growing separate-type termites are the most prosperous (Wood and Sands 1978; Abe and Matsumoto 1979; Brian 1983; Josens 1985).

Sociality of Termites

The Isoptera (termites) and certain groups of Hymenoptera (bees, ants, and wasps) are the major groups of social insects, constituting a large part of the world of social insects. The Isoptera have a diplo-diploid reproduction system, while the Hymenoptera have a haplo-diploid reproduction system. A colony is initiated by a reproductive pair (a queen and a king) in the Isoptera (Fig. 6.5), but only by a queen among the Hymenoptera. Soldiers and workers may either be male or female in the Isoptera, but they are only female in the Hymenoptera. Although eusociality has evolved in both insect orders, the Isoptera and the Hymenoptera, the hemimetabolous termites, which can be neotenous, have a very different caste system from social Hymenoptera, which are metabolous and thus cannot show neoteny. In termites, workers are not imagoes and imagoes are restricted to reproductives, while in social Hymenoptera both workers and reproductives are imagoes. The soldiers of termites, which have no equivalent in other social insects and are unique in their postembryonic development and their exclusive defense functions, appeared very early in the evolution of the Isoptera and are likely to have arisen only once (Noirot and Pasteels 1987; Myles 1988; Noirot 1990).

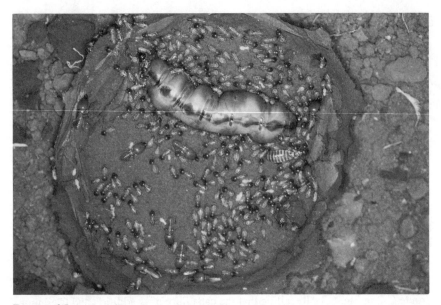

FIGURE 6.5. A termite colony consisting of a queen, a king, workers, and soldiers of *Macrotermes michaelseni* in the grassland of Kenya.

Termites have a large number of occasional or regular predators, among which ants are the most important (Longhurst et al. 1978; Collins 1981b; Abe and Darlington 1985). The defense of termites lies essentially in the defense of their nests. Soldiers and workers take part in this task by fighting against predators and by building the nest, respectively. All termites, with a few exceptions in higher termites, have a sterile soldier caste that is highly varied morphologically (Fig. 6.6). Among soldiers two types can be distinguished, mandibulate (mechanical) and nasute (chemical) types. Mandibulate soldiers occurring in all families rely on large prominent mandibles, and defense is achieved by biting or snapping actions with the mandibles. On the other hand, in nasute types, which occur only in Nasutitermitinae of the Termitidae (higher termites), the front of the head is modified for chemical warfare and the mandibles are in most cases reduced.

In spite of the presence of soldiers, most termite individuals are very vulnerable to predators on the ground surface outside their nests; exceptions are some groups of Nasutitermitinae with nasute soldiers and *Macrotermes* of Macrotermitinae with giant mandibulate soldiers. Termites of *Hospitalitermes* and *Longipeditermes* (in Nasutitermitinae), whose body color is black, form foraging columns from their nests to feeding sites on the ground surface in the tropical rain forests of South East Asia (Abe 1979). Their foraging distance is sometimes more than 40 m. Both sides of the foraging columns (consisting of workers) are guarded by many nasute soldiers. Those columns are in many cases free from predation. Thus, nasute soldiers seem to be successful in providing enemy-free foraging

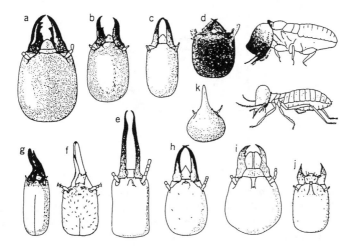

FIGURE 6.6a–k. The diversity in morphology of soldiers' heads in termites. (After Harris 1961.) (a) *Neotermes;* (b) *Odontotermes;* (c) *Reticulitermes;* (d) *Cryptotermes;* (e) *Termes;* (f) *Procapritermes;* (g) *Pericapritermes;* (h) *Coptotermes;* (i) *Schedorhinotermes;* (j) *Cornitermes;* (k) *Nasutitermes.*

space. The situation is similar in *Macrotermes carbonarius,* which is also black and forms foraging columns on the ground surface (Sugio 1995). The proportion of soldiers changes with colony growth (Maki and Abe 1986) and differs greatly among species, however; soldiers occupy 30% of all individuals in some species of Nasutitermitinae, while they occupy about 5% in most termites with mandibulate soldiers (Haverty 1977). In comparison with mandibulate soldiers, the nasute soldier is small and the proportion of soldiers in a colony is very high.

Contrary to the case of the soldier caste, not all termites have a sterile worker caste. The presence of a true (sterile) worker caste, the caste of workers that totally lose at an early stage of development the ability to reproduce, is polyphyletic at the family level; i.e., it is thought to be distributed over multiple branches in the phylogenetic tree of Isoptera (see Fig. 6.1). A clear trend can be identified in the correlation between food–nest separation and true worker evolution in termites: the separate-type termites possess a true worker caste and never have a false (non-sterile) worker caste, the caste of workers that preserve an ability to reproduce, while the one-piece-type termites possess no true workers but only the false worker caste (Abe 1991).

An evolutionary explanation for this correlation, based on a mathematical model, has been proposed: the food–nest separation increases nest stability, which in turn enhances the evolution of true workers (Higashi et al. 1991). Here, nest stability is defined as the ratio $(1 - p)/p$ with p representing the probability that either the queen or king dies or the nest collapses. This result was extended to generate a more comprehensive explanation (Higashi et al. 1992): the higher nest stability and the higher C–N balance capability of separate-type termites should together promote the evolution of a sterile worker caste (Fig. 6.7). Here, the C–N balance

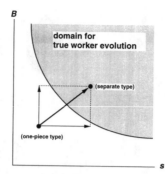

FIGURE 6.7. A summary of a theoretical result indicating that the higher the nest stability and the C–N balance capability, and thus increase in the contribution to the reproductive pair when a false worker changes into a true worker, the more likely is the evolution of true worker caste to occur.

capability of a termite species is defined as its capability of N addition and C elimination for balancing C and N (i.e., correcting the C–N imbalance) in the food, which sets an upper bound for food utilization, as was discussed previously. In the case of one-piece-type termites, a true worker cannot exploit a resource beyond a low upper bound imposed by the low C–N balance capability of one-piece-type termites, even if it is potentially more efficient in resource exploitation than a false worker. Thus, becoming a true worker should be less favored in the case of one-piece-type termites than in the case of separate-type termites. This conclusion can be combined with the previous result on the effect of nest stability to derive the more comprehensive explanation.

The Global Diversification of Termites

Introduction

We identify three major events of global diversification that the Isoptera as a whole order has experienced: (i) the evolutionary radiation of original termites, (ii) the expansion and diversification of separate-type termites and the diversification of one-piece-type termites driven by the expanding separate type termites into fragmented habitats, and (iii) the radiation of higher termites. By pulling together currently available empirical evidence with aid of logical inference, we develop here hypotheses that point to the conclusion that each of these global diversification events corresponds to, and was driven by, a new development in the sociality and symbiosis of termites.

The Evolutionary Radiation of Original Termites

The first event of global diversification in the Isoptera was the evolutionary radiation of original termites into a new habitat of wood over the tropical regions. We infer that this radiation was driven by the efficient utilization of wood as an abundant and stably supplied food resource and nest substrate, which became possible by the evolution of two symbioses for termites, i.e., cellulose digestion symbiosis with cellulolytic Protozoa and C–N balance symbiosis with nitrogen-fixing bacteria. We derive this inference next.

As we have already discussed, no termite that exists today depends on symbionts for cellulose digestion. There must, however, have existed a phase in which symbiotic Protozoa were necessary for cellulose digestion—because, if such a phase had not existed, the original termites with no cellulase activity would have had no reason to feed on a cellulose-rich resource and their cellulase production (or, acquisition of cellulase activity) would not have been favored by natural selection. It is likely that some Protozoa and bacteria eaten by ancestral termites happened to obtain resistance to digestion and became parasites, some of which further became symbionts (possibly through such a mechanism as suggested by Yamamura in Chapter 5, in this volume). The coevolution between the symbiotic partners would have enhanced the degree of mutual contribution to the extent that termites could utilize wood as the only food and nest substrate. Note that in this process the symbiosis with Protozoa for cellulose digestion and that with nitrogen-fixing bacteria for C–N balance must evolve together.

Once a termite obtains a means for cellulose digestion and C–N balance through association with microorganisms, then wood, which is superabundant but extremely hard to digest and has an extremely high C:N, becomes a "well-protected" food resource that can be monopolized by the termite. This event may be viewed as an example of niche opening, which often leads to adaptive radiation.

Effective competition for the wood habitat is presumed to promote the evolution of the false worker caste, because life in a piece of wood, which is that of today's one-piece-type termites, implies: (i) a greater need for concentrating nitrogen, (ii) a higher possibility of nest succession because of the longer duration of the nest relative to the longevity of reproductive individuals, and (iii) a lesser possibility to find an alternative nesting place because of the heterogeneity of resource (wood) distribution. The evolution of this subsociality should further have enhanced wood utilization and thus the radiation of termites.

Separate-Type Expansion and One-Piece-Type Diversification

The second major event of global diversification of termites occurred as separate-type termites, those that separate their feeding sites from their nest, appeared and, presumably, expanded quickly, replacing the existing one-piece-type termites through competitive exclusion. They may have gone through a diversification process in this expansion into diverse habitats that had been occupied by speciated one-piece-type termites, while some groups of one-piece-type termites that were driven out by the expanding separate-type termites into fragmented habitats should also have gone through a diversification process. These scenarios could be reconstructed using the existing peculiar distribution patterns of distinctive groups of one-piece-type termites and separate-type termites, as is summarized next.

The three groups of one-piece-type termites show quite different marginal distribution patterns (Fig. 6.8). Damp wood termites (all species of the Termopsidae), which nest in large damp fallen wood, show a clear amphitropical (bipolar) distribution that is characterized by a close relationship between the taxa in the north temperate (boreal) zone and those in the austral zone.

Three species of *Porotermes* of the Termopsidae are distributed in temperate regions of three southern continents: one species each in Chile, Australia, and South Africa (Fig. 6.9). *Stolotermes* of the Termopsidae is found in Australia, New Zealand, Tasmania, and South Africa. This discontinuous distribution of *Porotermes* and *Stolotermes* is similar to that of the tree genus *Nothofagus* (Humphries 1981). The other three genera of the Termopsidae are distributed in the temperate and subtropical north hemisphere: *Archotermopsis* in Kashmir

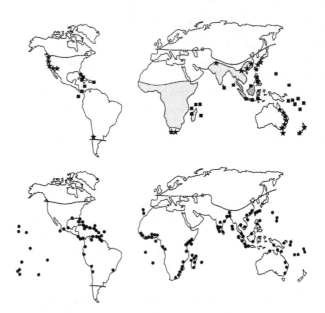

FIGURE 6.8. The distribution of three groups of one-piece-type species and that of a dominant group of separate-type species, subfamily Macrotermitinae (fungus-growing termites), of Termitidae. *Upper panel:* damp wood termites (*stars,* Termopsidae), ordinary wood termites (*squares, Prorhinotermes* of Rhinotermitidae), and Macrotermitinae (*stippled region*); *lower panel:* dry wood termites of *Cryptotermes* (*circles*). The distribution of *Cryptotermes* is from Chhotani (1970).

FIGURE 6.9. The distribution of three species of *Porotermes* (*closed circles*) and five species of *Stolotermes* (*open circles*) of the Termopsidae in temperate regions of three southern continents. Each circle corresponds to a species.

(India), *Zootermopsis* in North America, and *Hodotermopsis* in Japan and China.

Ordinary wood termites *(Prorhinotermes),* which are one-piece-type termites nesting in ordinary fallen wood, are widely distributed in tropical and subtropical regions but tend to be confined to islands (see Fig. 6.8). Dry wood termites (also one-piece-type termites) are widely distributed in the tropical and subtropical regions, but are confined to dry dead parts of standing trees. *Cryptotermes,* a dominant genus of the Kalotermitidae that is a typical dry wood termite, is widely distributed in the tropical region but tends to be confined to coastal forests and islands (Fig. 6.10) (Chhotani 1970). In contrast, separate-type termites, represented by the Termitidae, are widely distributed in tropical mainlands, the central areas of termite distribution. Intermediate-type termites, represented by Rhinotermitidae, are widely distributed from tropical to temperate regions. Thus the distribution of the one-piece type and the separate type is complementary, while that of the intermediate type covers the ranges of both types. What determines these distribution patterns?

For more than a century, biogeographers have been interested in the origin and cause of the similarity in biota among the highlands of tropical New Guinea, subtropical New Caledonia, and the temperate areas of South America, New Zealand, Tasmania, and Australia, sometimes including the southern tip of Africa (Darwin 1859; Darlington 1957; Humphries and Parenti 1986). Their further attention has been directed to amphitropical distribution (McDowell 1964; Rosen 1974; Humphries 1981). Many phylogenetically different groups show an amphitropical distribution and therefore some general explanation is expected. Reasoning based on plate tectonics (Nur and Ben-Avraham 1981; Humphries and Parenti 1986) explains the distribution of the Termopsidae (damp wood termites) in the northern hemisphere, *Zootermopsis* in the United States and Canada, and *Hodotermopsis* in the Ryukyu Islands and China, but does not explain well either the fossil record of

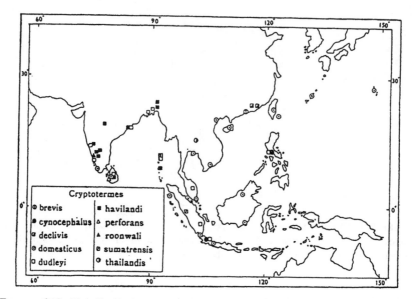

FIGURE 6.10. The distribution of *Cryptotermes,* a dominant genus of Kalotermitidae, which is a typical dry wood termite, at species level in the Indo-Malayan region (Chhotani, 1970).

the Termopsidae in Europe or the presence of *Archeotermopsis* in Kashimir. Alternative explanations are given on the basis of ecological interactions. Among predators of termites, ants are predominant, and the oldest fossil ants have been found in the Cretaceous, but it is difficult to find a clear correlation between the amphitropical distribution of termites and the distribution of ants. Therefore, we should seek interspecific competition as the best candidate for a major force determining the amphitropical distribution of the Termopsidae.

In fact, interspecific competition on the geographic scale may explain not only the amphitropical distribution of damp wood termites (Termopsidae) but also the overall distribution pattern of one-piece-type termites and thus the complimentary distribution of one-piece-type and separate-type termites. We argue that, although a full explanation of such distribution patterns requires other factors such as plate tectonics and global climate change to be taken into consideration, ecological mechanisms, specifically interspecific interactions and differential dispersal ability, have been a major force determining the range of distribution.

An assumption of this hypothesis is that the one-piece-type termite was widespread before the separate-type termite appeared. During the era of one-piece-type termites, the evolution of the false worker caste should have occurred, as we have already discussed. Note that once a false worker caste is developed, the separation of feeding (production site) from nest (reproduction site) becomes possible. Once separate-type termites appeared, their geographic expansion was driven by their superiority in resource utilization over one-piece-type termites under the conditions that are normally found in tropical central regions.

This asymmetrical competition between one-piece and separate types is a consequence of the fact that wood is food and nest for the one-piece type but only food for separate-type termites. This result has two components. First, the fact that one-piece-type termites use the same piece of wood as their house and food imposes a constraint on the size or quality of wood that they can utilize; i.e., one-piece-type termites can utilize only that wood available in their habitat (Fig. 6.11b) which has a large initial size or a decay rate slow enough to sustain them until the completion of their generation cycle (Fig. 6.11a). Second, when a colony of either the separate-type or the one-piece-type termites utilizes a piece of wood, it should impose more damage on other colonies of one-piece-type termites than on those of separate-type termites in the area; this results because the preempted wood can no longer be as valuable for other one-piece termites as a house–food resource while it can still be used as a food resource by separate-type termites (Fig. 6.11b).

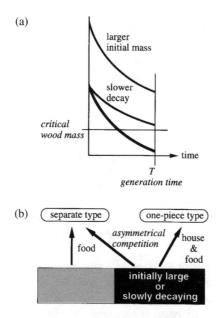

FIGURE 6.11. (a) A model for the wood decay process (*curved lines* on graph) resulting from consumption by a one-piece-type termite colony. It is assumed that there is a critical mass (or size) of wood (*horizontal line* across curved lines on graph) that is necessary to maintain a colony of one-piece-type termites. The larger initial size or the smaller decay rate the wood has, the longer it sustains the process. (b) Asymmetrical competition between one-piece-type and separate-type termites.

These two components of asymmetrical competition can be formulated most simply by the following Lotka–Volterra equations:

$$\frac{dx_o}{dt} = x_o(\varepsilon_o - a_o x_o - b_{os} x_s)$$

$$\frac{dx_s}{dt} = x_s(\varepsilon_s - a_s x_s - b_{so} x_o)$$

(1a)

with

$$\varepsilon_o = cR - d_o P$$

$$\varepsilon_s = R - d_s P$$

(1b)

where x_o and x_s, respectively, represent the colony number of one-piece-type termites and that of separate-type termites; it is assumed that the intraspecific growth rates of both colony populations are effectively determined by the resource supply rates, ε_o and ε_s, respectively; a_o and a_s, respectively, denote the intraspecific competition coefficients of the colony population of one-piece-type termites and that of the separate type; b_{os} and b_{so}, respectively, are the interspecific competition coefficients from separate type to one-piece type and vice versa. The resource supply rates are functions of predation pressure, denoted here by P, and the separate type, which forages out, is assumed to be more sensitive to predation pressure than the one-piece type, i.e., $d_s > d_o$. However, the resource supply rate under no predation pressure is reduced with the one-piece type by the rate of c ($0 < c < 1$), which reflects the constraint on the size and quality of wood that the one-piece type can utilize. The second component of asymmetrical competition between one-piece and separate types is represented by the condition:

$$\frac{b_{os}}{a_s} > 1 > \frac{b_{so}}{a_o}$$

(2)

Under this condition, if the resource supply rate of the one-piece type is small enough relative to that of the separate type (this condition is satisfied if the resource supply rate of the one-piece type is equal to, or less than, that of separate type), then the separate type outcompetes the one-piece type, resulting in separate-type dominance (Fig. 6.12a). Thus, once separate-type termites appeared, they competitively replace or drive out the original one-piece-type termites, expanding their geographic distribution.

If the resource supply rate of the one-piece type increases relative to that of the separate type (i.e., the ratio $\varepsilon_o/\varepsilon_s$ increases), however, the consequence of the dynamics will be reversed, resulting in one-piece-type dominance (Fig. 6.12a). This increase of the ratio may occur if c increases (i.e., the constraint is relaxed) under a given P value; the greater the predation pressure (P), the more easily this happens (Fig. 6.12b). For instance, as latitude increases, the decay rate of damp wood should decrease because of the temperature drop, which implies (according to the mechanism depicted in Fig. 6.11a) an increase in the range of wood available to the one-piece type (i.e., an increase of c). Likewise, as we approach the

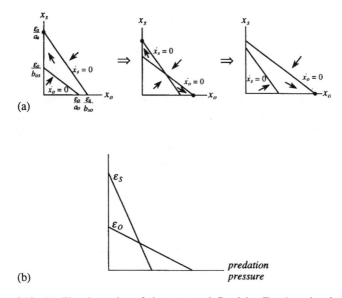

(a)

(b)

FIGURE 6.12. (a) The dynamics of the system defined by Eq. 1 under the condition

$$\frac{b_{OS}}{a_S} > 1 > \frac{b_{SO}}{a_O}.$$

Left, $\varepsilon_O/\varepsilon_S \leq 1$; center, $\varepsilon_O/\varepsilon_S$ gets larger; *right*, $\varepsilon_O/\varepsilon_S$ gets even larger. (b) The dependence on predation pressure (P) of the resource supply rates for one-piece type and separate type (ε_O and ε_S, respectively).

seacoast, the decay rate of dry wood should decrease because of the effect of wind, which implies an increase of c.

The main conclusions from the foregoing theoretical consideration are as follows. First, as the latitude (ocean) gets higher (closer), the resource supply rate of the one-piece type relative to that of the separate type, e_O/e_S, increases, because more wood is available to the one-piece type as the decay rate of damp (dry) wood is slower at higher latitude (seacoast) areas, under a given predation pressure. Thus, the one-piece type may compensate its disadvantage from its asymmetrical competition with the separate type and become dominant at higher latitude (seacoast) areas. This explains the questioned marginal distribution patterns of damp and dry wood termites (see Fig. 6.8). The distribution of ordinary wood termites confined to islands may simply be explained as a result of the escape from asymmetrical competition resulting from their dispersal ability (Fig. 6.8). Second, as a corollary, it is predicted that the one-piece-type termites that are dominant at higher latitudes should be damp wood termites, while those dominant at seacoast areas should be dry wood termites. This hypothesis is supported by empirical evidence (Fig. 6.8).

Separation of the feeding site from the nest (reproductive site) should enhance

C–N balance symbiosis by adding the option of selective carbon elimination and thus promote the evolution of a true worker caste, as we have already reviewed. These changes should further enhance the superiority of the separate type in resource utilization, thus driving the expansion and diversification of separate-type termites.

Among the one-piece-type termites driven out by the expanding separate-type termites, those groups driven into fragmented habitats along seacoasts and islands are expected to have gone through a diversification process. These groups of one-piece-type termites are dry wood and ordinary wood termites, and in fact both groups show high species diversity as a family or a subfamily.

The Radiation of Higher Termites

The third major event of global diversification of termites is the radiation of higher termites (family Termitidae), the termites that are free from Protozoa. We infer that this radiation of higher termites was driven by the advantage that a termite would have if it could digest cellulose efficiently enough to eliminate cellulolytic Protozoa, which became possible by attacking cellulose more directly in the absence of its association with lignin as lignocellulose, perhaps through the shift of the diet from wood to a food with less lignin. We discuss this next.

The present-day termites, including lower termites that contain cellulolytic Protozoa, have their own cellulase activity and do not need help from cellulolytic Protozoa for cellulose digestion. Then, are Protozoa in lower termites pure parasites competing with their host for cellulose, their common resource? It is well known, however, that lower termites, at least those species studied to date, exhibit a characteristic behavior in transmitting Protozoa immediately after molting to new workers or workers that lack Protozoa (Honigberg 1970), which could not be expected if the Protozoa were pure parasites. One possible resolution for this paradox is that, in spite of the cellulase activity of lower termites, their cellulose digestion is inefficient so that some residues are left for Protozoa in the hindgut to utilize; the Protozoa convert this residual cellulose, together with other microbial associates, eventually to acetate, providing an energy source for the host termites. Thus, cellulolytic Protozoa are still symbionts to lower termites, at least for energy supply, although they may no longer supply glucose.

Higher termites today are, then, those that have succeeded in increasing the efficiency of cellulose digestion to the extent that cellulolytic Protozoa are no longer useful for an energy supply either. Once this increase is achieved, the elimination of cellulolytic Protozoa should follow, resulting in the evolution of higher termites. To increase the efficiency of cellulose digestion, attacking cellulose more directly in the absence of its association with lignin as lignocellulose would be expected to be the most effective, because the association with lignin is the major obstacle to cellulose utilization. An obvious means to achieve this is to shift the diet from wood to a food with less lignin. Note that the nest–food separation makes possible the selection of food. Thus, it is expected that higher termites may arise only from separate-type termites, which is indeed the case.

Because the elimination of cellulolytic Protozoa saves much of the resource consumed by the Protozoa, higher termites are more efficient in resource utilization than lower termites; this is the driving force for their radiation. The elimination of Protozoa would make space to accommodate more symbionts (bacteria) for C–N balance, and an addition of C–N balance symbionts is favored by natural selection because it should increase the upper bound for the termite's resource exploitation rate and thus its fitness. Therefore, an increase of C–N balance symbionts should have occurred in higher termites, further enhancing their radiation.

It is known that pyruvate dehydrogenase is inactive in modern termites, including higher termites, so that they depend wholly on the acetate produced by microbial associates (bacteria in the case of higher termites) for an energy source (O'Brien and Breznak 1984; Slaytor 1992). This inactivity of pyruvate dehydrogenase in termites may suggest, and thus can be explained by the assumption, that the energy source for termites provided by microbial accetate production is great enough to make unnecessary energy generation by termites that is independent of microbial associates. The essential difference between lower and higher termites is that higher termites provide glucose for the microbial associates which produce acetate, the energy source for the termites, while in lower termites Protozoa generally do this job. [Note here that if pyruvate dehydrogenase in termites is inactive (thus there is no energy generation by termites that is independent of microbial associates), then termites will not make pyruvate, the source substance in energy generation by termites independent of microbial associates, and thus PEP carboxykinase, which acts on pyruvate to synthesize glucose, should also be inactive.]

Higher termites, including the four subfamilies Termitinae, Apicotermitinae, Nasutitermitinae, and Macrotermitinae, show three trends in diversification: (i) changing from wood to soil feeding, (ii) symbiosis with fungi, and (iii) production of powerful nasute soldiers that use chemical weapons. Each subfamily seems to have achieved a more direct attack on cellulose by shifting their diet from wood to a food with less lignin in its unique way.

The Evolution of Soil-Feeding Termites

Noirot (1992) pointed out the change in feeding habits from wood to soil (humus) feeding as a major evolutionary trend. Once a termite succeeded in utilizing as a food resource decomposed dead plant tissues, which have less lignin, it should have been able to eliminate cellulolytic Protozoa. The higher efficiency in cellulose utilization by a higher termite and the abundance and commoness of soil should together promote the expansion and diversification of the soil feeder. Furthermore, soil, composed of highly heterogeneous components, provides a diversity of feeding habitats for soil-feeding termites to choose. This heterogeneity in feeding habitats (or, trophic niches) may possibly enhance the diversification of soil-feeding termites. The morphological diversity in the gut (which is usually elongated) and the microbial composition of soil feeders (Bignell et al. 1983; Bignell 1994) are thought to be closely related to their adaptive diversification into heterogeneous habitats.

TABLE 6.1. Soil-feeding genera in the family Termitidae

Subfamily	Number of genera	Soil feeders	Percent of soil feeders
Macrotermitinae	13	0	0
Nasutitermitinae	78	34	44
Termitinae	76	56	74
Apicotermitinae	42	40	95
Total:	209	130	62

From Noirot (1992).

Soil feeding is recognized in three subfamilies of the Termitidae, i.e., the Nasutitermitinae, Termitinae, and Apicotermitinae, but not in the Macrotermitinae (fungus-growing termites) (Table 6.1). The Termitidae occupy 209 of 257 genera of the Isoptera, and 130 genera of the Termitidae are soil feeders. Therefore, approximately 50% of all termite genera and 62% of the genera in Termitidae have specialized in soil feeding. Soil feeding is supposed to have appeared at least twice in the Termitinae and several times in the Nasutitermitinae and as a primitive form in the Apicotermitinae (Noirot 1992). Table 6.2 shows the geographical localization of the genera in Termitidae. Comparison of Tables 6.1 and 6.2 is useful to detect endemicity: the total number of genera is similar in the two tables, indicating that endemicity at the genus level is very high in the Termitidae (94% compared to 51% for the lower termites). Endemicity is even higher (98.5%) among soil feeders. Apicotermitinae in Africa and the *Termes* complex in Australia are of special interest because they include species with the successive change from wood feeding to soil feeding (Miller 1992; Eggleton et al. 1995).

The Evolution of Fungus-Growing Termites

Many termites are dependent on fungi outside their nests in making wood and other dead plant material available to them as food resources (Collins 1983b). In contrast, the Macrotermitinae cultivate specific species of fungi in their nests, supplying suitable food, protecting them from competition with other species of fungi, and furthermore assisting the dispersal of the fungal spores (Darlington 1994). This makes it possible for the fungus-growing termites to utilize sound wood as well, which is hardly attacked by fungi and only slightly exploited by most soil animals in the tropical forest (Abe 1979). The nests of fungus-growing termites form a large-scale "super-decomposition factory" to which a large amount of dead plant material (including sound wood) is carried and processed into "stored food" (fungus comb) to be completely decomposed and consumed (Sands 1969; Batra and Batra 1979; Wood and Thomas 1989). Water produced by respiration of fungi maintains high humidity inside the nest, which makes possible the geographic expansion of fungus-growing termites into arid regions (Wood and Thomas 1989).

Although it is true that dead plant materials carried into the nests are rather

TABLE 6.2. Biogeography of Termitidae

| | Number of genera | | Percent of soil | Percent of region soil |
	Total	Soil feeders	feeders in region total	feeders in total soil feeders
Ethiopian	93	67	72	52
Neotropical	55	28	51	22
Oriental	57	28	49	22
Australian	15	6	40	4.5
Malagasy	9	3	33	2.5
Total:	229 (209)	132 (130)		

From Noirot (1992).

completely decomposed in fungus growers, the true nature of the relationship between termites and fungi is not clear. Three suggestions have been made on the fungal role: (i) fungi degrade lignin in the dead plant material (Rohrmann and Rossman 1980); (ii) fungi concentrate nitrogen and improve termite nutrition (Matsumoto 1976; Rohrmann 1978; Higashi et al. 1992); or (iii) termites ingest fungi as an acquired cellulolytic enzyme (Martin 1987). As for the first suggestion, we note that it is not necessary for lignin to be decomposed and digested by symbiotic fungi, but only to be broken down to the extent that termite access to cellulose is ensured. The second suggestion in essence concerns C–N balance, and fungi may contribute to C–N balance, as was discussed, through N addition by a local N-recycling (route N-2b in Fig. 6.3) and C elimination through a fungal respiration (route C-2 in Fig. 6.3). The third suggestion seems to have been refuted by Veivers et al. (1991), who showed that 90% of the glucose requirements of workers can be met by the action of endogenous intestinal cellulases.

About 300 species of Macrotermitinae are distributed throughout tropical Africa and parts of Arabia and Indomalaya, but the subfamily is not found in Australia and America. This distribution pattern suggests a fairly late evolution of the subfamily after the separation of the Gondwana continent, which might limit the global species number of this subfamily relative to Nasutitermitinae and Termitinae, whose species numbers exceed 500.

The fungus growers are abundant in savannas or somewhat dry forest, while soil feeders are abundant in the forest (Table 6.3).

The Evolution of Termites with Nasute Soldiers

The Nasutitermitinae, which contain more than 500 species, show a remarkable adaptive radiation, having the broadest distribution in terms of geographic regions, habitat range, and feeding habits, together with high species diversity, among the three groups of higher termites. They nest in various sites such as in dead wood, on tree trunks, on the ground, and in the soil, and they consume various dead plant materials such as dead wood, dead grasses, soil, dung, and also lichen. Therefore, they cover almost all food items and nesting sites utilized by termites as a whole (Fig. 6.13).

TABLE 6.3. Biomass (wet weight, g/m^2) of termites with reference to their feeding habits in a tropical rain forest and savanna ecosystem

| | Rain forest | | Savanna | |
Ecosystem	Malaysia	Nigeria	Ivory Coast	Senegal
Latitude	3° N	9° N	6° N	16.5° N
Annual rain fall (mm)	2000	1175	1270	375
Source	Abe (1979)	Wood et al. (1977)	Josens (1972)	Lepage (1974)
Feeding habits				
Lichen feeders	0–0.28	0	0	0
Soil feeders	2.17–2.86	0.66	0.16	0
Fungus growers	5.91–6.32	6.39	0.64	0.72
Others	0.61–0.67	3.54	0.93	0.24
Total:	8.69–10.13	10.59	1.73	0.96

Data after Wood and Sands (1978) and Abe (1979).

Most species of the other two groups of higher termites tend to nest in the soil. Their nesting habits make it difficult for them to cross oceans by rafting. However, many species of Nasutitermitinae, in the stage of young colonies, make their nests in tree trunks. This makes it possible for the Nasutitermitinae to cross the oceans by rafting. The Nasutitermtinae are found in the Krakatau Islands of Indonesia together with species of the one-piece type, while no species of the other two groups of higher termites are found in spite of their abundance in the west tip of Java (Abe 1984; Yamane et al. 1992). This difference is explained by differential dispersal ability across the oceans.

The adaptive radiation of the Nasutitermitinae seems to be based on the production of nasute soldiers, which have an elongated pear-shaped head (see Fig. 6.6k) and a small frontal pore that emerges at the tip of the long frontal tube (nasus). The mandibles of these soldiers are reduced. The nasute soldiers can project a sticky liquid that is irritating and toxic to arthropod predators (Deligne et al. 1981). The evolution of this new type of soldier with chemical weapons against ants, their major predators, made it possible for Nasutitermitinae to produce somewhat "enemy-free foraging space" and thus forage out freely and select food with less lignin cover. A typical example is *Hospitalitermes*, which forages actively for lichen.

Most species of termites forage cryptically, but termites of at least eight genera of Nasutitermitinae (*Nasutitermes, Fulleritermes, Syntermes, Tenuirostritermes, Longipeditermes, Hospitalitermes, Lacessititermes,* and *Trinervitermes*) forage in the open, although a few other groups also do so, i.e., harvester termites (*Drepanotermes* of Termitinae and *Hodotermes, Microhodotermes,* and *Anacanthotermes* of Hodotermitidae) and *Macrotermes carbonarious* of Macrotermitinae (Jander and Daumer 1974; Sugio 1995). In the tropical forests of East Asia *Hospitalitermes* spp., which consume lichen and have black bodies, march in enormous numbers on the ground. The distance from their nests to foraging site sometimes exceeds 100 m. Both sides of the foraging columns are guarded by soldiers and

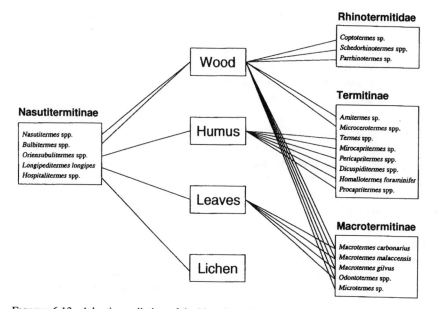

FIGURE 6.13. Adaptive radiation of the Nasutitermitinae. The feeding habits of Nasutitermitinae are compared with those of other termites in the tropical rain forest of West Malaysia (Pasoh Forest Reserve).

efficiently protected against occasional predatory ants. The proportion of soldiers in a colony of the Nasutitermitinae is extremely high, sometimes reaching 30%, while it is less than 10% in many other termites (Haverty 1977). This high figure in the Nasutitermitinae may partly compensate for the relatively small size of their soldiers.

Conclusions

The theoretical inference that has been developed on the basis of currently available empirical evidence leads us to the following main conclusion: three major global diversification events in Isoptera (termites) were ultimately driven by a new development in the sociality and symbiosis of termites. The first event, the evolutionary radiation of original termites, was driven by the evolution of symbiosis for cellulose digestion (with cellulolytic Protozoa) and of symbiosis for C–N balance by nitrogen addition. The second event, the expansion and diversification of separate-type termites and the diversification of one-piece-type termites, was ultimately driven by the evolution of the false worker caste, which makes possible the evolution of separate-type termites, and it was further driven by the evolution of symbiosis for C–N balance by selective carbon elimination and the evolution of a true worker caste, which were both promoted by food–nest separation. The third event, the radiation of higher termites, was ultimately driven by developments in sociality (the evolution of stronger soldiers, i.e., nasute-type soldiers) or in symbiosis (in particular, fungus growing), which provides termites with food with less lignin cover and thus promotes the elimination of cellulolytic Protozoa. In turn, this allows the termites to accommodate more C–N balance symbionts and thus increase their resource exploitation rate, or fitness, further enhancing their radiation.

We suspect that the main conclusion reached here should not be limited to the case of termites, but can be generalized to a wide range of higher animal and plant groups. The case of social Hymenoptera (ants, bees, and wasps) and that of corals have already been mentioned. The diversification of Dipterocarpaceae in tropical forests may also have been driven by the evolution of mycorrhizae, the symbiotic association of their root systems with fungi (Ashton 1982).

The case of termites should, however, serve as one of the best examples to articulate the point that symbiosis between higher organisms and microorganisms creates capabilities for exploiting new food resources, thus opening a new niche. It also illustrates the benefits from examining past and present spatial distribution and considering the dynamics of radiation and competition on the evolutionary time scale when we want to understand the causal mechanisms for the macroevolution of a higher taxonomic group.

Summary

1. Three major events of global diversification of Isoptera (termites) are identified, each of which corresponds to, and is driven by, a new development in its sociality and symbiosis. This hypothesis is derived by synthesizing available empirical evidence.
2. The first event is the radiation of termites into wood throughout the tropics. This dispersion was driven by an efficient utilization of wood as an abundant food resource and nest substrate, which became possible by the evolution of their cellulose digestion symbiosis with cellulolytic Protozoa and the C–N balance symbiosis with nitrogen-fixing bacteria. Full exploitation of the wood habitat should promote the evolution of a false worker caste, which should further have enhanced wood utilization and thus the radiation of termites.
3. The second event is the expansion and diversification of separate-type termites, which drove the one-piece type out into a marginal distribution. This was driven by the asymmetrical superiority of the separate type in resource utilization, which is based on the fact that wood is both food and nest for one-piece-type termites while it is only food for the separate type. Separation of feeding site from nest should enhance C–N balance symbiosis by adding the option of selective carbon elimination and promote the evolution of the true worker caste. These factors should have further enhanced the superiority of the separate type in resource utilization and thus driven the radiation of the separate type.

4. In marginal areas, however, the inferiority of the one-piece type may be compensated by the increase in resource supply caused by the predation effect combined with a reduction in the decay rate of wood, so that the one-piece type dominates over the separate type. Dry wood and ordinary wood termites, two groups of the one-piece type, show a high species diversity produced probably by a spatial isolation effect in coastal and island areas.

5. The third event is the radiation of higher termites, which are free from Protozoa. This was driven by the advantage that a termite would obtain when it could digest cellulose efficiently enough to eliminate cellulolytic Protozoa. This efficient digestion was made possible by the evolution of stronger soldiers or new symbionts, which provide the termite with food with less lignin cover and more efficiency in C–N balance.

Acknowledgments. We thank P. Eggleton, D.E. Bignell, M. Slator, and E.O. Wilson for careful reviews and helpful comments. This work is partly supported by a Japan Ministry of Education, Science and Culture Grant-in-Aid for Scientic Research on Priority Areas (No. 319), project "Symbiotic biosphere: an ecological interaction network promoting the coexistence of many species," and JMESC Grant-in-Aids Research Project numbers 07044193 and 07304052.

Literature Cited

Abe, T. 1979. Studies on the distribution and ecological role of termites in a lowland rain forest of West Malaysia. 2. Food and feeding habits of termites in Pasoh Forest Reserve. Japanese Journal of Ecology (Sendai) 29:121–135.

Abe, T. 1984. Colonization of the Krakatau Islands by termites (Insecta: Isoptera). Physiology and Ecology, Japan 21:63–88.

Abe, T. 1987. Evolution of life types in termites. In: S. Kawano, J.H. Connell, and T. Hidaka, eds. Evolution and Coadaptation in Biotic Communities, pp. 125–148. University of Tokyo Press, Tokyo.

Abe, T. 1991. Ecological factors associated with the evolution of workers and soldiers caste in termites. Annals of Entomology 9:101-107.

Abe, T. and J.P.E.C. Darlington. 1985. Distribution and abundance of a mound-building termite, *Macrotermes michaelseni,* with special reference to its subterranean colonies and ant predators. Physiology and Ecology, Japan 22:59–74.

Abe, T. and M. Higashi. 1991. Cellulose centered perspective on terrestrial community structure. Oikos 60:127–133.

Abe, T. and T. Matsumoto. 1979. Studies on the distribution and ecological role of termites in a lowland rain forest of West Malaysia. 3. Distribution and abundance of termites in Pasoh Forest Reserve. Japanese Journal of Ecology (Sendai) 29: 337–351.

Anderson, J.M. and P. Ineson. 1984. Interactions between microorganisms and soil invertebrates in nutrient flux pathways of forest ecosystems. In: J.M. Anderson, A.D.M. Rayner, and D.W.H Walton, eds. Invertebrate-Microbial Interactions, pp. 59-88. Cambridge University Press, Cambridge.

Ashton, P.S. 1982. Dipterocarpaceae, Flora Malesiana. Nijhoff, The Hague.

Batra, L.R. and S.W. T. Batra. 1979. Termite-fungus mutualism. In: L.R. Batra, ed. Insect Fungus Symbiosis, pp. 117–163. Allaheld, Osmun, Montclair, NJ.

Benemann, J.R. 1973. Nitrogen fixation in termites. Science 181:164–165.

Bignell, D.E. 1994. Soil-feeding and gut morphology in higher termites. In: J.H. Hunt and C.A. Nalepa, eds. Nourishment and Evolution in Insect Societies, pp. 131–158. Westview Press, Boulder, CO.

Bignell, D.E., H. Oskarsson, J.M. Anderson, P. Ineson, and T.G. Wood. 1983. Structure, microbial associations and function of the so-called mixed segment of the gut in two soil-feeding termites, *Procubitermes auriensis* and *Cubitermes serverus* (Termitidae, Termitinae). Journal of Zoology (London) 201:445–480.

Breznak, J.A. 1984. Biochemical aspects of symbiosis between termites and their intestinal microbiota. In: J.M. Anderson, A.D.M. Rayner, and D.W.H. Walton, eds. Invertebrate-Microbial Interactions, pp. 173–203. Cambridge University Press, Cambridge.

Breznak, J.A and A. Brune. 1994. Role of microorganism in the digestion of lignocellulose by termites. Annual Review of Entomology 39:453–487.

Breznak, J.A., W.J. Brill, J.W. Mertins, and H.C. Coppell. 1973. Nitrogen fixation in termites. Nature (London) 244:577–580.

Brian, M.V. 1983. Social Insects. Chapman & Hall, London.

Buckley, R. 1987. Ant–plant–homopteran interactions. Advances in Ecological Research 16:53–85.

Chhotani, O.B. 1970. Taxonomy, zoogeography and phylogeny of the genus Cryptotermes (Isoptera: Kalotermitidae) from Oriental region. Memoirs of the Zoological Survey of India, Calcutta 15:1–81.

Collins, N.M. 1981a. The role of termites in the decomposition of wood and leaf litter in the Southern Guinea Savanna of Nigeria. Oecologia 51:389–399.

Collins, N.M. 1981b. Populations, age structure and survivorship of colonies of *Macrotermes bellicosus* (Isoptera: Macrotermitinae). Journal of Animal Ecology 50:293–311.

Collins, N.M. 1983a. Termite populations and their role in litter removal in Malaysian rain forests. In: S.L. Sutton, T.C. Whimore, and A.C. Chadwick, eds. Tropical Rain Forest; Ecology and Management, pp. 311–325. Blackwell Scientific, Oxford.

Collins, N.M. 1983b. The utilization of nitrogen resources by termites (Isoptera). In: J.A Lee, S. McNeill, and I.H. Rorison, eds. Nitrogen As an Ecological Factor, pp. 381–412. Blackwell Scientific, Oxford.

Collins, N.M. and T.G. Wood. 1984. Termite and atmospheric gas production. Science, 224:84-86.

Cowling, E.B. and W. Merrill. 1966. Nitrogen in wood and its role in wood deterioration. Canadian Journal of Botany 44:1539–1554.

Darlington, J.P.E.C. 1990. Populations in nests of the termite *Macrotermes subhyalinus* in Kenya. Insectes Sociaux 37:158–168.

Darlington, J.P.E.C. 1994. Nutrition and evolution in fungus-growing termites. In: J.H. Hunt and C.A. Nalepa, eds. Nourishment and Evolution in Insect Societies, pp. 105–130. Westview Press, Boulder, CO.

Darlington, P.J. Jr. 1957. Zoogeography: The Geographical Distribution of Animals. Wiley, New York.

Darwin, C. 1859. On the Origin of Species by Means of Natural Selection, or the Preservation of Favoured Races in the Struggle for Life. John Murray, London.

Deligne, J., A.C. Quennedey, and M.S. Blum. 1981. The enemies and defence mechanisms of termites. In: H.R. Hermann, ed. Social Insects, Vol. 2, pp. 1–76. Academic Press, New York.

Eggleton, P., P.H. Williams, and K.J. Gaston. 1994. Explaining global termite diversity: productivity or history? Biodiversity and Conservation 3:318–330.

Eggleton, P., D.E. Bignell, W.A. Sands, B. Waite, T.G. Wood, and J.H. Lawton. 1995. The species richness of termites (Isoptera) under differing levels of forest disturbance in the Mbalmayo Forest Reserve, southern Cameroon. Journal of Tropical Ecology 11:85–98.

Emerson, A.E. 1955. Geographical origins and dispersions of termite genera. Fieldiana Zoology 3: 465–521.

Emerson, A.E. 1967. Cretaceous insects from Labrador 3. A new genus and species of termite (Isoptera: Hodotermitidae). Psyche (Cambridge) 74:276–289.

Fittkau, E.J and H. Klinge. 1973. On the biomass and trophic structure of the central Amazon rain forest ecosystem. Biotropica 5:2–14.

Glover, P.E., E.C. Trump, and L.E.D. Wateridge. 1964. Termitaria and vegetation pattern on the Loita Plains of Kenya. Journal of Ecology 52:367–377.

Grassé, P.-P. and C. Noirot. 1959. L'evolution de la symbiose chez les Isopteres. Experientia (Basel) 15:365–372.

Harris, W.C. 1961. Termites: Their Recognition and Control. Longman, London.

Haverty, M.I. 1977. The proportion of soldiers in termite colonies: a list and bibliography (Isoptera). Sociobiology 2:199–216.

Higashi, M., T. Abe, and T.P. Burns. 1992. Carbon-nitrogen balance and termite ecology. Proceedings of the Royal Society of London Series B, Biological Sciences 249: 303–308.

Higashi, M., N. Yamamura, T. Abe, and T. Burns. 1991. Why don't all termite species have a sterile worker caste? Proceedings of the Royal Society of London Series B, Biological Sciences 246:25–29.

Hölldobler, B. and E.O. Wilson. 1990. The Ants. Harvard University Press, Cambridge.

Honigberg, B.M. 1970. Termite protozoa and their role in digestion. In: K. Krishna and F.M.K. Weesner, eds. Biology of Termites, Vol. 2, pp. 1–36. Academic Press, New York.

Humphries, C.J. 1981. Biogeographical methods and the southern beeches (Fagaceae: *Nothofugus*). In: V.A. Funk and D.R. Brooks, eds. Advances in Cladistics: Proceedings of the First Meeting of the Willi Hennig Society, pp. 177–207. New York Botanical Garden, New York.

Humphries, C.J. and L.R. Parenti. 1986. Cladistic Biogeography. Clarendon Press, Oxford.

Jander, R. and K. Daumer. 1974. Guide-line and gravity orientation of blind termite foraging in the open (Termitidae: *Macrotermes, Hospitalitermes*). Insectes Sociaux 21:45–69.

Jarzembowski, E.A. 1981. An early Cretaceous termite from southern England (Isoptera: Hodotermitidae). Systematic Entomology 6:91–96.

Jones, C.G., J.H. Lawton, and M. Shachak. 1994. Organisms as ecosystem engineers. Oikos 69:373–386.

Josens, G. 1972. Etudes biologique et ecologiques des termites (Isoptera) de la savane de Lamto-Pakobo (Cote d'Ivoire). Doctoral thesis, Free University of Brussels, Brussels.

Josens, G. 1985. The soil fauna of tropical savanna. III. The termites. In: F. Bourliere, ed. Ecosystems of the World, Vol. 13. Tropical Savanna, pp. 505–524. Elsevier, Amsterdam.

Khalil, M.A.K., R.A. Rasmussen, J.R.J. French, and J.A. Holt. 1990. The influence of termites on atmospheric trace gases: CH_4, CO_2, $CHCl_3$, N_2O, CO, H_2 and light hydrocarbons. Journal of Geophysical Research 95(D4):3619–3634.

LaFage, J.P. and W.L. Nutting. 1978. Nutrient dynamics of termites. In: M.V. Brian, ed.

Production Ecology of Ants and Termites, pp. 165–244. Cambridge University Press, Cambridge.

Lal, R. 1987. Tropical Ecology and Physiological Edaphology. Wiley, Chichester.

Lawton, J.H. 1994. What do species do in ecosystems? Oikos 71:367–374.

Lee, K.E. and T.G. Wood. 1971. Termites and Soils. Academic Press, New York.

Lenz, M. 1994. Food resources, colony growth and caste development in wood-feeding termites. In: J.H. Hunt and C.A. Nalepa, eds. Nourishment and Evolution in Insect Societies, pp. 159–209. Westview Press, Boulder, CO.

Lepage, M. 1974. Les termites d'une savane sahelienne (Ferlo septentrional, Senegal): peuplement, populations, consommation, role dans l'ecosysteme. Doctoral thesis, University of Dijon, Dijon.

Lepage, M.G. 1981a. L'impact de populations recoltantes de Macrotermes michaelseni (Sjostedt) (Isoptera: Macrotermitinae) dans un ecosysteme semi-aride (Kajiado, Kenya). Insectes Sociaux 28: 297–308.

Lepage, M.G. 1981b. Etude de la predation de *Megaponera foetens* (F.) sur les populations recoltantes Macrotermitinae dans un ecosysteme semiaride (Kajiado-Kenya). Insectes Sociaux 28:247–262.

Longhurst, C., R.A. Johnson, and T.G. Wood. 1978. Predation by *Megaponera foetens* (Fabr.) (Hymenoptera: Formicidae) on termites in the Nigerian Southern Guinea savanna. Oecologia 32: 101–107.

Maki, K. and T. Abe. 1986. Proportion of soldiers in the colonies of a dry wood termite, *Neotermes koshunensis* (Kalotermitidae, Isoptera). Physiology and Ecology, Japan 23:109–117.

Mann, K.H. 1973. Seaweeds: their productivity and strategy for growth. Science 182:975–981.

Margulis, L. and R. Fester, eds. 1991. Symbiosis As a Source of Evolutionary Innovation: Speciation and Morphogenesis. MIT Press, Cambridge.

Martin, M.M. 1987. Invertebrate-Microbial Interactions. Ingested Fungal Enzymes in Arthropod Biology. Cornell University Press, Ithaca.

Martius, C. 1994. Diversity and ecology of termites in Amazonian forests. Pedobiologia 38:407–428.

Matsumoto, T. 1976. The role of termites in an equatorial rain forest ecosystem of west Malaysia. 1. Population density, biomass, carbon, nitrogen and calorific content and respiration rate. Oecologia 22:153–178.

Matsumoto, T. and T. Abe. 1979. The role of termites in an equatorial rain forest ecosystem of west Malaysia. 2. Leaf litter consumption on the forest floor. Oecologia 38: 261–274.

McDowall, R.M. 1964. The affinities and derivation of the New Zealand fresh-water fish fauna. Tuatara 12:59–67.

Messer, A.C. and M.J. Lee. 1989. Effect of chemical treatment on methane emission by the hindgut microbiota in the termite *Zootermoposis angusticolli*. Microbial Ecology 18:275–284.

Miller, L.R. 1991. A revision of the *Termes-Capritermes* branch of the Termitinae in Australia (Isoptera: Termitidae). Invertebrate Taxonomy 4:1147–1282.

Myles, T.G. 1988. Resource inheritance in social evolution from termites to man. In: C.N. Slobodchikoff, ed. The Ecology of Social Behaviour, pp. 379–423. Academic Press, San Diego.

Nalepa, C.A. 1994. Nourishment and the origin of termite eusociality. In: J.H. Hunt and C.A. Nalepa, eds. Nourishment and Evolution in Insect Societies, pp. 57–104. Westview Press, Boulder, CO.

Noirot, C. 1990. Social structure in termite societies. Ethology Ecology & Evolution 1:1–17.

Noirot, C. 1992. From wood to humus feeding: an important trend in termite evolution. In: J. Billen, ed. Biology and Evolution of Social Insects, pp. 107–119. Leuven University Press, Leuven, Belgium.

Noirot, C. and J.M. Pasteels. 1987. Ontogenic development and evolution of worker caste in termites. Experientia (Basel) 43:851–852.

Nur, A. and Z. Ben Avraham. 1981. Lost Pacifica continent: a mobilistic speculation. In: G. Nelson and D.E. Rosen, eds. Vicariance Biogeography; a Critique, pp. 341–358. Columbia University Press, New York.

O'Brien, R.W. and R.W. Breznak. 1984. Enzymes of acetate and glucose metabolism in termites. Insect Biochemistry 14:639–643.

O'Brien, R.W. and M. Slaytor. 1982. Role of microorganisms in the metabolism of termites. Australian Journal of Biological Science 35:239–262.

Potrikus, C.J. and J.A. Breznak. 1981. Gut bacterial recycle uric acid nitrogen in termites: a strategy for nutrient conservation. Proceedings of the National Academy of Sciences of the United States of America 78:4601–4605.

Prestwich, G.D. and B.L. Bentley. 1981. Nitrogen fixation by intact colonies of the termite *Nasutitermes corniger*. Oecologia 49:249–251.

Prestwich, G.D., B.L. Bentley, and E.J. Carpenter. 1980. Nitrogen sources for neotropical nasute termite, fixation and selective foraging. Oecologia 46:379–401.

Rohrmann, G.F. 1978. The origin, structure and nutritional importance of the comb in two species of Macrotermitinae (Insecta: Isoptera). Pedobiologia 18:89–98.

Rohrmann, G.F. and A.Y. Rossman. 1980. Nutrient strategies of *Macrotermes ukuzii* (Isoptera: Termitidae). Pedobiologia 20:61–73.

Rosen, D.E. 1974. The phylogeny and zoogeography of salmoniform fishes and the relationships of *Lepidogalaxias salamandroides*. Bulletin of the American Museum of Natural History 153:265–326.

Roubik, D.W. 1989. Ecology and Natural History of Tropical Bees. Cambridge University Press, Cambridge.

Sands, W.A. 1969. The association of termites and fungi. In: K. Krishna and F.M. Weesner, eds. Biology of Termites, Vol. 1, pp. 495–524. Academic Press, New York.

Sinclair, A.R.E. and N. Norton-Griffiths, eds. 1979. Serengeti: Dynamics of an Ecosystem. Chicago University Press, Chicago.

Slaytor, M. 1992. Cellulose digestion in termites and cockroach: what role do symbionts play? Comparative Biochemistry and Physiology B 103:775–784.

Slaytor, M. and D.J. Chappell. 1994. Nitrogen metabolism in termites. Comparative Biochemistry and Physiology B 107:1–10.

Sugio, K. 1995. Trunk trail foraging of the fungus-growing termite *Macrotermes carbonarius* (Hagen) in southeastern Thailand. Tropics 4:211–222.

Swift, M.J., O.W. Heal, and J.M. Anderson. 1979. Decomposition in Terrestrial Ecosystem. Blackwell Scientific, Oxford.

Tayasu, I., A. Sugimoto, E. Wada, and T. Abe. 1994. Xylophagous termites depending on atmospheric nitrogen. Naturwissenschaften 81:229–231.

Veivers, P.C., R. Muhlenmann, M. Slaytor, R.H. Leuthold, and D.E. Bignell. 1991. Digestion, diet and polyethism in two fungus-growing termites: *Macrotermes subhyalinus* Rambur and *M. michaelseni* Sjostedt. Journal of Insect Physiology 37:675–682.

Waller, D.A. and J.P. LaFage. 1987. Nutritional ecology of termites. In: F. Slanky and J.G. Rodriguez, eds. Nutritional Ecology of Insects, Mites and Spiders, pp. 487–531. Wiley, New York.

Wilson, E.O. 1971. The Insect Societies. Belknap Press of Harvard University Press, Cambridge.

Wilson, E.O. 1975. Sociobiology: The New Synthesis. Belknap Press of Harvard University Press, Cambridge.

Wilson, E.O. 1990. Success and Dominance in Ecosystems: The Case of the Social Insects. Ecology Institute, Oldendorf/Luhe, Germany.

Wood, T.G. 1978. Food and feeding habits of termites. In: M.V. Brian, ed. Production Ecology of Ants and Termites, pp. 55–80. Cambridge University Press. Cambridge.

Wood, T.G. 1988. Termites and the soil environment. Biology and Fertility of Soils 6:228–236.

Wood, T.G. and R.A. Johnson. 1986. The biology, physiology and ecology of termites. In: S.B. Vinson, ed. Economic Impact and Control of Social Insects, pp. 1–67. Praeger, Greenwood Press, Westport, CT.

Wood, T.G. and W.A. Sands. 1978. The role of termites in ecosystems. In: M.V. Brian, ed. Production Ecology of Ants and Termites, pp. 245–292. Cambridge University Press. Cambridge.

Wood, T.G. and R.J. Thomas. 1989. The mutualistic association between Macrotermitinae and *Termitomyces*. In: N. Wilding, N.M. Collins, P.M. Hammond, and J.F. Webber, eds. Insect-Fungus Interactions, pp. 69–92. Academic Press, London.

Wood. T.G., R.A. Johnson, C.E. Ohiagu, N.M. Collins, and C. Longhurst. 1977. Ecology and importance of termites in crops and pastures in northern Nigeria. Project Report, 1973–1976. COPR, London.

Yamane, S., T. Abe, and J. Yukawa. 1992. Recolonization of the Krakataus by Hymenoptera and Isoptera (Insecta). Geojournal 28:213–318.

Zimmerman, P.R., J.P. Greenberg, S.O. Wandiga, and P.J. Crutzen. 1982. Termites: a potentially large source of atmospheric methane, carbon dioxide, and molecular hydrogen. Science 218:563–565.

III

Biodiversity and Ecological Complexity

7

Plant-Mediated Interactions Between Herbivorous Insects

Takayuki Ohgushi

Introduction

Traditional arguments on species interactions have focused mainly on classification of the interactions into several different types, such as competition, predation, and mutualism, through negative or positive impact of one species on the fitness of another species (Pianka 1983). The classification largely rests on an averaged effect of the first species on size or growth rate of the population, or individual fitness, in terms of survivorship and reproduction of the second species. If these properties of interacting species change in time and space, an outcome of the interaction could change accordingly. Commonly, impacts of the first species on the second species have the potential to change through changes in exogenous factors in local communities or endogenous factors in interacting species (Gilbert 1983; Thompson 1988a; Cushman and Whitham 1991; Singer and Parmesan 1993). Involved are population density and variability, and structure in terms of genotype or phenotype of interacting populations; environmental variables, and direct or indirect effects of another species on the interactions. For instance, population density may substantially change the intensity of the interaction by means of the probability of encounter with the interacting species (Ohgushi 1992a). When the interacting species approaches a high population density, the pairwise interaction becomes more apparent. Conversely, the interaction becomes less apparent when the population declines to a low density.

Increasing evidence of insect–plant interactions has provided rich sources of information indicating the variability and flexibility of species interactions (Price et al. 1980; Haukioja and Neuvonen 1987; Schultz 1988; Faeth 1987, 1991; Karban and Myers 1989; Ohgushi 1992b), confounded by temporal and spatial variation in secondary chemistry of the plant, abiotic and biotic factors, and other plant features, such as genotype, age, phenology, and architecture, that may interact with variation in secondary chemistry (Schultz 1988; Faeth 1991; Damman 1993). The variability and flexibility of these species interactions are of crucial importance for improving biodiversity in two ways. First, they could provide a new arena to maintain a wide variety of species interactions. Spatiotemporal variability and flexibility of species interactions have the potential to lead to more opportunities for noninteracting species to participate in the associations, compared to tight species associations. Second, insect–plant associations could increase species diversity in a local community by enhancing speciation through a process of coevolution (Ehrlich and Raven 1964; Futuyma and Slatkin 1983; Futuyma and Keese 1992).

In this context, reciprocal changes in plant chemical defenses and counteradaptation of specialist herbivores have been considered as a coevolutionary arms race (Thompson 1986; Berenbaum 1988; Caprio and Tabashnik 1992). Furthermore, resistance of plants against herbivores has been linked to particular chemical substances that have a genetic basis (Kennedy and Barbour 1992; Fritz 1992; Berenbaum and Zangerl 1992), suggesting the view of heritable genetic variation in plant resistance to herbivores (Edmunds and Alstad 1978; Whitham et al. 1984). Thus, herbivory may provide a selective force enhancing genetic diversity associated with chemical defenses within plant populations.

This chapter throws light on the dynamic aspects of insect–plant interactions. The particular focus is on how interspecific interactions among herbivorous insects vary through direct and indirect effects, mediated by changes in quality or quantity of a plant or plant parts.

Plant-Mediated Species Interactions

It has been generally thought that herbivorous insects are unlikely to compete with each other for host plants, because plants are rarely defoliated by associated herbivorous insects (Hairston et al. 1960; Lawton and Strong 1981). However, interest in interspecific competition among herbivorous insects has recently resurged with accumulating evidence that herbivory by one species can induce changes in quality or quantity in host plants that alter resource availability to other species which feed concurrently or subsequently on the same host plant (Schultz 1988; Karban and Myers 1989; Denno et al. 1995). In particular, induced changes in chemical, morphological, and phenological properties of plants are thought to deter herbivores by reducing their survival, growth rate, and fecundity.

Plant Responses to Herbivory

A wide variety of secondary metabolites in plants can have a significant role in protecting them from herbivore attack. Feeny (1976) distinguished two categories of defensive chemical substances: qualitative toxins (e.g., alkaloids, terpenoids, and hydrogen cyanide) and quantitative digestion inhibitors (e.g., tannins, lignins, and phenols). Qualitative chemicals are effective at low concentrations against nonadapted insects, while quantitative chemicals, operating in a dose-dependent manner, inhibit the digestive process of a wide range of insects. Several studies have reported negative correlations between larval performance and quantitative substances (Feeny 1976; Johnson et al. 1984; Bryant et al. 1987; Lindroth and

Peterson 1988; Karowe 1989) or qualitative substances (Erickson and Feeny 1974; Miller and Feeny 1989; Cates et al. 1987).

Injuries to plant tissues cause a wide array of plant responses. Physical damage by herbivores to leaves of plants induces changes in their secondary chemistry and nutritional quality (Haukioja et al. 1985; Edwards et al. 1986; Neuvonen et al. 1987; Silkstone 1987; Gibberd et al. 1988; Rossiter et al. 1988; Raupp and Sadof 1991). Some of these insect-inducible changes in plant quality have been considered as a defense strategy against herbivores, because even low levels of herbivory often reduce plant fitness considerably (Marquis 1984; Crawley 1985). For instance, proteinase inhibitors were immediately induced in leaves of potatoes and tomatoes when they were attacked by Colorado potato beetles (Green and Ryan 1972). Also, leaf damage by the autumnal moth *Epirrita autumnata* increased the total phenol content of a mountain birch, *Betula pubescens,* which reduced the growth rate and pupal weight of the moth, and the effect of the induced defense remained over subsequent years (Haukioja and Neuvonen 1985; Haukioja et al. 1985). Likewise, red oak trees that received heavy herbivory in the previous year exhibited significantly higher concentrations of tannins and phenols, and these chemicals had negative effects on larval growth of the gypsy moth *Lymantria dispar* (Schultz and Baldwin 1982). Herbivory also influences the morphology of their host plants by increasing the density of prickles, spines, and hairs, or by affecting the phenology of plant processes such as leaf abscission (Karban and Myers 1989; Myers and Bazely 1991).

It should be noted that many agents of environmental stress can affect or even override variation in secondary metabolites. Thus, plant defenses show enormous intra- and interspecific variation in response to the environmental light or soil nutrition (Coley et al. 1985; Bazzaz et al. 1987; Bryant et al. 1987; Shure and Wilson 1993). For example, slow-growing plants adapted to resource-limited habitats have higher replacement costs for damaged leaves and hence higher levels of defense. On the other hand, fast-growing plants adapted to high resource habitats invest very little in defense, but compensate rapidly for damage by rapid regrowth (Coley et al. 1985).

Interactions Among Herbivore Guild Species

Janzen (1973) first postulated the possibility that herbivores feeding on different parts of a plant, or at different times, are still involved in significant interactions. There is recently increasing evidence that herbivorous insects sharing the same host plant often substantially affect each other by mediating host plant quality even at low levels of herbivory (Faeth 1987, 1991; Karban and Myers 1989; Damman 1993). Interactions among herbivore guild species involve the following processes: (1) insect 1 feeds on a plant, (2) the plant changes chemical composition or morphology in response to the herbivory, and (3) these changes of the plant, in turn, affect survival and reproduction of insect 2 (Fig. 7.1).

Interactions Between Temporally Separated Species

Plants damaged by herbivores often change nutrient levels and composition of secondary compounds, which are of substantial importance in defining the suitability of food to subsequent herbivores. In this context, feeders early in the season can indirectly influence the abundance and survival of subsequent feeders by altering the quantity or suitability of a common plant resource (Karban and Carey 1984; Harrison and Karban 1986; Faeth 1985, 1986, 1987, 1988; Hunter 1987, 1992; Leather 1993). In return, if the effects of late feeders are carried over to the next growing season, the late feeders may have an impact on early feeders in the next season (Faeth 1988). These plant-mediated interactions among temporally separated species could occur commonly in nature under even low levels of herbivory.

Leaves of the emory oak *Quercus emoryi* are attacked by several species of insect chewers including Lepidoptera, Coleoptera, and Orthoptera from April through May, and then by leaf-mining species including Lepidoptera from May through July. Faeth (1986) found that survival of the leaf miners in leaves previously damaged by the spring chewers was significantly lower than that in intact leaves because of increased parasitism in damaged leaves. Also, early-season herbivory resulted in higher levels of condensed tannins and lower levels of protein in damaged leaves, which reduced the size of the leaf miner, probably affecting

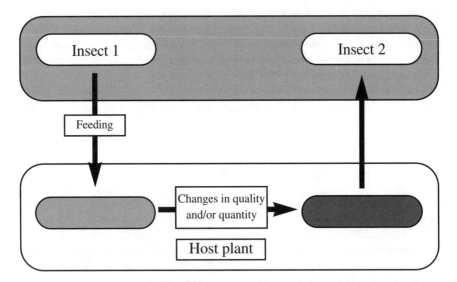

FIGURE 7.1. Interactions among herbivore guild species (*arrows*), mediated by changes in a host plant. Plant-mediated interactions consist of the following processes: (1) insect 1 feeds on a plant; (2) the plant changes chemical composition or morphology in response to the herbivory; and (3) these changes of the plant, in turn, affect survival and reproduction of insect 2. The reverse is also true for the effect of insect 2 on insect 1.

fecundity (Faeth 1988). Parasitism was enhanced on damaged leaves, probably because parasitoids used physical and chemical changes as cues to locate leaf miners (Faeth and Bultman 1986; Faeth 1991), or because exposure of leaf miners to parasitoids was prolonged.

In a similar way, within a tree of the pedunculate oak *Quercus robur,* survival to pupation of a leaf miner, *Phyllonorycter,* a late-season feeder, was significantly poorer on leaves that previously had been damaged by lepidopteran larvae in the spring (Hunter 1992). This interaction seems to be mediated by direct chemical changes rather than by increased parasitism. In an artificial herbivory experiment on the bird cherry to mimic leaf damage by the ermine moth *Yponomeuta evonymellus* in spring, Leather (1993) found that defoliated trees had significantly less damage from the bird cherry-oat aphid *Rhopalosiphum padi* than those trees which were undefoliated or suffered only slight defoliation. Defoliation also induced significant changes in leaf chemistry at the time of aphid colonization, increasing nitrogen but decreasing calcium. Larvae of the ranchman's tiger moth *Platyprepia virginalis* and the western tussock moth *Orgyia vetusta* both feed on the bush lupine *Lupinus arboreus.* The former appears from February through April and the latter from May through July. Harrison and Karban (1986) demonstrated that feeding by the tiger moth larvae in early spring affected the suitability of the host plant for the tussock moth larvae feeding late in the season. Growth rates of first-instar larvae of the tussock moth were significantly reduced when they grew on damaged branches. Also, spring feeding by the tiger moth significantly reduced the pupal weight of the tussock moth and thus its fecundity. It is suggested that early herbivory by the tiger moth has the potential to change nitrogen content in subsequently emerging leaves of the host plant, which may reduce performance of the tussock moth.

However, early-season herbivory does not always affect late-emerging insects negatively. For instance, the fall webworm *Hyphantria cunea,* feeding on leaves of the red alder from August through September, had a larger pupal size and higher pupation success on trees that had been previously damaged by the western tent caterpillar *Malacosoma californicum pluviale* (Williams and Myers 1984). The heavier pupae on damaged trees actually could result in a 12.5% improvement in fecundity over pupae on undamaged trees.

Interactions Between Spatially Separated Species

Recent studies have begun to focus on interactions between spatially separated insects that share one plant but utilize different parts of it. Studies have demonstrated that insects interact significantly with each other through changes in the way energy is allocated within the plant (Gange and Brown 1989; Moran and Whitham 1990; Masters and Brown 1992; Masters et al. 1993). These plant-mediated interactions among spatially separated species stem from the fact that tissues of an individual plant depend on a common resource budget.

Moran and Whitham (1990) demonstrated a host-mediated interaction between two aphid species that share *Chenopodium album.* One aphid species, *Hayhurstia atriplicis,* makes leaf galls, while the other aphid, *Pemphigus betae,* feeds underground on roots. They evaluated effects of one aphid species on another, using plants both resistant and susceptible to the leaf galler. The root feeder had,no significant effects on its host. On the other hand, the leaf feeder largely reduced root biomass. As a result, numbers of the root feeder significantly decreased, often being eliminated entirely. In contrast to the negative impact of the leaf galler, the root feeder had little effect on the numbers of the leaf galler. Thus, the interaction between the two aphid species was highly asymmetrical.

Conversely, the garden chafer *Phyllopertha horticola,* feeding on the roots of an annual herb, *Capsella bursa-pastoris,* improves performance of the sap-sucking aphid *Aphis fabae* (Gange and Brown 1989). The root feeder induced water stress to the host plant by marked reduction in vegetative biomass, resulting in an increase in nitrogen. The enhanced host quality improved the growth rate and longevity and thus the fecundity of the aphid. The aphid produced more offspring in the presence of the root feeder. On the other hand, the aphid had no detectable effects on either the host plant or the garden chafer. However, the garden chafer interacts with the dipteran leaf miner *Chromatomyia syngenesiae* in quite a different way on the host plant *Sonchus oleraceus* (Masters and Brown 1992). Root herbivory increased the pupal weight of the leaf miner and thus its fecundity, probably because of changes in host quality by root feeding. In contrast, leaf herbivory reduced the growth rate of the root feeder, because herbivory by the leaf miner reduced root biomass considerably. The interaction is symmetrical, but exhibited effects in opposing directions; the leaf miner benefits from the root herbivory but performance of the root feeder is reduced by the leaf miner.

Masters et al. (1993) suggested a possible mechanism for plant-mediated interactions between above- and belowground insect herbivores. Root herbivory reduces the biomass of the root, which is important in taking up water and nutrients essential for plant growth. Consequently, the root herbivores may induce a stress response within the host plant that leads to the accumulation of soluble amino acids and carbohydrates in the foliage. The improved host quality may, in turn, lead to increased performance of the aboveground insect herbivores. On the other hand, foliar feeding has a negative effect on the belowground herbivore by reducing the root biomass available to the root feeder.

Interactions Between Taxonomically Separated Species

Highly unrelated organisms sharing a host plant may interact strongly. Karban et al. (1987) demonstrated the plant-mediated interaction between a spider mite, *Tetranychus urticae,* and a vascular wilt fungus, *Verticillium dahliae,* on cotton. When cotton seedlings were previously damaged by the spider mite, the severity of symptoms by the wilt fungus and the probability of fungal infection were highly suppressed. On the other hand, spider mite population growth was significantly reduced on cotton seedlings that were infected with the fungus. The negative effect is likely to result from the reduction in amount of leaf tissue caused by fungal infection. Thus, the fungus and the herbivorous mite can strongly and negatively affect each other.

Regrowth of plants following herbivory can change plant architecture by increasing the biomass of vegetative and reproductive parts (Paige and Whitham 1987; Mopper et al. 1991). Such architectural modification by one herbivore may indirectly enhance resources available to another species. When browsed heavily by mule deer or elk in spring, a scarlet gilia, *Ipomopsis arizonica,* increases florescences and stalks. The increased inflorescences, in turn, maintain a larger population of a noctuid caterpillar that feeds exclusively on the fruits of the scarlet gilia (Mopper et al. 1991).

Interactions Between Co-Occurring Species

Both a calendula plume moth, *Platyptilia williamsii,* and a meadow spittlebug, *Philaenus spumarius,* feed in the terminal shoot of the rosette of the seaside daisy *Erigeron glaucus.* Karban (1986) found that the spittlebugs without the moth caterpillars were consistently nearly 40% larger than spittlebugs that co-occurred with the moth caterpillars. Because the moth larvae consumed the entire terminal bud and greatly reduced new leaf production, the negative effect of the plume moths on spittlebug persistence may be duplicated by making the vegetative bud and newly expanded leaves physically unavailable to spittlebug nymphs. On the other hand, the spittlebug had little effect on the performance of the moth larvae. Thus, the interaction between the co-occurring herbivorous insects was highly asymmetrical.

However, in a system of two species sharing a host plant, one species does not always have negative effects on another. Damman (1989) demonstrated a facilitative interaction between two lepidopteran species sharing the pawpaw *Asimina* spp. When the pawpaw was highly damaged by the pyralid moth larvae, it produced a number of young shoots by a compensating response. The increased new leaves resulted in increased larval abundance of the swallowtail butterfly *Eurytides marcellus,* which fed exclusively on young leaves. Coupled with regular and severe defoliation by the moth, lack of a defensive response to leaf damage by the plant is more likely to cause a positive effect on population size of the butterfly larva. It should be also noted that herbivorous insects which share a same plant often benefit from other species by changes in host plant characteristics following herbivory (see Damman 1993 for a review).

The evidence reviewed here clearly illustrates that substantial interactions occur among temporally, spatially, and even taxonomically separated species through changes in resource availability, which long has been ignored in earlier arguments of species interactions in community ecology. The traditional view suggests that interactions should be most prevalent among closely related species within guilds and among species that utilize a host plant simultaneously. If interactions commonly occur between species that are highly separated in terms of time, space, and phylogeny, ecologists must revise the importance of these interspecific effects on the community structure of organisms that share a host plant.

Three-Trophic-Level Interactions

Three-trophic-level interactions have recently received much attention in insect–plant interactions. Changes in host plant quality may directly affect survivorship and reproduction of herbivorous insects through increases in plant toxins or decreases in nutrition, but may also indirectly increase vulnerability of insect hosts to predators and parasitoids (Fig. 7.2) (Vinson 1976; Price et al. 1980; Schultz 1983). This view addresses a significant role of the third trophic level (natural enemies) as part of a plant's battery of defenses against herbivores. Plants have evolved to manipulate natural enemies of herbivorous insects through induced responses as another line of defense against herbivores (Price et al. 1980). Induced defenses via natural enemies are thought to operate in several ways: (1) natural enemies can use induced chemical and physical cues associated with damage to locate their hosts (Vinson 1976; Faeth 1986); (2) avoidance of damaged leaves may cause increased movement of insects searching for suitable feeding sites such that exposure and apparency to natural enemies are increased (Edwards and Wratten 1983; Schultz 1983); and (3) induced chemicals such as tannins may prolong development time so that probability of discovery by natural enemies is enhanced (Price et al. 1980; Faeth 1985, 1986).

As mentioned earlier, Faeth (1986) examined experimentally whether leaf herbivory increased parasitism of leaf miners feeding on the emory oak *Quercus emoryi*. Mortality of leaf miners on damaged leaves was significantly higher than that on intact leaves, mainly because of enhanced parasitism in response to higher levels of condensed tannins in damaged leaves. It is likely that leaf herbivory induced these chemical changes in the oak leaves, thus providing cues for parasitic wasps to locate the leaf miners, or prolonged the exposure of leaf miners available to parasitoids.

Induced changes in plant quality, however, may interfere with the operation of natural enemies. Larvae of the gypsy moth *Lymantria dispar* feeding on leaves of the red oak *Quercus rubra* benefit from induced changes in host quality because of reduced mortality from the nuclear polyhedrosis virus that usually causes high larval mortality. Hunter and Schultz (1993) demonstrated that the gypsy moth larvae were less susceptible to the virus on foliage with higher levels of defoliation. Foliar defoliation induced higher concentrations of gallotannin, which increased inhibition of the virus. In other words, the gypsy moth can alter the leaf quality of a host plant and gain protection from its own virus. The result would be a balance between direct deleterious forces acting on the herbivore and indirect beneficial forces acting through inhibition of the herbivore's pathogens.

Three-trophic-level interactions mediated by plant quality are also seen in ant–homopteran insect mutualisms. Host plants can indirectly affect the fitness of many homopterans by mediating the outcome of their interactions with ants. Having examined a mutualism in a membracid–ant association, Cushman (1991) emphasized that differences in quality of the reward (honeydew) causes a significant variation in the strength and sign of herbivore–ant interactions. He suggested the possibility that ants act as predators on the herbivores providing rewards to

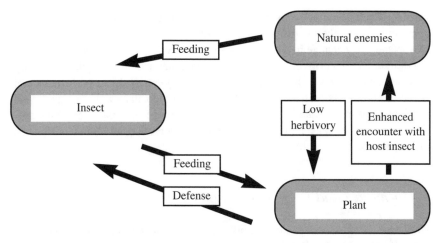

FIGURE 7.2. Three-trophic-level interactions (*arrows*) consisting of a plant, insect herbivores, and natural enemies of host insects. Induced chemical substances caused by herbivore attack directly affect survivorship and reproduction of insects but may also indirectly enhance vulnerability of insect hosts to predators or parasitoids.

ants. Ant attendance on the herbivores will be determined in part by the nutritional content and quality of rewards, which are largely dependent on the quality of host plants of the homopterans. Thus, homopterans feeding on high-quality host plants are well guarded by ants because high-quality rewards are presented. On the other hand, homopterans on low-quality host plants produce secretions low in nutrition and with low nitrogen. In this case, ants may benefit more from directly attacking the insects than from getting poor honeydew rewards.

Future Directions

The chemical or morphological plant responses following herbivory have significant consequences for intensity and variability of direct or indirect effects on the interacting species that share a host plant. For further understanding of the plant-mediated species interactions, I draw some future directions that are chosen for their relative importance and lack of adequate coverage in the literature on insect–plant interactions.

Field Manipulations to Separate Direct and Indirect Effects

Indirect effects undoubtedly play a significant role in enhancing the variation in outcomes of plant-mediated interactions (Price et al. 1980; Faeth 1987; Strauss 1991; Masters et al. 1993). The outcomes of interactions between any two species can be affected indirectly by additional interactions with other species within communities. In other words, strength and often direction of interactions between two species may change in the presence of others. Thus, a more exact understanding of indirect effects involved in plant-mediated interactions is required to clarify how induced changes of host plants directly and indirectly affect performance of herbivorous insects and their natural enemies. To do this, well-designed field manipulations to separate indirect and direct effects are essential, employing artificial herbivory or cage experiments in adding or removing additional herbivores or their natural enemies in a system.

Performance and Preference of Adult Herbivores

Until now, most studies evaluating effects by one insect species on another have focused mainly on the larval performance of the insect species in terms of survival, development, and body weight. In contrast, we know little about how induced changes in host plants affect adult performance, such as fecundity, reproductive life span, and mating success, which may substantially determine lifetime fitness. For example, poor resource availability in a larval period could have a negative impact on fecundity or adult survival in the reproductive season because of reduced adult size. As components of lifetime fitness are not always positively correlated with each other (Thompson 1988b), when estimated fitness covers only a larval period partial analyses may fail to reveal important consequences of induced changes for fitness components through lifetime. Thus, without monitoring adult performance, misleading conclusions may be reached. For instance, a shorter development period in the larval stage, which is thought to enhance larval performance, may result in reduction of fecundity or reproductive life span for oviposition.

For many herbivorous insects, the searching abilities of larvae are poor relative to those of adults, and thus oviposition behavior is of paramount importance in selecting suitable host plants or plant parts for their offspring (Thompson 1988b; Thompson and Pellmyr 1991). Hence, oviposition behavior has the potential to determine offspring performance when larval survival and development are largely dependent on conditions of individual plants. Because ovipositing females of many herbivorous insects use chemical stimuli from the host plant to locate it (Renwick and Chew 1994), oviposition behavior could be affected indirectly by other insect species that change host plant quality and cues for ovipositing females. Host plant selection by ovipositing females may, in turn, change outcomes of interactions among insects in subsequent generations by distributing offspring on different individual plants associated with different herbivore species (Maddox and Root 1990). Nevertheless, little is known about how changes in host plant characteristics imposed by other herbivores affect the process of host plant selection by ovipositing females. In addition, reproductive performance including fecundity, reproductive life span, and mating success provides physiological constraints to females in choice of oviposition sites. We should therefore pay much more attention to the effects of induced changes in host plants on both performance and oviposition preference of adult females, with equal emphasis on larval performance.

Long-Term Consequences of Induced Changes

Induced chemical changes in response to herbivory are highly variable in initiation and persistence. They are detectable within a few hours, days, or weeks, and last a few hours, days, weeks, or even years (Haukioja and Neuvonen 1987; Schultz 1988). If induced defense persists across several years, it could have substantial effects on the interactions and population dynamics of insect herbivores in subsequent years, by changing not only resource availability but also the probability of encounter with different guild species because of changes in population size (Rhoades 1983; Haukioja and Neuvonen 1987; Myers 1988). Also, even the responses of a plant to the same type or same amount of herbivory may differ among different years partly because of different physiological conditions (Haukioja and Neuvonen 1987). Through changes in resource allocation within a plant, induced chemical changes in a given year could alter plant responses to herbivory in subsequent years. Therefore, we need long-term monitoring of changes in host plant characteristics induced by herbivory and the consequences for plant-mediated interactions.

Population-Based Approach

A long-term population-based approach, using life table analysis and mark-recapture experiments, will greatly contribute to understanding the dynamic aspects of plant-mediated species interactions. Detailed information on the survival and reproductive processes of herbivorous insects relevant to the interaction is necessary to evaluate the intensity and variability of effects of induced changes in the plant on lifetime performance and oviposition preference. In particular, the life table was originally designed to estimate demographic parameters of survivorship and reproduction over the lifetime.

Mark-recapture techniques applied to adult females provide a powerful tool to evaluate detailed processes of oviposition site selection by individual females (Ohgushi 1992b). Also, long-term changes in size and structure of populations of insects and host plants will provide much insight into how demographic features of interacting species affect outcomes of the plant-mediated interaction. Coupled with well-designed field manipulations, a long-term population study will reveal the underlying mechanisms responsible for temporal and spatial variations in the plant-mediated interactions and thereby teach us much about the dynamic features of species interactions.

Summary

Recent studies on insect–plant interactions have revealed that interspecific inter-actions are often indirect, asymmetrical, and subtle, and that morphological, phe-nological, and chemical changes in the host plant alter the success of predation or parasitism by natural enemies. Even insects feeding at different times or on differ-ent parts of a plant may have a substantial effect on the quality or quantity of resources available to one another. Three-trophic-level interactions have recently received much attention in insect–plant interactions. This approach addresses a significant role of the third trophic level (natural enemies) as part of a plant's bat-tery of defenses against herbivores. It also recognizes that changes in host plant quality directly or indirectly affect the efficacy of parasitoids or predators by alter-ing insect host location or vulnerability. Interactions between temporally sepa-rated guilds may be critical in distribution and survivorship and thus in population dynamics of insect herbivores. In this context, recent increasing evidence supports the view that one species attacking the host plant early in the season can change the performance or abundance of another species attacking late in the season, mediated by changes in host quality. Likewise, this notion is applicable to the sit-uation of spatially separated guilds that utilize different parts of the shared host plant in very different manners; for example, one attacking leaves and another the roots. Such interactions among temporally or spatially separated guild members sharing the same host plant are often asymmetrical.

Acknowledgments. I thank Peter Price and four anonymous reviewers for their valuable comments on the earlier version of this chapter. Financial support was provided by a Japan Ministry of Education, Science and Culture Grant-in-Aid for Scientific Research on Priority Areas (#319).

Literature Cited

Bazzaz, F.A., N.R. Chiariello, P.D. Coley, and L.F. Pitelka. 1987. Allocating resources to reproduction and defense. Bioscience 37:58–67.

Berenbaum, M.R. 1988. Allelochemicals in insect-microbe-plant interactions; agents provocateurs in the coevolutionary arms race. In: P. Barbosa and D.K. Letourneau, eds. Novel Aspects of Insect–Plant Interactions, pp. 97–123. Wiley, New York.

Berenbaum, M.R. and A.R. Zangerl. 1992. Genetics of secondary metabolism and herbi-vore resistance in plants. In: G.A. Rosenthal and M.R. Berenbaum, eds. Herbivores: Their Interactions with Secondary Plant Metabolites, 2nd Ed., pp. 415–438. Academic Press, San Diego.

Bryant, J.P., T.P. Clausen, P.B. Reichardt, M.C. McCarthy, and R.A. Werner. 1987. Effect of nitrogen fertilization upon the secondary chemistry and nutritional value of quaking aspen (*Populus tremuloides* Michx.) leaves for the large aspen tortrix (*Choristoneura conflictana* (Walker)). Oecologia 73:513–517.

Caprio, M.A. and B.E. Tabashnik. 1992. Evolution of resistance to plant defensive chemi-cals in insects. In: B.D. Roitberg and M.B. Isman, eds. Insect Chemical Ecology: An Evolutionary Approach, pp. 179–215. Chapman & Hall, New York.

Cates, R.G., C.B. Henderson, and R.A. Redak. 1987. Responses of the western spruce bud-worm to varying levels of nitrogen and terpenes. Oecologia 73:312–316.

Coley, P.D., J.P. Bryant, and F.S.I. Chapin. 1985. Resource availability and plant antiher-bivore defense. Science 230:895–899.

Crawley, M.J. 1985. Reduction of oak fecundity by low-density herbivore populations. Nature 314:163–164.

Cushman, J.H. 1991. Host-plant mediation of insect mutualisms: variable outcomes in her-bivore-ant interactions. Oikos 61:138–144.

Cushman, J.H. and T.G. Whitham. 1991. Competition mediating the outcome of a mutual-ism: protective services of ants as a limiting resource for membracids. American Natu-ralist 138:851–865.

Damman, H. 1989. Facilitative interactions between two lepidopteran herbivores of *Asim-ina*. Oecologia 78:214–219.

Damman, H. 1993. Patterns of interaction among herbivore species. In: N.E. Stamp and T.M. Casey, eds. Caterpillars, pp. 132–169. Chapman & Hall, New York.

Denno, R.F., M.S. McClure, and J.R. Ott. 1995. Interspecific interactions in phytophagous insects: competition reexamined and resurrected. Annual Review of Entomology 40:297–331.

Edmunds, G.F.J. and D.N. Alstad. 1978. Coevolution in insect herbivores and conifers. Science 199:941–945.

Edwards, P.J., and S.D. Wratten. 1983. Wound-induced defences in plants and their conse-quences for patterns of insect grazing. Oecologia 59:88–93.

Edwards, P.J., S.D. Wratten, and S. Greenwood. 1986. Palatability of British trees to insects: constitutive and induced defences. Oecologia 69:316–319.

Ehrlich, P.R., and P.H. Raven. 1964. Butterflies and plants: a study in coevolution. Evolu-tion 18:586–608.

Erickson, J.M. and P. Feeny. 1974. Sinigrin: a chemical barrier to the black swallowtail butterfly, *Papilio polyxenes*. Ecology 55:103–111.

Faeth, S.H. 1985. Host leaf selection by leaf miners: interactions among three trophic lev-els. Ecology 66:870–875.

Faeth, S.H. 1986. Indirect interactions between temporally separated herbivores mediated by the host plant. Ecology 67:479–494.

Faeth, S.H. 1987. Community structure and folivorous insect outbreaks: the roles of verti-cal and horizontal interactions. In: P. Barbosa and J.C. Schultz, eds. Insect Outbreaks, pp. 137–171. Academic Press, San Diego.

Faeth, S.H. 1988. Plant-mediated interactions between seasonal herbivores: enough for evolution or coevolution? In: K.C. Spencer, ed. Chemical Mediation of Coevolution, pp. 391–414. Academic Press, New York.

Faeth, S.H. 1991. Variable induced responses: direct and indirect effects on oak folivores. In: D.W. Tallamy and M.J. Raupp, eds. Phytochemical Induction by Herbivores, pp. 293–323. Wiley, New York.

Faeth, S.H. and T.L. Bultman. 1986. Interacting effects of increased tannin levels on leaf-mining insects. Entomologia Experimentalis et Applicata 40:297–300.

Feeny, P. 1976. Plant apparency and chemical defense. Recent Advances in Phytochem-istry 10:1–40.

Fritz, R.S. 1992. Community structure and species interactions of phytophagous insects on resistant and susceptible host plants. In: R.S. Fritz and E.L. Simms, eds. Plant Resis-tance to Herbivores and Pathogens, pp. 240–277. University of Chicago Press, Chicago.

Futuyma, D.J. and M.C. Keese. 1992. Evolution and coevolution of plants and phy-tophagous arthropods. In: G.A. Rosenthal and M.R. Berenbaum, eds. Herbivores: Their

Interactions with Secondary Plant Metabolites, 2nd Ed., pp. 439–475. Academic Press, San Diego.

Futuyma, D.J. and M. Slatkin. 1983. Coevolution. Sinauer, Sunderland, MA.

Gange, A.C. and V.K. Brown. 1989. Effects of root herbivory by an insect on a foliar-feeding species, mediated through changes in the host plant. Oecologia 81:38–42.

Gibberd, R., P.J. Edwards, and S.D. Wratten. 1988. Wound-induced changes in the acceptability of tree-foliage to Lepidoptera: within-leaf effects. Oikos 51:43–47.

Gilbert, L.E. 1983. Coevolution and mimicry. In: D.J. Futuyma and M. Slatkin, eds. Coevolution, pp. 263–281. Sinauer, Sunderland, MA.

Green, T.R. and C.A. Ryan. 1972. Wound-induced proteinase inhibitor in plant leaves: a possible defense mechanism against insects. Science 175:776–777.

Hairston, N.G., F.E. Smith, and L.B. Slobodkin. 1960. Community structure, population control, and competition. American Naturalist 94:421–425.

Harrison, S. and R. Karban. 1986. Effects of an early-season folivorous moth on the success of a later-season species, mediated by a change in the quality of the shared host, *Lupinus arboreus* Sims. Oecologia 69:354–359.

Haukioja, E. and S. Neuvonen. 1985. Induced long-term resistance of birch foliage against defoliators: defensive or incidental? Ecology 66:1303–1308.

Haukioja, E. and S. Neuvonen. 1987. Insect population dynamics and induction of plant resistance: the testing of hypotheses. In: P. Barbosa and J.C. Schultz, eds. Insect Outbreaks, pp. 411–432. Academic Press, San Diego.

Haukioja, E., P. Niemela, and S. Siren. 1985. Foliage phenols and nitrogen in relation to growth, insect damage, and ability to recover after defoliation, in the mountain birch *Betula pubescens* ssp. *tortuosa.* Oecologia 65:214–222.

Hunter, M.D. 1987. Opposing effects of spring defoliation on late season oak caterpillars. Ecological Entomology 12:373–382.

Hunter, M.D. 1992. Interactions within herbivore communities mediated by the host plant: the keystone herbivore concept. In: M.D. Hunter, T. Ohgushi, and P.W. Price, eds. Effects of Resource Distribution on Animal–Plant Interactions, pp. 287–325. Academic Press, San Diego.

Hunter, M.D. and J.C. Schultz. 1993. Induced plant defenses breached? Phytochemical induction protects an herbivore from disease. Oecologia 94:195–203.

Janzen, D.H. 1973. Host plants as islands. II. Competition in evolutionary and contemporary time. American Naturalist 107:786–790.

Johnson, N.D., C.C. Chu, P.R. Ehrlich, and H.A. Mooney. 1984. The seasonal dynamics of leaf resin, nitrogen, and herbivore damage in *Eriodictyon californicum* and their parallels in *Diplacus aurantiacus.* Oecologia 61:398–402.

Karban, R. 1986. Interspecific competition between folivorous insects on *Erigeron glaucus.* Ecology 67:1063–1072.

Karban, R. and J.R. Carey. 1984. Induced resistance of cotton seedlings to mites. Science 225:53–54.

Karban, R. and J.H. Myers. 1989. Induced plant responses to herbivory. Annual Review of Ecology and Systematics 20:331–348.

Karban, R., R. Adamchak, and W.C. Schnathorst. 1987. Induced resistance and interspecific competition between spider mites and a vascular wilt fungus. Science 235:678–680.

Karowe, D.N. 1989. Differential effect of tannic acid on two tree-feeding Lepidoptera: implications for theories of plant anti-herbivore chemistry. Oecologia 80:507–512.

Kennedy, G.G. and J.D. Barbour. 1992. Resistance variation in natural and managed systems. In: R.S. Fritz and E.L. Simms, eds. Plant Resistance to Herbivores and Pathogens, pp. 13–41. University of Chicago Press, Chicago.

Lawton, J.H. and D.R. Strong Jr. 1981. Community patterns and competition in folivorous insects. American Naturalist 118:317–338.

Leather, S.R. 1993. Early season defoliation of bird cherry influences autumn colonization by the bird cherry aphid, *Rhopalosiphum padi*. Oikos 66:43–47.

Lindroth, R.L. and S.S. Peterson. 1988. Effects of plant phenols on performance of southern armyworm larvae. Oecologia 75:185–189.

Maddox, G.D. and R.B. Root. 1990. Structure of the encounter between goldenrod (*Solidago altissima*) and its diverse insect fauna. Ecology 71:2115–2124.

Marquis, R.J. 1984. Leaf herbivores decrease fitness of a tropical plant. Science 226:537–539.

Masters, G.J. and V.K. Brown. 1992. Plant-mediated interactions between two spatially separated insects. Functional Ecology 6:175–179.

Masters, G.J., V.K. Brown, and A.C. Gange. 1993. Plant-mediated interactions between above- and below-ground insect herbivores. Oikos 66:148–151.

Miller, J.S. and P.P. Feeny. 1989. Interspecific differences among swallowtail larvae (Lepidoptera: Papilionidae) in susceptibility to aristolochic acids and berberine. Ecological Entomology 14:287–296.

Mopper, S., J. Maschinski, N. Cobb, and T.G. Whitham. 1991. A new look at habitat structure: consequences of herbivore-modified plant architecture. In: S.S. Bell, E.D. McCoy, and H.R. Mushinsky, eds. Habitat Structure, pp. 260–280. Chapman & Hall, London.

Moran, N.A. and T.G. Whitham. 1990. Interspecific competition between root-feeding and leaf-galling aphids mediated by host-plant resistance. Ecology 71:1050–1058.

Myers, J.H. 1988. Can a general hypothesis explain population cycles of forest Lepidoptera? Advances in Ecological Research 18:179–242.

Myers, J.H. and D. Bazely. 1991. Thorns, spines, prickles, and hairs: are they stimulated by herbivory and do they deter herbivores? In: D.W. Tallamy and M.J. Raupp, eds. Phytochemical Induction by Herbivores, pp. 325–344. Wiley, New York.

Neuvonen, S., E. Haukioja, and A. Molarius. 1987. Delayed inducible resistance against a leaf-chewing insect in four deciduous tree species. Oecologia 74:363–369.

Ohgushi, T. 1992a. From population to species interaction. In: M. Higashi and T. Abe, eds. What Is Symbiotic Biosphere? pp. 200–217. Heibon-sha, Tokyo.

Ohgushi, T. 1992b. Resource limitation on insect herbivore populations. In: M.D. Hunter, T. Ohgushi, and P.W. Price, eds. Effects of Resource Distribution on Animal–Plant Interactions, pp. 199–241. Academic Press, San Diego.

Paige, K.N. and T.G. Whitham. 1987. Overcompensation in response to mammalian herbivory: the advantage of being eaten. American Naturalist 129:407–416.

Pianka, E.R. 1983. Evolutionary Ecology, 3rd Ed. Harper & Row, New York.

Price, P.W., C.E. Bouton, P. Gross, B.A. McPheron, J.N. Thompson, and A.E. Weis. 1980. Interactions among three trophic levels: influence of plants on interactions between insect herbivores and natural enemies. Annual Review of Ecology and Systematics 11:41–65.

Raupp, M.J. and C.S. Sadof. 1991. Responses of leaf beetles to injury-related changes in their salicaceous hosts. In: D.W. Tallamy and M.J. Raupp, eds. Phytochemical Induction by Herbivores, pp. 183–204. Wiley, New York.

Renwick, J.A.A. and F.S. Chew. 1994. Oviposition behavior in lepidoptera. Annual Review of Entomology 39:377–400.

Rhoades, D.F. 1983. Herbivore population dynamics and plant chemistry. In: R.F. Denno and M.S. McClure, eds. Variable Plants and Herbivores in Natural and Managed Systems, pp. 155–220. Academic Press, New York.

Rossiter, M.C., J.C. Schultz, and I.T. Baldwin. 1988. Relationships among defoliation, red oak phenolics, and gypsy moth growth and reproduction. Ecology 69:267–277.

Schultz, J.C. 1983. Impact of variable plant defensive chemistry on susceptibility of insects to natural enemies. In: P. Hedin, ed. Plant Resistance to Insects, pp. 37–54. American Chemical Society, Washington, DC.

Schultz, J.C. 1988. Plant responses induced by herbivores. Trends in Ecology & Evolution 3:45–49.

Schultz, J.C. and I.T. Baldwin. 1982. Oak leaf quality declines in response to defoliation by gypsy moth larvae. Science 217:149–151.

Shure, D.J. and L.A. Wilson. 1993. Patch-size effects on plant phenolics in successional openings of the Southern Appalachians. Ecology 74:55–67.

Silkstone, B.E. 1987. The consequences of leaf damage for subsequent insect grazing on birch (*Betula* spp.). A field experiment. Oecologia 74:149–152.

Singer, M.C. and C. Parmesan. 1993. Sources of variations in patterns of plant-insect association. Nature 361:251–253.

Strauss, S.Y. 1991. Indirect effects in community ecology: their definition, study and importance. Trends in Ecology & Evolution 6:206–210.

Thompson, J.N. 1986. Constraints on arms races in coevolution. Trends in Ecology & Evolution 1:105–107.

Thompson, J.N. 1988a. Variation in interspecific interactions. Annual Review of Ecology and Systematics 19:65–87.

Thompson, J.N. 1988b. Evolutionary ecology of the relationship between oviposition preference and performance of offspring in phytophagous insects. Entomologia Experimentalis et Applicata 47:3–14.

Thompson, J.N. and O. Pellmyr. 1991. Evolution of oviposition behavior and host preference in lepidoptera. Annual Review of Entomology 36:65–89.

Vinson, S.B. 1976. Host selection by insect parasitoids. Annual Review of Entomology 21:109-133.

Whitham, T.G., A.G. Williams, and A.M. Robinson. 1984. The variation principle: individual plants as temporal and spatial mosaics of resistance to rapidly evolving pests. In: P.W. Price, C.N. Slobodchikoff, and W.S. Gaud, eds. A New Ecology: Novel Approaches to Interactive Systems, pp. 15–51. Wiley, New York.

Williams, K.S. and J.H. Myers. 1984. Previous herbivore attack of red alder may improve food quality for fall webworm larvae. Oecologia 63:166–170.

8

Herbivore-Induced Plant Volatiles with Multifunctional Effects in Ecosystems: A Complex Pattern of Biotic Interactions

JUNJI TAKABAYASHI AND MARCEL DICKE

Introduction

Plants defend themselves against herbivores directly in various ways. They may do so chemically by producing toxins, digestibility reducers, repellents, etc., or physically by constructing a hard structure or spines, secreting viscous materials, etc. (Schoonhoven 1981; Bernays and Chapman 1994). In addition to these various direct defenses, plants may defend themselves indirectly by promoting the effectiveness of the natural enemies of herbivores (Price et al. 1980).

It has been found that plants respond to herbivore damage by changing the blend of volatiles that indicates the presence and nature of infesting herbivores to the natural enemies (Dicke et al. 1990a; Dicke 1994; Takabayashi et al. 1994a, 1995; Turlings et al. 1991). This change may be effected by (1) increasing the amounts of some volatile compounds or (2) producing new volatile compounds. In both cases, we call such compounds herbivore-induced plant volatiles (HIPVs).

Besides their attractive function to natural enemies, HIPVs have been found to mediate interactions between several trophic levels. For example, in the tritrophic system consisting of plants, the two-spotted spider mite *Tetranychus urticae*, and the predatory mite *Phytoseiulus persimilis* (Fig. 8.1), HIPVs that attract predatory mites were also found to mediate interaction between (1) spider mites and predatory mites (Sabelis and van de Baan 1983); (2) spider mites and other spider mites of the same species (Dicke 1986); and (3) plants and other plants (Bruin et al. 1991, 1992; Dicke et al. 1993b). This chapter reviews these multifunctional aspects of the HIPVs of plants infested by *T. urticae*.

Different plant–herbivore combinations result in the emission of different HIPVs that mediate different interactions between plant and carnivore (Sabelis and van de Baan 1983). This may also be true in the HIPV-mediated interactions between several trophic levels. Thus, in ecosystems, there must be a vast number of HIPV network systems functioning in different ways. Intra- and interspecific variation in the composition of the HIPVs is discussed here also.

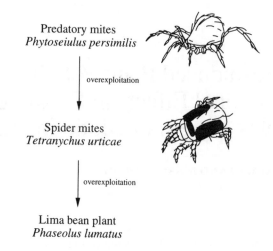

Predatory mites
Phytoseiulus persimilis

overexploitation

Spider mites
Tetranychus urticae

overexploitation

Lima bean plant
Phaseolus lumatus

FIGURE 8.1. The tritrophic interactions of the lima bean plant, two-spotted spider mites, and predatory mite.

Interaction Between Plants and Predatory Mites Mediated by Herbivore-Induced Plant Volatiles

Production of Predator Attractants by a Plant Infested by Herbivores

Sabelis and van de Baan (1983) first showed the olfactory response of the predatory mites *Phytoseiulus persimilis* toward volatiles of lima bean leaves infested by spider mites using a Y-tube olfactometer. The predatory mite preferred the odor of infested lima bean leaves to the odor of uninfested leaves. When they removed all the spider mites and visible products such as feces, exuviae, and webbings from the infested leaves, the predatory mites still preferred the infested to uninfested leaves, at least for a few hours (Sabelis et al. 1984). The removed spider mites, exuviae, and webbing were not attractive, while feces were slightly attractive (Sabelis et al. 1984). The predatory mites preferred the odor of uninfested lima bean leaves to clean air (Takabayashi et al. 1991a) but did not prefer the odor of mechanically damaged lima bean leaves (imitating spider mite damage) to the odor of undamaged leaves (M. Dicke, unpublished data).

Chemical Analysis

Lima bean leaves respond to *Tetranychus urticae* damage by emitting four terpenoid attractants [(3*E*)-4,8-dimethyl-1,3,7-nonatriene, linalool, (*E*)-ß-ocimene, and methyl salicylate] as HIPVs for the predatory mites *Phytoseiulus persimilis*

(Fig. 8.2) (Dicke et al. 1990a). Methyl salicylate was not found at all in the volatiles of uninfested or artificially damaged lima bean leaves; (3E)-4,8-dimethyl-1,3,7-nonatriene, linalool, and (E)-ß-ocimene were found in minute amounts in the headspace of both uninfested and artificially damaged leaves (Takabayashi et al. 1994a).

These data suggest two possibilities: (1) the leaves may produce predatory mite attractants in response to herbivore damage, or (2) the attractants may be plant chemicals that are modified after uptake by the spider mites and deposited on the leaf surface. The second possibility seems highly unlikely because Takabayashi et al. (1991a) reported that both the infested and uninfested parts of a lima bean leaf emit two predatory mite attractants [(3E)-4,8-dimethyl-1,3,7-nonatriene and (E)-ß-ocimene]. The emission of the two attractants by the uninfested part could reflect adsorption of the two attractants produced in the infested part onto the uninfested part, from which secondhand emission occurs. However, this seems unlikely, because attempts to adsorb synthetic (E)-ß-ocimene to uninfested lima bean leaves by placing the chemical next to the uninfested leaves were not successful (J. Takabayashi and M. Dicke, unpublished data). It seems more likely that the uninfested part of the infested leaf produces the two terpenoids.

Mechanical damage by carborundum, simulating spider mite infestation, did not result in the production of the predator attractants. Thus, it is the spider mite per se that causes the plant to produce more of the attractants (more than 100-fold). We suspect that the saliva of the two-spotted spider mite is responsible for triggering the production of the predatory mite attractants. However, it has been impossible to collect spider mite saliva because of the small size (~0.8 mm) of the mite.

Systemic Production

As described, the uninfested part of the infested lima bean leaf starts to emit two predator attractants (Takabayashi et al. 1991a). In addition, Dicke et al. (1990b) showed that spider mite damage also affected undamaged leaves of the infested

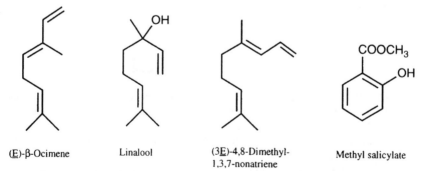

(E)-β-Ocimene   Linalool   (3E)-4,8-Dimethyl-   Methyl salicylate
                                          1,3,7-nonatriene

FIGURE 8.2. The predator attractants emitted by lima bean leaves that are infested by two-spotted spider mites.

plant. Uninfested leaves next to an infested leaf are more attractive to *P. persim-ilis* than uninfested leaves of uninfested plants (Fig. 8.3). Adsorption of the volatiles emitted from the infested leaves onto the uninfested leaves was pre-cluded by removing HIPVs from a climate room.

The data show the systemic production of the predator attractants: not only the infested part of the infested leaf but also the uninfested part of the infested leaf and the leaves next to the infested leaf produced the predator attractants on spider mite infestation. Takabayashi et al. (1991a) and Dicke et al. (1993a) reported the presence of the endogenous elicitor(s) mediating the production of the predatory mite attractants in uninfested leaves. The presence of this elicitor will be partly responsible for this systemic production.

Function of HIPVs as Predator Attractants

The plants may be overexploited by the two-spotted spider mites in the absence of predatory mites such as *P. persimilis* because of their high capacity of population

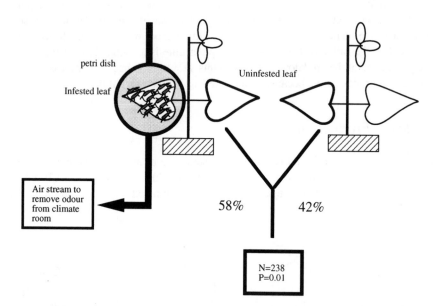

FIGURE 8.3. Response of satiated *Phytoseiulus persimilis* females in a Y-tube olfactometer when offered uninfested leaves from infested lima bean plants vs. uninfested leaves from uninfested lima bean plants. Volatiles emitted from infested leaves may adsorb to unin-fested leaves. Volatiles emitted from infested leaves are led to the vacuum system of the building (*left*) to prevent adsorption to uninfested leaves. *n*, Number of predators tested; *p*, critical level determined with significance test for differences from 50:50 distribution of predators over the two arms of the device. 58% of the predators preferred the uninfested leaves of infested plants to the uninfested leaves of uninfested plants.

increase. However, the predatory mites have an even higher capacity of population increase, which results in the overexploitation of their prey. Thus, the invasion of the predatory mites of the spider mite colony is adaptive to the infested plant. For the predatory mites, it is also adaptive to respond to the attractants emitted by a plant infested by their prey. Thus, the predatory mite attractants emitted by plants on spider mite damage are adaptive to both the infested plant (emitting signaler) and the predatory mite (receiver). When allelochemicals are adaptive to both signaler and receiver, they are classified as synomones (Dicke and Sabelis 1988). The predator attractants in this tritrophic context are, therefore, termed herbivore-induced synomones (Vet and Dicke 1992).

Interaction Between Spider Mites and Predatory Mites Mediated by Herbivore-Induced Plant Volatiles

The interaction between plants and predatory mites through a herbivore-induced synomone can also be studied from the point of view of the interaction between predatory mites and two-spotted spider mites. When we focus on this interaction, the two-spotted spider mite is the inducer of the signal (inducing signaler) and the predatory mite is the receiver. For a two-spotted spider mite, the induction of predator attractants is clearly maladaptive. The allelochemical that is adaptive to a receiver but not to the signaler is called a kairomone (Dicke and Sabelis 1988). Thus, the herbivore-induced predator attractants are classified as spider mite kairomones in the interaction between the spider mite and the predatory mite (Sabelis and van de Baan 1983) .

As described, the predatory mite *Phytoseiulus persimilis* is a ravenous predator that overexploits the prey population. Thus, there will be a strong selective pressure for the spider mites to be able to decrease the invasion of the predatory mites to their colony. When considering the interaction between the spider mite and the predatory mite, an important question is "why do spider mites induce the plant's production of the predator attractants that are maladaptive to the spider mites themselves?"

To discuss this, we next examine the function of HIPVs from the point of view of two-spotted spider mites.

Interaction Between Conspecific Spider Mites Through Herbivore-Induced Plant Volatiles

One possible answer to the foregoing question is that there is no adaptive function of HIPVs for the spider mites, but that the spider mite's elicitor(s) for the volatile production have an original function that is essential to mite fitness (case 1). In this case, the mite is not able to stop producing the elicitor(s).

Another possible answer to this question is that the HIPVs have another function(s) that are adaptively favorable to the spider mite (case 2). If so, there are two cases for the production of HIPVs:

1. This adaptive function may be relatively more important than the function that is maladaptive to the spider mite (attracting the predatory mite). In this case, the spider mites continue to induce the volatiles.
2. The adaptive function for the spider mites may be relatively less important than the function that is maladaptive to the spider mite. In this case, the character of the spider mite to induce volatiles will be selected against, but it will last longer, because of their relative importance, than in the case in which the HIPVs have no adaptive function at all for the spider mites.

Here, we consider one of the adaptive functions of HIPVs for the spider mites. The response of the two-spotted spider mites toward the volatiles emitted from the plants infested by conspecific mites was studied by Dicke (1986) with a vertical olfactometer. In clean air, the adult female spider mite did not walk or walked with increased turning movements. When volatiles of lima bean leaves infested by the spider mites were offered to a conspecific adult female spider mite in the olfactometer, the mite usually walked straight, away from the odor source. Artificially damaged leaf volatiles elicited the same response as uninfested leaf volatiles did. On the other hand, when volatiles of uninfested lima bean leaves were offered, the mites walked with increased turning and stayed within the odor source. These data show that volatiles emitted by lima bean leaves in response to spider mite damage function as a two-spotted spider mite-dispersing pheromone. So far, one of the HIPVs, linalool, is identified as a component of the spider mite-dispersing pheromone. In addition, the volatiles emitted by uninfested lima bean leaves function as a plant kairomone to enable the mites to track the volatiles from uninfested leaves (Dicke 1986). The dispersing stage of the two-spotted spider mite is the adult female of the oviposition phase and the nymph. To them, it is adaptive to choose an uninfested leaf rather than a leaf already used by conspecifics as a food resource and oviposition site.

Dicke (1986) further studied the response of the spider mites to a mixture of clean and infested leaves in the vertical olfactometer. At a low ratio of infested leaf (plant kairomone) to uninfested leaf (spider mite pheromone), spider mite response did not differ significantly from the response to uninfested leaves, and no aggregative response was observed in two-spotted spider mites. However, at a high ratio of infested leaf to uninfested leaf, the mite showed an aggregative response. This behavior was called spaced-out gregariousness (Kennedy and Crawley 1967).

The spaced-out gregariousness of spider mite response toward a mixture of infested leaf and uninfested leaf volatiles would restrict the habitat structure of the spider mites as well as that of the predatory mites in the habitat. For example, two-spotted spider mites have a tendency to make a new colony next to their parental colony on the plant, and consequently a cluster of colonies arises. This colonizing pattern may partly be explained with this spaced-out gregariousness. The predator foraging behavior will also be affected by this spaced-out gregariousness because the predatory mites *Phytoseiulus persimilis* respond to infested lima bean leaf volatiles more than to uninfested conspecific leaf volatiles, and to the uninfested leaf volatiles more than to clean air (see Production of Predator Attractants by a Plant Infested by Herbivores).

However, it is not clear whether the function of the volatile as dispersing pheromone is relatively more or less important than the kairomone function (the predatory mite attractant). Further study is necessary on the adaptive function of the two-spotted spider mite-induced plant volatiles for the mites.

Interaction Between Conspecific Plants Through Herbivore-Induced Plant Volatiles

Function of Herbivore-Induced Plant Volatiles as Plant Pheromone

Spider mites disperse from a plant to another plant on wind currents. Thus, when considering an uninfested plant that is located downwind from the plant infested by two-spotted spider mites, the uninfested plant is in danger of being infested by spider mites. They are, however, also exposed to the herbivore-induced plant volatiles of the upwind plant. This raises the question whether the downwind plant responds to this chemical information. Bruin and his co-workers have been focusing on this and found effects of lima bean plants and cotton plants infested by the spider mites on downwind uninfested conspecific plants in the tritrophic context (Bruin et al. 1991, 1992; Dicke et al. 1990b, 1993b).

Bruin and co-workers used a three-compartment wind tunnel: the compartments were connected to each other with a tube covered with gauze to prevent mite transfers (Fig. 8.4). In this wind tunnel, they put the uninfested cotton (*Gossypium hirsutum*) or lima bean plants in the first compartment (upwind compartment), the conspecific plants infested by two-spotted spider mites in the second, and uninfested conspecific plants in the third (downwind compartment). The uninfested plants in the downwind compartment were exposed to the odors of upwind infested plants while the uninfested plants in the upwind compartment were exposed to clean air. After 5 days of exposure, they tested the attraction of leaves exposed to infested plant volatiles (sample leaves) and of leaves exposed to clean air (control leaves) in the Y-tube olfactometer and found that the sample leaves were more attractive to the predatory mites *Phytoseiulus persimilis* than control leaves (Bruin et al. 1991; Dicke et al. 1990b). Further elaborate experiments were carried out in cotton plants (Bruin et al. 1991, 1992; Dicke et al. 1990b). Because an infochemical that mediates an interaction between organisms of the same species is called a pheromone (Dicke and Sabelis 1988), HIPVs from cotton or lima bean plants are plant pheromones in this interaction between infested and uninfested plants.

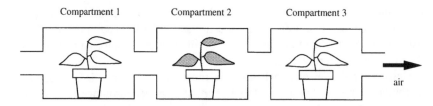

Compartment 1   Compartment 2   Compartment 3

air

FIGURE 8.4. Three-compartment wind tunnel. Uninfested lima bean plants (*unstippled leaves*) were placed in compartment 1 and compartment 3; lima bean plants infested with *Tetranychus urticae* (*stippled leaves*) were placed in compartment 2. The air was sent from compartment 1 to compartment 3.

Chemical Analysis

Preliminary chemical investigations revealed that the composition of the blend emitted by these exposed plants is different from the blend emitted by control plants (J. Takabayashi, M.A. Posthumus, and M. Dicke, unpublished data). The headspace of the exposed lima bean plants contains relatively larger amounts of (*E*)-ß-ocimene and (3*E*)-4,8-dimethyl-1,3,7-nonatriene than unexposed uninfested control leaves. For instance, in the headspace of undamaged plants that were never exposed to two-spotted spider mite-infested plants, peak areas of (*E*)-ß-ocimene and (3*E*)-4,8-dimethyl-1,3,7-nonatriene are much smaller than that of (*Z*)-3-hexenyl acetate (one of the common green leaf odors). In contrast, in the headspace of plants exposed to two-spotted spider mite-infested plants, those peak areas are as much as 5 fold [(3E)-4,8-dimethyl-1,3,7-nonatriene] and up to 26 fold [(*E*)-ß-ocimene] larger than that of (*Z*)-3-hexenyl acetate. These data may be explained by adsorption of (*E*)-ß-ocimene and (3*E*)-4,8-dimethyl-1,3,7-nona-triene emitted from the infested leaves onto the exposed leaves. However, this seems highly unlikely because only trace amounts of (*E*)-ß-ocimene were found in the headspace of plants that were exposed to synthetic (*E*)-ß-ocimene vapor. An alternative exploration is that plants start producing the two predator attractants by themselves by receiving information from nearby plants infested by the spider mites.

Interaction Between Spider Mites and the Uninfested Plant That Is Exposed to Herbivore-Induced Plant Volatiles

Using the same experimental set-up as used for the plant–plant interactions, Bruin et al. (1992) studied the rate of oviposition of spider mites on uninfested cotton plants that received the herbivore-induced plant volatiles of a conspecific plant. The rate was reduced significantly on leaves exposed to volatiles from infested plants when compared with the rate on uninfested leaves exposed to volatiles from the uninfested plant. This will also influence the spacing pattern of the spider mite colonies and the distribution pattern of associating predatory mites.

Interspecific Variation in Composition of Herbivore-Induced Plant Volatiles

Plant Species

Different plant species emit different blends of HIPVs. We studied HIPVs in four species: lima bean, cucumber, apple, and nightshade (*Solanum luteum*). *Phytoseiulus persimilis* preferred infested leaves of the four plant species to uninfested leaves of the respective plant species (Dicke and Sabelis 1988; Takabayashi and Dicke 1993). However, there are large qualitative and quantitative differences in HIPV composition (Table 8.1) (Takabayashi et al. 1994a).

Of the four plant species listed in Table 8.1, only *Solanum luteum* has an effective direct defense mechanism against the spider mites. The *S. luteum* leaf and stem have glandular hairs on its surface from which the plant secretes viscous materials. Such glandular hairs hamper mites (van Haren et al. 1987). When we introduced *Tetranychus urticae* on *S. luteum* leaves, however, momentary severe damage occurred. Interestingly, such infested leaves emitted the simplest blend of herbivore-induced synomone in the four plant species (Table 8.1). We hypothesize here that plants with effective direct defense invest less effort for the production of HIPVs that attract natural enemies of spider mites than the plant without such defense. To test this, further studies on plants with effective direct defense mechanisms against spider mites are necessary.

Herbivore Species

Several reports have shown that different herbivore species elicit different herbivore-induced synomone blends. For example, the predatory mites *Amblyseius andersoni* and *A. finlandicus* responded to apple leaves infested by the European red spider mite *Phytoseiulus ulmi*, which is a suitable prey species (Sabelis and van de Baan 1983). However, they did not respond to apple leaves infested by the spider mite *T. urticae*, which is not a very suitable prey species for these predators (Sabelis and van de Baan 1983). In contrast, satiated female *P. persimilis* responded to volatiles of apple leaves (Cox Orange Pippin) infested by *T. urticae*, but not to volatiles of apple leaves infested by *P. ulmi* (Sabelis and van de Baan 1983). Starved females of *P. persimilis*, however, responded to apple leaves infested by both *T. urticae* and by *P. ulmi* compared to uninfested apple leaves. Such differential responses may be based on the chemical differences recorded by Takabayashi et al.: the composition of the volatile blends emitted by apple leaves when infested by *T. urticae* or *P. ulmi* showed many quantitative differences (Fig. 8.5) (Takabayashi et al. 1991b).

TABLE 8.1. Herbivore-induced plant volatiles extracted from infested leaf volatiles[a]

	Lima bean	Cucumber	Apple	*Solanum luteum*
Terpenoids				
Linalool	3.6			
Limonene			<0.1	
(*E*)-ß-Ocimene	15.3	24.9	23.8	1.1
(*Z*)-ß-Ocimene		0.3	<0.1	
(3*E*)-4,8-Dimethyl-1,3,7-nonatriene	12.9	53.8	8.5	59.2
(3*E*,7*E*)-4,8,12-Trimethyl-1,3,7,11-tridecatetraene	5.7	3.4	0.2	
(*E*,*E*)-α-Farnesene			4.0	
Germacrene-D			1.9	
Esters				
Methyl salicylate	2.7	<0.1	<0.2[b]	
Ethyl benzoate			2.5	
Nitriles				
2-Methylpropanenitrile		0.1		
2-Methylbutanenitrile		0.8	<0.6	
3-Methylbutanenitrile		2.5	<1.3	
Phenylacetonitrile			<1.2	
Oximes				
2-Methylbutanal *O*-methyloxime		0.8		
3-Methylbutanal *O*-methyloxime		3.8	<0.3	
Unknown oxime		0.1		

Data are relative amount in mean percent.
[a]Constitutively produced volatiles are not listed in the table.
[b]Maximum value. Compounds were not found in all samples.
From Takabayashi et al. (1994b).

Intraspecific Variation in Composition of Herbivore-Induced Plant Volatiles

Individual Plants

We studied the production of herbivore-induced synomone in cucumber leaves of different developmental stages and found that the predatory mite *P. persimilis* responded to young *T. urticae*-infested leaves but not to old *T. urticae*-infested leaves in a Y-tube olfactometer (Takabayashi et al. 1994a). We also showed that the infested old cucumber leaf volatiles mask the attraction of young infested cucumber leaves to *P. persimilis* (Takabayashi et al. 1994a).

A chemical analysis showed that both young and old infested cucumber leaves emitted the herbivore-induced predator attractants [(3*E*)-4,8-dimethyl-1,3,7-nonatriene and (*E*)-ß-ocimene] among the major spider mite-induced plant volatiles. In addition, we found that two oximes (3-methylbutanal *O*-methyloxime and an unknown oxime) were much more abundant in infested old cucumber leaves than

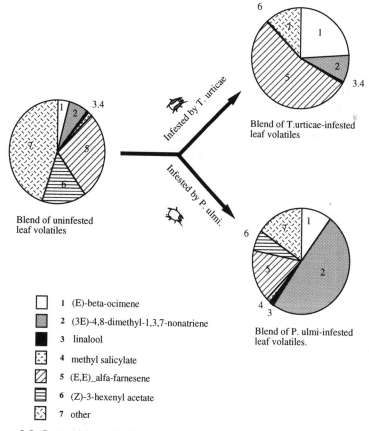

1 (E)-beta-ocimene

2 (3E)-4,8-dimethyl-1,3,7-nonatriene

3 linalool

4 methyl salicylate

5 (E,E)_alfa-farnesene

6 (Z)-3-hexenyl acetate

7 other

FIGURE 8.5. Composition of volatile blend emitted by leaves of apple cultivar Summer Red when uninfested or infested by the spider mite *T. urticae* or *Panonychus ulmi*. (After data from Takabayashi et al. 1991b.)

in infested young cucumber leaves. These oximes probably mask the two predator attractants emitted by old cucumber leaves (Takabayashi et al. 1994a).

This difference in attractiveness of old and young *T. urticae*-infested leaves reflects the different value of young and old leaves to the plant; young leaves at the growing tip of the plant are of high value (Edwards et al. 1992). Therefore, it is highly adaptive for a plant to ensure that predatory mites are directed to the growing points.

Cultivar

The blends of HIPVs emitted by leaves of two different apple cultivars (Summer Red and Cox Orange Pippin) infested by *T. urticae* varied in their composition

(Takabayashi et al. 1991b). When *T. urticae* infested leaves of apple cv. Summer Red, the relative percentages of two HIPVs, (E)-ß-ocimene and $(3E)$-4,8-dimethyl-1,3,7-nonatriene, were higher than when this spider mite infested apple leaves of cv. Cox Orange Pippin. The reverse is the case of relative percentages of one of the HIPVs, phenylacetnitrile; it was higher in the volatiles of infested Cox Orange leaves than those of infested Summer Red leaves. We also observed that the constitutively produced volatile compounds 2-hexenal, (Z)-3-hexen-1-yl acetate, 2-hexen-1-yl acetate, and 2-hexene-1-ol were present in larger proportions in the blend emitted by infested Cox Orange Pippin leaves.

The intraspecific variation in HIPVs in different cultivars may be the result of the different genotypes of two apple cultivars. Sabelis and de Jong (1988) predicted that there is a rather wide range of conditions for synomone-producing plants to coexist with conspecific plants that spend their energy in other ways. The result of intraspecific variation between cultivars supports their prediction.

Summary and Future Perspective

It has long been considered that a plant infested by herbivores can only defend itself directly. However, recent studies of the tritrophic interaction between plants, herbivores, and carnivores show that the plants actively function as a source of chemical information in the form of HIPVs, while the carnivores are involved in the plant's indirect defense. Furthermore, several interactions of the same/different trophic levels are involved and consequently create an information network system.

As one example of such a source of information, we have looked at the tritrophic interaction between lima bean plants, two-spotted spider mites, and the predatory mite *P. persimilis*. The volatiles emitted by lima bean leaves in response to spider mite damage serve the following four functions. (1) In the plant–predatory mite interaction, HIPVs that attract predatory mites function as an herbivore-induced plant synomone; (2) in the spider mite–predatory mite interaction, HIPVs that attract predatory mites function as a herbivore kairomone; (3) in spider mite–spider mite interaction, HIPVs function as dispersing pheromone; and (4) in infested plant–uninfested plant interaction, HIPVs function as plant pheromones that make the uninfested plant more attractive to the predatory mite and possibly less suitable for the spider mites.

In addition, the uninfested leaf volatiles of the lima bean plant have been found to have the following two functions: (1) in the uninfested plant–spider mite interaction, the uninfested leaf volatiles function as a plant kairomone (Dicke 1986); and (2) in the uninfested plant–predatory mite interaction, the uninfested leaf volatiles function as a plant synomone (Takabayashi and Dicke 1992). However, the attraction of uninfested leaves is weaker than that of infested leaves and equal to artificially damaged leaves. Thus, in the tritrophic interaction between lima bean plants, two-spotted spider mites, and predatory mites we can see a complex

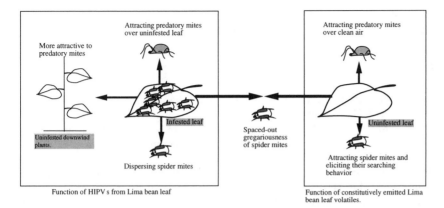

Function of HIPV s from Lima bean leaf

Function of constitutively emitted Lima bean leaf volatiles.

FIGURE 8.6. Function of plant volatiles in the tritrophic system consisting of lima bean plants, two-spotted spider mites, and predatory mites. *HIPVs*, Herbivore-induced plant voltiles.

pattern of biotic interactions involving different tritrophic levels and ecological conditions (Fig. 8.6).

We also review here the following inter- and intraspecific variation in composition of HIPVs (Takabayashi et al. 1994b). (1) Different plant species emit different blends of HIPVs on damage by the same herbivore species. (2) Apple leaves infested by *P. ulmi* or *T. urticae* attracted the respective natural enemies by emitting different blends of HIPVs. (3) In the plant developmental stage, we showed the different attractiveness of an infested leaf within an individual cucumber plant according to its leaf developmental stage. (4) We also found intraspecific variation in HIPVs in different cultivars (genotypes) of apple plants.

The function of HIPVs was first found to be as attractants of carnivorous natural enemies of herbivores (Dicke and Sabelis 1988). However, the function of HIPVs is not so simple as it first appears. For example, as seen in Figure 8.6, HIPVs mediate several interactions at the same and at different trophic levels. In addition, the quality and quantity of HIPVs vary because of biotic and abiotic effects. In the ecosystem, therefore, there will be a number of infested plants that emit their own specific HIPVs with multifunctional purposes. So far, more than 15 plant species, 10 herbivore species, and 10 carnivore species have been used in different combinations in studies on HIPVs (Dicke 1994; Dicke et al. 1990b; Takabayashi et al. 1994a, 1995; Turlings et al. 1993a; Vet and Dicke 1992). However, we do not know yet whether the multiple functions of HIPVs play an important role to maintain ecosystems and biodiversity, or whether the functions are only secondary in the ecosystems. Further studies on HIPVs using seminatural conditions such as the Ecotron are necessary to elucidate the ecological significance of HIPVs in time and space in ecosystems.

Acknowledgments. This study was partly supported by a Japan Ministry of Education, Science and Culture Grant-in-Aid for Scientific Research on Priority

Areas (#319), project "Symbiotic Biosphere: An Ecological Interaction Network Promoting the Coexistence of Many Species."

Literature Cited

Bernays, E.A. and R.F. Chapman. 1994. Host Plant Selection by Phytophagous Insects. Chapman & Hall, London.

Bruin, J., M. Dicke, and M.W. Sabelis. 1992. Plants are better protected against spider-mites after exposure to volatiles from infested conspecifics. Experientia (Basel) 48:525–529.

Bruin, J., M.W. Sabelis, J. Takabayashi, and M. Dicke. 1991. Uninfested plants profit from their infested neighbors. In: M.J. Sommeijer and J. van de Brom, eds. Proceedings of the Section Experimental and Applied Entomology of the Netherlands Entomological Society 2:103–108.

Dicke, M. 1986. Volatile spider-mite pheromone and host-plant kairomone, involved in spaced-out gregariousness in the spider mite *Tetranychus urticae*. Physiological Entomology 11:251–256.

Dicke, M. 1994. Local and systemic production of volatile herbivore-induced terpenoids: their role in plant-carnivore mutualism. Journal of Plant Physiology 143:465–472.

Dicke, M. and A. Groeneveld. 1986. Hierarchical structure in kairomone preference of the predatory mite *Amblyseius potentillae*: dietary component indispensable for diapause induction affects prey location behavior. Ecological Entomology 11:131–138.

Dicke, M. and M.W. Sabelis. 1988. Infochemical terminology: should it be based on cost-benefit analysis rather than origin of compounds? Functional Ecology 2:131–138.

Dicke, M., T.A. van Beek, M.A. Posthumus, N. Ben Dom, H. van Bokhoven, and A.E. de Groot. 1990a. Isolation and identification of a volatile kairomone that affects acarine predator-prey interaction: involvement of host plant in its production. Journal of Chemical Ecology 16:381–396.

Dicke, M., M.W. Sabelis, J. Takabayashi, J. Bruin, and M.A. Posthumus. 1990b. Plant strategies of manipulating predator-prey interactions through allelochemicals: prospects for application in pest control. Journal of Chemical Ecology 16:3091–3118

Dicke, M., P. van Baarlen, B. Wessels, and H. Dijkman. 1993a. Herbivory induces systemic production of plant volatiles that attract predators of the herbivore: extraction of endogenous elicitor. Journal of Chemical Ecology 19:581–599.

Dicke, M., J. Bruin, and M.W. Sabelis. 1993b. Herbivore-induced plant volatiles mediate plant-carnivore, plant-herbivore and plant-plant interactions: talking plant revisited. In: J.C. Schultz and I. Paskin, eds. Plant Signals in Interactions with Other Organisms, Vol. 11. Current Topics in Plant Physiology, pp. 182–196. American Society of Plant Physiologists, Rockville, MD.

Edwards, P.J., S.D. Wratten, and E.A. Parker. 1992. The ecological significance of rapid wound-induced changes in plants: insect grazing and plant competition. Oecologia 91:266–272.

Kennedy, J.S. and L. Crawley. 1967. Spaced-out gregariousness in sycamore aphids *Drepanosiphum platanoides* (Schrank) (Hemiptera: Callaphicidae). Journal of Animal Ecology 36:147–163.

Price, P.E., C.E. Bouton, P. Gross, B.A. McPheron, J.N. Thompson, and A.E. Weis. 1980. Interaction among three trophic levels: influence of plant interactions between insect herbivores and natural enemies. Annual Review of Ecology and Systematics 11:41–65.

Sabelis, M.W. and H.E. van de Baan. 1983. Location of distant spider mite colonies by phytoseiid predators: demonstration of specific kairomones emitted by *Tetranychus urticae* and *Panonychus ulmi*. Entomologia Experimentalis et Applicata 33:303–314.

Sabelis, M.W. and M.C.M. de Jong. 1988. Should all plants recruit bodyguards? Condition for a polymorphic ESS of synomone production in plants. Oikos 53:247–252.

Sabelis, M.W., B.P. Afman, and P.J. Slim. 1984. Location of distant spider mite colonies by *Phytoseiulus persimilis*: localization and extraction of a kairomone. Acarology VI(1):431–440.

Schoonhoven, L.M. 1981. Chemical mediators between plant and phytophagous insects. In: D.A. Nordlund, R.L. Jones, and W.J. Lewis, eds. Semiochemicals. Their Role in Pest Control, pp. 31–50. Wiley, New York.

Takabayashi, J. and M. Dicke. 1992. Response of predatory mites with different rearing histories to volatiles of uninfested plants. Entomologia Experimentalis et Applicata 64:187–193.

Takabayashi, J. and M. Dicke. 1993. Volatile allelochemicals that mediate interactions in a tritrophic system consisting of predatory mites, spider mites and plants. In: H. Kawanabe, J.E. Cohen, and K. Iwasaki, eds. Mutualism and Community Organization. Behavioural, Theoretical and Food Web Approaches, pp. 280–295. Oxford University Press, London.

Takabayashi, J., M. Dicke, and M.A. Posthumus. 1991a. Induction of indirect defense against spider-mites in uninfested lima bean leaves. Phytochemistry (Oxford) 30:1459–1462.

Takabayashi, J., M. Dicke, and M.A. Posthumus. 1991b. Variation in composition of predator-attracting allelochemicals emitted by herbivore-infested plants: relative influence of plant and herbivore. Chemoecology 2:1–6.

Takabayashi, J., M. Dicke, S. Takahashi, M.A. Posthumus, and T.A. van Beek. 1994a. Leaf age affects composition of herbivore-induced synomones and attraction of predatory mites. Journal of Chemical Ecology 20:373–386.

Takabayashi, J., M. Dicke and M.A. Posthumus. 1994b. Volatile herbivore-induced terpenoids in plant-mite interactions: variation caused by biotic and abiotic factors. Journal of Chemical Ecology 20:1329-1354.

Takabayashi, J., S. Takahashi, M. Dicke, and M.A. Posthumus. 1995. Developmental stage of the herbivore *Pseudaletia separata* affects production of herbivore-induced synomone by corn plants. Journal of Chemical Ecology 21:273–287.

Turlings, T.C.J., J.H. Tumlinson, R.R. Heath, A.T. Proveaus, and R.E. Doolittle. 1991. Isolation and identification of allelochemicals that attract the larval parasitoid, *Cotesia marginiventris* (CRESSON), to the microhabitat of one of its hosts. Journal of Chemical Ecology 17:2235–2251.

Turlings, T.C.J., P.J. McCall, H.T. Alborn, and J.H. Tumlinson. 1993a. An elicitor in caterpillar oral secretions that induces corn seedlings to emit chemical signals attractive to parasitic wasps. Journal of Chemical Ecology 19:411–425.

Turlings, T.C.J., F.L. Wackers, L.E.M. Vet, W.J. Lewis, and J.H. Tumlinson. 1993b. Learning of host-hiding cues by hymenopterous parasitoids. In: D.R. Papaj and A.C. Lewis, eds. Insect Learning, pp. 51–78. Chapman & Hall, New York.

van Haren, R.J.F., M.M. Steenhuis, M.W. Sabelis, and O.B.M. de Ponti. 1987. Tomato stem trichomes and dispersal success of *Phytoseiulus persimilis* relative to its prey *Tetranychus urticae*. Experimental & Applied Acarology 3:115–121.

Vet, L.E.M. and M. Dicke. 1992. Ecology of infochemical use by natural enemies in a tritrophic context. Annual Review of Entomology 37:141–172.

9

How a Butterfly Copes with the Problem of Biological Diversity

Daniel R. Papaj

Introduction

In a volume such as this one, it is a given that biological diversity is treated as something of great value. Justifiable though this perspective seems to biologists and other human beings, it is worth noting that, for other animal species, biological diversity can be either a boon or a bane. Among insects that feed on plants, for example, some species clearly benefit from diversifying their diet. Mixing of different plants in the diet of some grasshoppers tends to promote growth rates and survival (Bernays et al. 1992). For such species, a nutritionally ideal habitat is a botanically diverse one. In contrast, other insect herbivores feed on only a few or perhaps even a single plant species. For such insects, botanical diversity may actually be a nuisance. The practice of intercropping, for example, wherein various agricultural crop species are intermingled in a single field, often reduces pest damage by making it difficult for specialist insects to move from host to host (Andow 1991). In this chapter, an example in which botanical diversity poses a problem for an herbivorous insect is reviewed. The nature of the insect's predicament is described and behavioral strategies employed to handle it are discussed. Four levels of diversity with which the insect must contend are enumerated, and an hypothesis is proposed for a mechanism by which the insect copes adaptively and simultaneously with these different levels. Finally, this work is related to findings on other herbivorous insects, and recommendations are made for directions in research.

Natural History and Description of Behavior

The pipevine swallowtail, *Battus philenor,* is distributed across much of the continental United States. Everywhere it occurs, females lay eggs on and larvae feed on a single genus of plants, the genus *Aristolochia* in the family Aristolochiaceae. Despite its extreme specialization in the larval food plant, the butterfly occupies a remarkable diversity of habitats including longleaf pine uplands (eastern Texas), riparian oak woodlands (California), mixed deciduous and coniferous montane forest (Appalachian Mountains), and mesquite grasslands and Sonoran desert washes (Arizona).

Perhaps the major problem to be solved by a female of this species in the 2 weeks or so in which she can be expected to live is the selection of plants on which to lay eggs. This process of host selection is essentially a simple form of maternal care in which the female places young of limited mobility directly in contact with their preferred resource. Wherever the species is found, the behavior of females looking for host plants on which to lay eggs is remarkably similar. Females adopt a slow, fluttering flight pattern that is close to the vegetation and characterized by frequent sharp turns. Periodically, they alight on vegetation. If the leaf landed upon belongs to an *Aristolochia* plant, females may or may not lay a clutch of eggs of 1 to 20 eggs. If the leaf does not belong to an *Aristolochia* plant, females virtually never lay eggs but resume search flight.

Problems Posed by Diversity

Level of Diversity Type 1: Host Species Versus Nonhost Species

Most herbivorous insects must detect hosts from a complex vegetative background of nonhost plants. The significance of nonhosts in host selection is illustrated by behavioral responses of herbivores to nonhost deterrent chemicals (Bernays and Chapman 1994; Renwick and Huang 1995) and by the impact of variation in nonhost background on herbivore attack (Rausher 1981a; Andow 1991; Kostal and Finch 1994). A female pipevine swallowtail butterfly must identify *Aristolochia* plants in a habitat containing possibly hundreds of other plant species.

Perhaps the principal criteria used by the butterfly to discriminate *Aristolochia* plants from all other plants are chemosensory in nature. After landing, a female tastes a leaf with chemoreceptors on her foretarsi. As demonstrated recently by Paul Feeny and colleagues, tarsal contact with a mixture of inositols, acids, and a lipid from the host species, *Aristolochia macrophylla,* suffices to elicit egglaying by *Battus philenor* (Papaj et al. 1992; Sachdev-Gupta et al. 1993) (Fig. 9.1). All three classes of compounds must be represented in the mixture. Given the presence of these compounds, however, the butterfly will lay eggs on all manner of artificial substrates (Papaj 1986a). These chemical constituents thus constitute the key in a lock-and-key mechanism governing egglaying behavior. This system of host recognition is highly effective: only a single oviposition "mistake" by this butterfly on a nonhost plant was observed by myself, my colleague Mark Rausher, and assistants in hundreds of person-hours of observation at field sites in eastern Texas (Papaj 1986b).

A butterfly's use of its "taste buds," while a virtually error-free means of identifying plants as hosts, is yet not by itself a very efficient way to select them. In

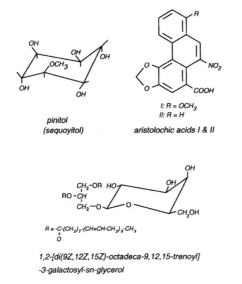

pinitol
(sequoyitol)

I: R = OCH₃
II: R = H

aristolochic acids I & II

R = -C-(CH₂)₇-(CH=CH-CH₂)₃-CH₃
       ‖
       O

1,2-[di(9Z,12Z,15Z)-octadeca-9,12,15-trenoyl]
-3-galactosyl-sn-glycerol

FIGURE 9.1. Compounds recovered from *Aristolochia macrophylla* leaf tissue that act as contact oviposition stimulants for the pipevine swallowtail butterfly, *Battus philenor*. Sequoyitol (not pictured) is an isomer of pinitol.

most habitats, there is a great deal of non-*Aristolochia* vegetation upon which a butterfly might land and taste; in fact, nonhost vegetation is usually far more abundant than host vegetation. If landing on nonhost plants takes time, the rate at which hosts are discovered might be reduced as a consequence. That nonhost vegetation impedes discovery of hosts by *B. philenor* is clear. Rauscher (1981a) reported that a seasonal increase in the height and density of surrounding vegetation was accompanied by a seasonal decrease in the rate at which *Aristolochia reticulata* plants were discovered by butterflies at sites in eastern Texas. While Rauscher provided experimental evidence that nonhost vegetation reduced host discovery by masking hosts physically, the seasonal decline in rate of host discovery could be caused at least in part by increase in time wasted by butterflies in sampling nonhost vegetation.

Exactly how much even small amounts of wasted time can reduce rates of host discovery was formalized using a two-prey modification of the well-known Holling disk equation for prey capture rate (Holling 1959; Papaj 1990). In this formulation, hosts and nonhosts were treated as alternative "prey," although of course only one of the prey (i.e., the host) has any value to the insect. The model assumed that it took some finite time to land on a nonhost and reject that plant, a time represented in the model as the "handling time" of the nonhost. In calculations summarized in Figure 9.2a, even very small values for time wasted in sampling a nonhost plant (e.g., 0.1 s) had profound effects on the rate at which a butterfly finds and lays eggs on its host. Estimates of the actual time wasted handling a nonhost are unavailable. Given that a butterfly's forward velocity fre-

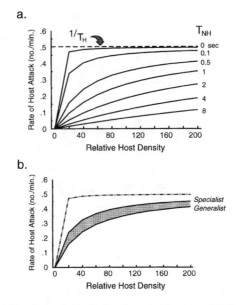

FIGURE 9.2a,b. Results of calculations using the modification of Holling disk equation for two-prey case. (a) The effect of time wasted in rejecting a single nonhost (T_{NH}) on the rate of host attack at different host densities relative to nonhost density. *Dashed line* indicates the asymptotic value of rate of host attack, which is equal to $1/T_{H}$ where T_{H} is the time required to evaluate a host plant. (b) The effect of specializing on leaf shape in a habitat containing a single host species with a single leaf shape. *Dashed line,* rate of host attack when no time is wasted on nonhosts; *shaded area,* advantage of leaf-shape specialist. In (b) half of nonhosts are assumed to have broad leaves and half narrow leaves. In both (a) and (b), time T_{H} is assumed to be 2 min; nonhost density is set at 10,000 plants; the parameter a, which describes the rate of successful search, was set at 0.4. All parameters fall within estimates for butterflies foraging under field conditions in an eastern Texas population. (See Papaj 1990 for further details.)

quently decreases greatly while landing on a leaf, it would not be surprising if values were frequently in the neighborhood of 0.5 s or more.

Any solution to the problem of wasted time would have to involve evaluations of vegetation while "on the wing." If a butterfly could distinguish between the odors of host and nonhost species, for example, she might be able to reduce time spent wasted by landing on nonhost plants. Although a related papilionid, *Papilio polyxenes,* discriminates between host and nonhost odors in laboratory assays (Feeny et al. 1989), the extent to which this cue is used by pipevine swallowtail butterflies is not known. However, pipevine swallowtail butterflies do use a visual cue, specifically shape, to distinguish at least crudely between hosts and nonhosts before landing. In field studies in eastern Texas, Rausher (1978) discovered that individual butterflies were biased as to the shapes of nonhost leaves on which they landed. Some butterflies alighted mostly on broad leaves, others mostly on narrow leaves; very few butterflies failed to show a bias. Moreover, females showed a parallel bias in which of two local host species they discovered. Females that preferred broad nonhost leaves tended to discover a broad-leaved host, *A. reticulata*. In contrast, females that preferred narrow nonhost leaves tended to discover a narrow-leaved host, *A. serpentaria*.

In both field and field-enclosure observations, butterflies that specialized on leaf shape generally found hosts at higher rates than butterflies that did not (Rausher 1978; Papaj 1990). Exactly how selective orientation to shape enhances the rate at which host are found could not be determined from these studies, but it is conceivable that it does so by reducing time wasted in sampling nonhost plants. Considering for simplicity's sake a habitat containing a single host species with a single leaf shape, a response to even the general shape (broad versus narrow) of that host's leaves would permit the butterfly to exclude from tarsal sampling any nonhost bearing leaves of the "wrong" shape. Because such nonhosts are generally very numerous (usually more numerous than the hosts themselves), significant time savings and a corresponding increase in rate of host attack could be realized with even a rudimentary assessment of shape (Papaj 1990) (Fig. 9.2b).

Level of Diversity Type 2: Good Host Species Versus Bad Host Species

The manner in which feeding and ovipositing insects discriminate among host species has been studied relatively intensively (Nylin 1988; Thompson 1988; Howard and Bernays 1991; Bowers and Stamp 1992; Nylin and Janz 1993; Thompson 1993; Bernays and Chapman 1994; Singer 1994). Such discrimination is often, although not always, functional (Thompson 1988; Fox and Eisenbach 1992; Nylin and Janz 1993 and references within; Bernays and Chapman 1994). In *B. philenor,* too, discrimination at this level of botanical diversity probably reflects both adaptation and constraint. Where more than one host species with different leaf shapes are found in the same habitat (as in eastern Texas), for example, the butterfly has a potential dilemma. On one hand, specializing on shape may save time wasted on nonhosts; on the other hand, such specialization may cause the insect to "overlook" hosts with the wrong shape. In theory, the trade-off will depend on the relative abundance of hosts and nonhosts with particular leaf shapes (Papaj 1990). In reality, host species preference probably depends as well on the relative suitability of alternative species for larval growth and survival.

In eastern Texas, there is a consistent seasonal shift in the host species on which eggs are laid (Rausher 1980; Papaj 1986c). Early in the season, females specialize on the broad-leaved *A. reticulata;* late in the season, females specialize on the narrow-leaved *A. serpentaria.* Paralleling this shift is a shift in the dominant nonhost leaf-shape preference in the population. Early in the season, females alight mainly on broad-shaped nonhost leaves; late in the season, females alight mainly on narrow-shaped nonhost leaves.

In mechanistic terms, the shift in host species preference is mediated by a shift in response to leaf shape. In functional terms, the shift in host species preference is related to seasonal patterns in host abundance and suitability for larval consumption (Rausher 1980, 1981b). Early in the season, both host species are equally suitable for larval growth and survival. However, *A. reticulata* is far more abundant than *A. serpentaria*. Given the discrepancy in abundance, it apparently is advantageous for the butterfly to specialize on broad leaves: the decrement in overall host discovery rate associated with overlooking the rare *A. serpentaria* is perhaps more than compensated by the increment in host discovery rate associated with saving time that would be otherwise wasted in sampling the narrow-leaved nonhost community (Papaj 1990). Later in the season, the discrepancy in host abundance has diminished somewhat. Perhaps more importantly, the leaves of the now phenologically mature *A. reticulata* are largely unsuitable for larval growth and survival, while those of *A. serpentaria* remain highly suitable. For this reason, late-season females search preferentially for *A. serpentaria*.

The seasonal shift in leaf-shape preference at the population level is mediated by learning at the level of individuals. When naive butterflies in a large field enclosure were permitted to forage for hosts of a particular species (and therefore a particular leaf shape), they made mistakes according to the shape of the leaf of the host species. Females given experience with the broad-leaved *A. reticulata* landed mainly on broad nonhost leaves; females given experience with the narrow-leaved *A. serpentaria* landed mainly on narrow nonhost leaves. Females in one search mode could be induced to switch to the other search mode by giving them experience with a host species with a new leaf shape (Papaj 1986a). Field observations supported a learning hypothesis. Older butterflies presumed to have more experience exhibited more extreme shape preferences than younger ones (Papaj 1986c). In addition, both Rausher (1978) and Papaj (1986c) witnessed numerous instances in which females in one search mode switched to the other search mode after "accidentally" landing on the host plant with the "wrong" leaf shape.

Finally, learning is associative. Using host extracts applied to nonhost plants, Papaj (1986a) was able to show that shape learning involves an association between shape and contact chemosensory stimuli. Interestingly, egglaying is not required for learning to take place, although learning appears to be stronger if eggs are actually deposited (Papaj 1986c).

Level of Diversity Type 3: Good Conspecific Hosts Versus Bad Conspecific Hosts

A variety of herbivorous insects discriminate among plants within a given host species (Mackay 1985; Fitt 1987; Price 1991; Bingaman and Hart 1992; Larsson and Strong 1992; Pettersson 1992; Zangerl and Berenbaum 1992; Hamilton and Zalucki 1993; Zangerl and Berenbaum 1993). Pipevine swallowtail butterflies likewise make choices among individual plants within a host species, and these choices are functional. In field manipulations involving *A. reticulata* and *A. ser-*

pentaria, survival inferred for larvae placed experimentally on plants on which females laid eggs was markedly higher than for larvae placed on plants rejected after landing by females (Rausher and Papaj 1983). These differences in fitness measures were associated with morphological differences: accepted *A. reticulata* plants were smaller and accepted *A. serpentaria* plants larger than their rejected congeners. For *A. reticulata,* smaller size connotes plants that are phenologically younger (Papaj and Rausher 1987) and therefore more highly suitable for larval growth than others (Rausher 1981b); for *A. serpentaria,* larger size may confer better larval performance by providing young with more food to eat before they must disperse in search of the next host plant.

Level of Diversity Type 4: Good Host Part Versus Bad Host Part

Herbivorous insects are known to evaluate at least one more aspect of botanical diversity, that is where specifically on the host plant to feed or place their eggs (Price 1991; Brody 1992; Lalonde and Roitberg 1992). Pipevine swallowtail butterflies are no different in this regard. Analogous to their preference for conspecific plants, females seem to prefer the part of the plant that is relatively phenologically young. In laboratory assays using *A. fimbriata,* for example, significantly more eggs were laid per unit area on leaves above node 1 on the vine than on leaves at node 2 or below (Table 9.1) (M.L. Henneman, D.R. Papaj, and R. Jensen, unpublished data). This intraplant preference is based at least in part on a response to chemosensory cues: ethanolic extracts of foliage above the first expanded leaf node were more likely to elicit egglaying by individual females than were extracts of foliage below the fourth leaf node (Table 9.1). HPLC analysis of extracts of foliage above the first node and below the fourth node indicated that levels of two oviposition stimulants, pinitol and sequoyitol (see Fig. 9.1), were significantly higher in above-first-node tissue than in below-fourth-node tis-

TABLE 9.1. Distribution of eggs along *Aristolochia fimbriata* stem and response of female *Battus philenor* to host extracts.[a]

Node	Mean eggs/cm^2 (± 1 SE)	Acceptance (%)
1	2.50 (0.57)a	83a
2	0.31 (0.22)b	—
3	0.22 (0.08)b	—
4+	0.00 (—)b	42b

[a] *n* for egg distribution experiment, 7 females; *n* for extract assay, 12 females. Values within a column followed by the same letter are not significantly different at $p < .05$ using either pairwise Wilcoxon signed rank tests (mean eggs/cm^2) or two-way G-test with Williams correction (% acceptance).

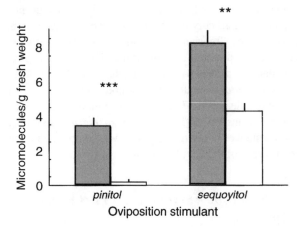

FIGURE 9.3. Levels of pinitol and sequoyitol in ethanolic extracts of above-first-node and below-fourth-node leaves of *Aristolochia fimbriata*. *Shaded bars,* above-first-node results; *open bars,* below-fourth-node results. *Line above bar,* 1 SE; *asterisks,* significance according to Mann–Whitney U-test: ***, $p < .0005$; **, $p < .005$.

sue (Fig. 9.3). Pending analysis of the other two classes of stimulants, it appears that females assess the suitability of a particular part of a host plant according to the levels of one or both cyclitols. Cyclitol levels may be good indicators of leaf water content (Richter and Popp 1992), which is a correlate of larval performance in *B. philenor* (Rausher 1981b).

Levels of Diversity and the Concept of Shared Features

A Template Model for Host Acceptance

In the literature on plant–insect interactions, we commonly refer to a hierarchical series of usually dichotomous categories into which we expect a successful insect to divide its botanical universe: host versus nonhost, good host species versus bad host species, good conspecific individuals versus bad conspecifics, and good host part versus bad part. It is important to appreciate that while this hierarchical system of categorization helps us to construct and test hypotheses about the function of behavior, an herbivorous insect probably views that universe neither as polarized (i.e., good versus bad) nor as consisting of categories arranged in hierarchical fashion. Rather, I propose that the insect uses a template of the best egglaying substrate available, a template composed of a variety of stimuli such as odor, color, and shape. The insect compares a plant encountered in its search path with this multisensory template and feeds or lays eggs according to the degree to which the plant matches this template.

I further propose that the template consists disproportionately of features of host tissue that span the levels of discrimination defined earlier, features that I term *shared features*. It is my supposition that insects have (in an evolutionary sense) both motive and opportunity to give greater weight to shared features than to other features useful for selection of suitable host tissue at a particular level of discrimination. In terms of motive, physiological limitations on sensory reception and information processing demand an efficient mechanism by which to select hosts. This point was driven home recently by Bernays and Wcislo (1994), who argued that neural constraints underlie patterns of host specialization in herbivorous insects. In terms of opportunity, characteristics that are useful in discrimination at one level of botanical diversity can often be expected to be useful in discrimination at one or more other levels (Papaj 1986d). I next expand on these ideas with reference to several plant characteristics that may serve as shared features in host selection by the pipevine swallowtail butterfly.

Chemosensory Stimuli As a Shared Feature

On one hand, pinitol and possibly sequoyitol (along with the aristolochic acids and galactolipid) permit the butterfly to discriminate host from nonhost; on the other, they may be used to distinguish phenologically young host tissue from phenologically mature host tissue on the same plant. This dual function embodies what is meant by a shared feature. Given that choices among conspecific host individuals are also based on phenological age (Papaj and Rausher 1987), it seems likely that these stimulants facilitate discrimination among conspecific plants as well. The degree to which chemosensory stimuli serve as shared features will require more complete chemical analysis and further behavioral assay.

Chemosensory stimuli are known to be involved at all levels of host discrimination in herbivorous insects (Pasteels and Rowell-Rahier 1992; Rank 1992; Larsen et al. 1992; Hanks et al. 1993; Huang and Renwick 1993; Kostal 1993; Bernays and Chapman 1994; Renwick and Huang 1995). However, few analyses of the phytochemical basis of herbivore behavior have evaluated more than one level of discrimination simultaneously. Possibly owing to keen interest in coevolution between plants and insects at the level of species and higher taxa (and therefore special emphasis on host species differences in chemistry), the chemistry of conspecific host and intrahost preference is particularly poorly known (but see Zangerl and Berenbaum 1993 for a beginning toward this end). In short, we lack almost completely the data with which to evaluate the shared features model with respect to plant chemistry.

Foliar Color As a Shared Feature

Recently we endeavored to determine if butterflies are capable of associating another cue, namely color, with an egglaying reward (M.R. Weiss and D.R. Papaj, in preparation). We constructed paper models to which ethanolic host extracts were applied and on which butterflies would readily lay eggs. Individual butterflies were given the opportunity to lay eggs on treated models of a particular color (blue, green, yellow, or red) and then permitted to search in an array containing a mixture of equal numbers of untreated models of each color. The number of landings made by a test female on the different colors were counted and a female's preference for her training color estimated.

Our measure of training color preference was simply the proportion of all landings that were made on models of the training color. As there were four colors, the value of the training preference expected in the absence of color learning is 0.25. As shown in Figure 9.4a, the mean training color preference was significantly greater than 0.25 regardless of training color used (pairwise single-sample t-tests, $p < .001$).

Their excellent performance on all colors notwithstanding, females performed particularly well when given experience with green. Despite relatively small sample sizes, the difference in training color preference between green and each of the other colors was significant (pairwise t-tests, $p < .01$). This green bias could result from either an innate disposition to land on green substrates or a greater capacity for learning green in association with a reward. Examination of a colorwise distribution of mistakes for all females trained to any color but green (i.e., females trained to red, yellow, and blue) supports the former mechanism (Fig. 9.4b). Green mistakes were by far most common. Because even females that had no experience laying eggs on green models showed a bias toward landing on green, it follows that butterflies have an innate disposition to land on green substrates.

At this time, we do not know if butterflies are learning hue (the dominant reflected wavelength) or brightness (the intensity of reflected light). True color learning is usually thought to involve learning of hue. However, evidence to that effect is not easily gained (Scherer and Kolb 1987). We also do not know if females can learn to discriminate finer shades of particular colors, notably green. Finally, we currently have no idea if color learning is a factor in host selection in nature.

These unknowns aside, foliar color is an ideal shared feature. First, foliar color might be used to distinguish host from nonhost. While most nonhost foliage is green, for example, some *Aristolochia* species (e.g., *A. serpentaria*) have a purplish tinge. Second, foliar color could be used to discriminate a preferred host species from other host species.; *A. serpentaria* has a purplish cast, for instance, but its congener in east Texas, *A. reticulata,* does not. Third, foliar color could provide a means of distinguishing a good conspecific individual from a bad one. *A. reticulata* hosts preferred by our butterflies consistently appear a brighter green than those less preferred (Papaj and Rausher 1987). Individuals of the host species native to Arizona, *A. watsoni,* vary even more dramatically in color. Some individuals bear leaves that appear dark purple; other individuals bear leaves that appear a deep green. If these morphs vary in suitability for larval growth and survival, butterflies could use foliar color in choosing among conspecific individuals. Finally, foliar color could permit butterflies to distinguish good host tissue from

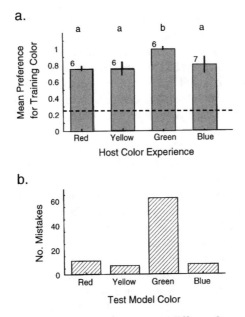

FIGURE 9.4a,b. Results of host color learning assay. (a) Effects of experience with extract-treated model of particular color. *Dashed line,* preference expected if butterfly lands on color in proportion to its representation in the array; *number above bar,* number of butterflies tested; *line over bar,* 1 SE. *Letters in row above bars that differ from one another* indicate mean preferences that are significantly different according to pairwise Mann–Whitney U-tests at $p < .05$. (b) Distribution of mistakes on test colors pooled over butterflies trained on any color except green.

bad host tissue. The highly preferred growing tips of *A. fimbriata, A. reticulata,* and other species are to our eye a much brighter green than are the less preferred mature leaves. It is not beyond reckoning that a searching preference for a single foliar color could in certain situations permit butterflies to evaluate plant tissue at all levels of botanical diversity simultaneously.

Foliar color is known to be important in host-finding by herbivorous insects ranging from aphids to butterflies (Prokopy 1968; Harris and Miller 1983, 1988; Scherer and Kolb 1987; Roessingh and Städler 1990; Pittara and Katsoyannos 1992; Harris et al. 1993) Despite the large body of literature on the subject of color and host selection, I could find no studies that attempt to examine if and how foliar color acts in host discrimination at different levels.

Visual Aspects of Plant Architecture As a Shared Feature

Early in the flight season in eastern Texas when phenologically young *A. reticulata* plants are preferred hosts, butterflies were found to form search modes for leaf buds, landing on nonhost buds in excess of their occurrence (relative to nonhost leaves) in the environment (Papaj 1986d). Moreover, the leaf buds on which females landed belonged primarily to broad-leaved nonhosts. It seems unlikely that, in cognitive terms, butterflies are representing bits and pieces of host architecture such as leaf shape and bud presence; rather, they are probably representing a composite image of a suitable host. Such photograph-like representation of visual patterns in the central nervous system has been postulated for other insects (e.g, bees; Gould 1993) and is intuitively consistent with the notion of template-matching.

Plant architecture is arguably an ideal shared feature. In the case of the broad-leaved, large-budded image possibly employed by early-season butterflies in eastern Texas, such information provides for at least partial discrimination (1) between hosts and nonhosts (i.e., those nonhosts without leaf buds); (2) among conspecifics differing in phenological age; and, because females tend to place eggs on the terminal bud of *A. reticulata,* (3) among parts of a host. Discrimination at each of these levels would be functional.

Use of a host image is not without its drawbacks, particularly if insects cannot search efficiently for more than one image at a time. For example, a butterfly foraging for the image of an *A. reticulata* leaf bud early in the flight season is likely to find fewer of the rare, narrow-leaved *A. serpentaria* as a consequence. Its reduced tendency to find *A. serpentaria* amounts to host discrimination, but discrimination that is not functional. Rather, the failure of butterflies searching for broad-leaved plants with a prominent leaf bud to find narrow-leaved, small-budded *A. serpentaria* plants probably reflects a fundamental trade-off under which females are foraging. With respect to *A. serpentaria* at this particular time of the season, one presumes that the benefits of omitting nonhosts and *A. reticulata* hosts with the wrong architecture from chemosensory consideration simply outweigh the cost of overlooking the rare host.

Generally, plant architecture has been studied less than other cues used in host selection by herbivorous insects. There is nevertheless evidence that aspects of plant architecture, notably shape and size, are used as criteria in host selection by insects other than *B. philenor* (Prokopy 1968; Harris and Miller 1984, 1988; Mackay and Jones 1989; Roessingh and Städler 1990; Pittara and Katsoyannos 1992; Kostal 1993). Moreover, there is strong evidence that insects in general are capable of reasonably sophisticated image processing. This evidence notwithstanding, I know of no systematic efforts to examine whether aspects of a plant's architecture are used as shared features over more than one level of host discrimination.

Closing Remarks

Clearly, we require more information both in terms of identification of the stimuli involved in host recognition and in terms of the levels of discrimination represented in the host selection behavior of the full spectrum of herbivorous insects. With respect to stimulus identification, we need not only to examine each possible kind of stimulus in turn but also to determine how detection and processing of these stimuli are integrated and translated to behavior. While I have addressed candidates for shared features of a putative host template separately, it is important to stress that insects integrate different stimuli in complex ways (Harris and Foster 1995). It is probably as inappropriate to study these cues in isolation as it is to study levels of discrimination in isolation.

For their own part, the various levels of host discrimination have not only not been studied in concert, but most levels continue to go more or less unstudied. The tremendous emphasis on host species preference as it relates to the coevolution of plants and insects seems almost to have suppressed the study of preference at other levels. The oversight is particularly ironic in that evolutionary studies of host use at the level of host species almost certainly depend critically on consideration of other levels of host discrimination (Rausher and Papaj 1983). This point seems particularly to have been ignored in the recent spate of phylogenetic analyses of plant–herbivore evolution.

Finally, the shared-features hypothesis will be difficult to evaluate until and unless studies of the function of host discrimination are wedded to studies of its mechanism. To date, the separate types of studies have typically proceeded almost as though in ignorance of one another. Certainly, because different studies often involve different study organisms, it can be difficult to relate the results of one kind of study to the other. For this reason, we need to develop model systems in which both types of analyses are carried out more or less simultaneously for model systems. The pipevine swallowtail butterfly would seem to be such a model system, but there are others. Onion flies, pierid butterflies, grasshoppers, and tephritid fruitflies are all worthy candidates for such studies.

Summary

Botanical diversity can pose a problem for a specialist herbivorous insect. The predicament and its solutions are described for one species, the pipevine swallowtail butterfly *Battus philenor*. In searching for *Aristolochia* host plants on which to lay its eggs, the butterfly contends with at least four levels of botanical diversity. I propose here that the insect uses shared features, i.e., features of plants that cut across these levels as a way of selecting hosts in an ideal way. Finally, I assert that success in efforts to maintain biological diversity depends in part on how much we know about how the organisms themselves assess patterns in diversity of relevance to their reproduction. Recognition that botanical diversity represents a problem in certain respects for a butterfly or other herbivorous insect should obviously not be construed as sentiment against the preservation of diversity. To the contrary, enumeration of the levels of diversity of apparent demographic importance to this butterfly should encourage in us an awareness that biological diversity is about much more than the richness and evenness of species. To borrow from Janzen (1979), life does not view life in terms of Latin binomials. The maintenance of even a fraction of existing biological diversity will require the continued analysis of that diversity from the perspective of other forms of life.

Literature Cited

Andow, D. 1991. Vegetational diversity and arthropod population response. Annual Review of Entomology 36:561–586.

Bernays, E.A. and R.F. Chapman. 1994. Host-Plant Selection by Phytophagous Insects. Chapman & Hall, New York. Bernays, E.A. and W.T. Wcislo. 1994. Sensory capabilities, information processing and resource specialization. Quarterly Review of Biology 69:187–204.

Bernays, E.A., K. Bright, J.J. Howard, D. Raubenheimer, and D. Champagne. 1992. Variety is the spice of life: frequent switching between foods in the polyphagous grasshopper *Taeniopoda eques* Burmeister (Orthoptera: Acrididae). Animal Behavior 44:721–732.

Bingaman, B.R. and E.R. Hart. 1992. Feeding and oviposition preferences of adult cottonwood leaf beetles (Coleoptera, Chrysomelidae) among *Populus* clones and leaf age classes. Environmental Entomology 21:508–517.

Bowers, M.D. and N.E. Stamp. 1992. Early stage of host range expansion by a specialist herbivore, *Euphydryas phaeton* (Nymphalidae). Ecology 73:526–536.

Brody, A.K. 1992. Oviposition choices by a predispersal seed predator (*Hylemya* sp.). 2. A positive association between female choice and fruit set. Oecologia 91:63–67.

Feeny, P.P., E. Städler, I. Ahman, and M. Carter. 1989. Effects of plant odor on oviposition by the black swallowtail butterfly, *Papilio polyxenes* (Lepidoptera: Papilionidae). Journal of Insect Behavior 2:803–827.

Fitt, G.P. 1987. Ovipositional responses of *Heliothis* spp. to host plant variation in cotton (*Gossypium hirsutum*). In: V. Labeyrie, G. Fabres, and D. Lachaise, eds. Insect-Plants: Proceedings on the Sixth International Symposium on Insect-Plant Relationships. Dordrecht, The Netherlands, Junk Publishers.

Fox, L.R. and J. Eisenbach. 1992. Contrary choices: possible exploitation of enemy-free space by herbivorous insects in cultivated vs. wild crucifers. Oecologia 89:574–579.

Gould, J.L. 1993. Ethological and comparative perspectives on honey bee learning. In: D.R. Papaj and A.C. Lewis, eds. Insect Learning: Ecological and Evolutionary Perspectives. Chapman & Hall, New York.

Hamilton, J.G. and M.P. Zalucki. 1993. Interactions between a specialist herbivore, *Crocidosema plebejana,* and its host plants *Malva parviflora* and cotton *Gossypium hirsutum* oviposition preference. Entomologia Experimentalis et Applicata 66:207–212.

Hanks, L.M., T.D. Paine, and J.G. Millar. 1993. Host species preference and larval performance in the wood-boring beetle *Phoracantha semipunctata* F. Oecologia 95:22–29.

Harris, M.O. and S.P. Foster. 1995. Behavior and integration. In: R.T. Carde and W.J. Bell, eds. Chemical Ecology of Insects 2. Chapman & Hall, New York.

Harris, M.O. and J.R. Miller. 1983. Color stimuli and oviposition behavior of the onion fly, *Delia antiqua* (Meigen) (Diptera: Anthomyiidae). Annals of the Entomological Society of America 76:766–771.

Harris, M.O. and J.R. Miller. 1984. Foliar form influences ovipositional behaviour of the onion fly. Physiological Entomology 9:145–155.

Harris, M.O. and J.R. Miller. 1988. Host-acceptance behaviour in an herbivorous fly, *Delia antiqua.* Journal of Insect Physiology 34:179–190.

Harris, M.O., S. Rose, and P. Malsch. 1993. The role of vision in the host plant-finding behavior of the Hessian fly. Physiological Entomology 18:31–42.

Holling, C.S. 1959. The components of predation as revealed by a study of small mammal predation of the European pine sawfly. Canadian Entomologist 91:293–320.

Howard, J.J. and E.A. Bernays. 1991. Effects of experience on palatability hierarchies of novel plants in the polyphagous grasshopper *Schistocerca americana*. Oecologia 87:424-428.

Huang, X.P. and J.A.A. Renwick. 1993. Differential selection of host plants by two *Pieris* species: the role of oviposition stimulants and deterrents. Entomologia Experimentalis et Applicata 68:59-69.

Janzen, D. 1979. New horizons in the biology of plant defenses. In: G.A. Rosenthal and D.H. Janzen, eds. Herbivores: Their Interactions with Secondary Plant Metabolites. Academic Press, New York.

Kostal, V. 1993. Physical and chemical factors influencing landing and oviposition by the cabbage root fly on host-plant models. Entomologia Experimentalis et Applicata 66:109-118.

Kostal, V. and S. Finch. 1994. Influence of background on host-plant selection and subsequent oviposition by the cabbage root fly (*Delia radicum*). Entomologia Experimentalis et Applicata 70:153-163.

Lalonde, R.G. and B.D. Roitberg. 1992. Field studies of seed predation in an introduced weedy thistle. Oecologia 65:363-370.

Larsen, L.M., J.K. Nielsen, and H. Sorenson. 1992. Host plant recognition in monophagous weevils: specialization of *Ceutorhynchus inaffectatus* to glucosinolates from its host plant *Hesperis matronalis*. Entomologia Experimentalis et Applicata 64:49-55.

Larsson, S. and D.R. Strong. 1992. Oviposition choice and larval survival of *Dasineura marginemtorquens* (Diptera, Cecidomyiidae) on resistant and susceptible *Salix viminalis*. Ecological Entomology 17:227-232.

Mackay, D.A. 1985. Conspecific host discrimination by ovipositing *Euphydryas editha* butterflies and its consequences for offspring survivorship. Researches on Population Ecology (Kyoto) 27:87-98.

Mackay, D.A. and R.E. Jones. 1989. Leaf shape and the host-finding behavior of two ovipositing monophagous butterfly species. Ecological Entomology 14:423-431.

Nylin, S. 1988. Host plant specialization and seasonality in a polyphagous butterfly, *Polygonia c-album* (Nymphalidae). Oikos 53:381-386.

Nylin, S. and N. Janz. 1993. Oviposition preference and larval performance in *Polygonia c-album* (Lepidoptera: Nymphalidae): the choice between bad and worse. Ecological Entomology 18:394-398.

Papaj, D.R. 1986a. Conditioning of leaf-shape discrimination by chemical cues in the butterfly, *Battus philenor*. Animal Behaviour 34:1281-1288.

Papaj, D.R. 1986b. An oviposition "mistake" by *Battus philenor*. Journal of the Lepidopterists' Society 40:348-349.

Papaj, D.R. 1986c. Shifts in foraging behavior by a *Battus philenor* population: field evidence for switching by individual butterflies. Behavioral Ecology and Sociobiology 19:31-39.

Papaj, D.R. 1986d. Leaf buds: a factor in host selection by *Battus philenor* butterflies. Ecological Entomology 11:301-307.

Papaj, D.R. 1990. Interference with learning in pipevine swallowtail butterflies: selective constraint or possible adaptation? Symposia Biologica Hungarica 30:89-101.

Papaj, D.R. 1994. Optimizing learning and evolutionary change in behavior. In: L. Real, ed. Behavioral Mechanisms in Evolutionary Ecology. University of Chicago, Chicago.

Papaj, D.R. and M.D. Rausher. 1987. Components of conspecific host plant discrimination by *Battus philenor* (Papilionidae). Ecology 68:245-253.

Papaj, D.R., P. Feeny, K. Sachdev, and L. Rosenberry. 1992. D-(+)-Pinitol, an oviposition stimulant for the pipevine swallowtail butterfly (*Battus philenor*). Journal of Chemical Ecology 18:799–815.

Pasteels, J.M. and M. Rowell-Rahier. 1992. The chemical ecology of herbivory on willows. Proceedings of the Royal Society of Edinburgh, B Biological Sciences 98:63–73.

Pettersson, M.W. 1992. Taking a chance on moths: oviposition by *Delia flavifrons* (Diptera: Anthomyiidae) on the flowers of bladder campion, *Silene vulgaris* (Caryophllaceae). Ecological Entomology 17:57–62.

Pittara, I.S. and B.I. Katsoyannos. 1992. Effect of shape, size and color on selection of oviposition sites by *Chaetorellia australis*. Entomologia Experimentalis et Applicata 63:105–113.

Price, P.W. 1991. The plant vigor hypothesis and herbivore attack. Oikos 62:244–251.

Prokopy, R.J. 1968. Visual responses of apple maggot flies, *Rhagoletis pomonella* (Diptera: Tephritidae): orchard studies. Entomologia Experimentalis et Applicata 11:403–422.

Rank, N.E. 1992. Host plant preference based on salicylate chemistry in a willow leaf beetle (*Chrysomela aeneicollis*). Oecologia 90:95–101.

Rausher, M.D. 1978. Search image for leaf shape in a butterfly. Science 200:1071–1073.

Rausher, M.D. 1980. Host abundance, juvenile survival and oviposition preference in *Battus philenor*. Evolution 34:342–355.

Rausher, M.D. 1981a. The effect of native vegetation on the susceptibility of *Aristolochia reticulata* (Aristolochiaceae) to herbivore attack. Ecology 62:1187–1195.

Rausher, M.D. 1981b. Host plant selection by *Battus philenor* butterflies: the roles of predation, nutrition, and plant chemistry. Ecological Monographs 51:1–20.

Rausher, M.D. and D.R. Papaj. 1983. Demographic consequences of conspecific host discrimination by *Battus philenor* butterflies. Ecology 64:1402–1410.

Renwick, J.A. and X. Huang. 1995. Interacting chemical stimuli mediating oviposition by Lepidoptera. In: T.N. Ananthakrishnan, ed. Functional Dynamics of Phytophagous Insects. Oxford & IBH, New Delhi.

Richter, A. and M. Popp. 1992. The physiological importance of cyclitols in *Viscum album* L. New Phytologist 121:431–438.

Roessingh, P. and E. Städler. 1990. Foliar form, color and surface characteristics influence oviposition behaviour in the cabbage root fly *Delia radicum*. Entomologia Experimentalis et Applicata 53:103–109.

Sachdev-Gupta, K., P.P. Feeny, and M. Carter. 1993. Oviposition stimulants for the pipevine swallowtail butterfly, *Battus philenor* (Papilionidae), from an *Aristolochia* host plant: synergism between inositols, aristolochic acids and a monogalactosyl diglyceride. Chemoecology 4:19–28.

Scherer, C. and G. Kolb. 1987. Behavioral experiments on the visual processing of color stimuli in *Pieris brassicae*. Journal of Comparative Physiology A 160:645–656.

Singer, M.C. 1994. Behavioral constraints on the evolutionary expansion of insect diet: a case history from checkerspot butterflies. In: L. Real, ed. Behavioral Mechanisms in Evolutionary Ecology, pp. 279–296. University of Chicago Press, Chicago.

Thompson, J.N. 1988. Evolutionary ecology of the relationship between oviposition preference and performance of offspring in phytophagous insects. Entomologia Experimentalis et Applicata 47:3–14.

Thompson, J.N. 1993. Preference hierarchies and the origin of geographic specialization in host use in swallowtail butterflies. Evolution 47:1585–1594.

Zangerl, A.R. and M.R. Berenbaum. 1992. Oviposition patterns and host plant suitability: parsnip webworms and wild parsnip. American Midland Naturalist 128:292–298.

Zangerl, A.R. and M.R. Berenbaum. 1993. Plant chemistry, insect adaptations to plant chemistry, and host plant utilization patterns. Ecology 74:47–54.

IV

Biodiversity and Ecological Function

10

Successional Development, Energetics and Diversity in Planktonic Communities

COLIN S. REYNOLDS

Introduction

This chapter addresses issues for discussion at the International Symposium on "Symbiosphere; Ecological Complexity for Promoting Biodiversity." It is concerned almost exclusively with processes relating to the development of population and communities in the pelagic (open water) of freshwaters, although many of the same principles apply equally to those in the sea. Moreover, these principles also provide convenient analogs to events in terrestrial ecosystems in everything except their diminutive time scales. This makes them very amenable to study, to useful experiment, and to the validation of ecological hypotheses.

My own work on small lakes in the United Kingdom, on experimental enclosures, and on particular systems in other countries, together with some success in computer simulations of the dynamics of mixed populations, has allowed me to build toward this present synthesis over a number of years. It has been subject to the experiences and to the influences of others. Indeed, I claim very little of the originality of the ideas presented; some barely amplify what is already classical theory. Their assembly is a personal one, however, and the framework is one that I have used previously (Reynolds 1992a, 1993a), that of ecological succession. It examines how phytoplankton populations increase through time, how morphological and behavioral adaptations are differentially selected, and how communities are assembled. It considers the various outcomes of successional sequences as they move toward their respective climactic, steady-state conditions. It considers the failure of many planktonic systems ever to achieve a self-regulated steady state and argues for the efficacy of quantifiable, counter-organizational, external influences in prolonging weak community organization and in maintaining high biological productivity and high species diversity.

The Establishment of Populations

In the best traditions of island biogeography (as promulgated by MacArthur and Wilson 1967) but with the places of land and water reversed, it is convenient to think of a body of water isolated in a terrestrial ocean. It is less realistic to think of these bodies as being devoid of life, although a rain puddle, a freshly flushed mountain tarn, or a newly constructed reservoir are fairly close to anyone's concept of a completely open and exploitable habitat. Nevertheless, instances abound where, in the wake of a storm or a flood or simply with the advance of a temperate-latitude winter to spring, a pelagic system finds itself sparsely populated in relation to its carrying capacity (Reynolds 1988a, 1992b). The subsequent dynamics of phytoplankton development demonstrate several principles of population establishment.

Two criteria require to be satisfied so that a planktonic species might be successful at a given location: (a) that it should have arrived there; and (b) that the minimum environmental requirements for its survival and increase are met. Much could be said on the first criterion, with allusions to the large numbers of species of planktonic algae, their cosmopolitan distribution (at least at the generic level), and the fact that large numbers of species at any one location are relatively rare, pending an opportunity either to expand or to recall a previous extant phase (Padisák 1992). These attributes suggest a high efficiency of dispersal, but it has to be stated that considerable uncertainties surround the means of transmission to hydraulically isolated water bodies. Some phytoplankters produce resistant cysts that might survive at least partial desiccation and airborne transport as dust; others may truly remain dependent on the proverbial "bird's foot" for carriage between waters, although it now seems that the feces of birds, as well as the bodies of certain insects, might be at least as important (Atkinson 1980). Certainly some species of coccoid and flagellate chlorophytes appear very quickly in new rain pools, or in a few milliliters of rinse water standing in an abandoned glass bottle. Other species, especially colonial cyanobacteria, may take several years to spread from one lake to a neighbor that is a few kilometers distant but which is supposedly chemically similar (Reynolds and Lund 1988). These differences in behavior are presumed to be related to the size of the organisms and the frequency with which they produce perennating propagules.

On the second criterion, some algae will develop in particular locations, so apparently spontaneously and with such spectacular fidelity as to have gained colloquial names. *Haematococcus,* a calcicolous chlamydomonad with a propensity to turn red with accumulated hematochrome pigment in the later stages of its growth cycle, is a sufficiently common colonist of water-retaining cavities in rough concrete to have earned it the name "bird-bath alga." Several species of Chrysophyceae and their allies have been shown to be unable to survive in environments lacking free carbon dioxide (e.g., in waters of high pH; Saxby 1990) but to be tolerant, therefore, of acidic or nonproductive waters. Conversely, many cyanobacteria have a relatively low half-saturation requirement for carbon uptake, and their growth can be maintained at quite high pH values (Shapiro 1990).

Our own work has distinguished among the growth requirements and susceptibilities of functionally distinct groups of planktonic photoautotrophs (Reynolds 1988b, 1993b). Three broad categories of response have been detected. Algae are either (i) initially fast growing but become susceptible to being eaten by organisms at higher trophic levels; or (ii) initially fast growing but vulnerable to sinking losses; or (iii) slow growing, both relatively and absolutely at low water temperatures, requiring lengthened time periods of appropriate conditions under which to develop populations. Some of the observational data collected for representative species of each category are summarized in Figure 10.1. Moreover, these behaviors are closely coherent with morphological distinctions: they are, respectively, (i) small unicells or simple coenobia of small cells, characterized by a large cell surface area/volume ratio; (ii) larger, often more rapidly moving unicells or simple coenobia but with distortions of form that can maintain a high surface

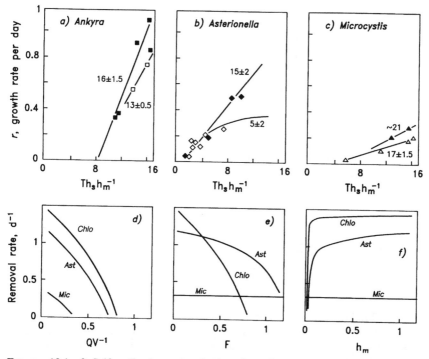

FIGURE 10.1a–f. Self-replication rates of selected species (a, *Ankyra judayi;* b, *Asterionella formosa;* c, *Microcystis aeruginosa*) in limnetic mesocosms, plotted against effective day length $Th_s h_m^{-1}$, where T is the hours of daylight between sunrise and sunset and h_s/h_m is the ratio of the Secchi disk depth and mixed-layer depth; note that when $h_s > h_m$, $h_s h_m^{-1}$ is evaluated at 1 and grouped according to ambient temperature. The effect of removal on net population increase rate, either through (d) dilution of the volume, V, by a diluting flow, Q, up to one complete flushing per day, (e) consumption by filter-feeding zooplankton filtering up to F = 1 entire volume of water per day and making certain assumptions about the filterability of the algae, or (f) sinking out from a mixed layer whose depth is expressed as a fraction of a 10-m water column and making certain assumptions about the sinking rates of the algae concerned or their abilities to regulate their own vertical movements. (Data were assembled over numerous individual studies and collected and reviewed in Reynolds 1986.)

area/volume ratio and which, additionally, may well contribute to the ability of the cells to "harvest" light in environments or layers in which it is deficient; and (iii) slow growing, mostly much larger organisms having low surface area/volume ratios that are poor light antennae but have well-developed abilities to control their own movements. These latter algae overcome sinking out and are also resistant to common types of planktonic grazers, including rotifers and crustacea (Fig. 10.2).

I considered this coherence to be not merely coincidental but a major factor in distinguishing among the ecological traits of the functional groups (Reynolds 1984a). I later went so far as to equate the distinctions with Grime's (1979) separa-

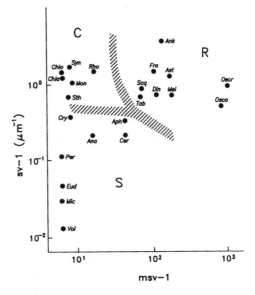

FIGURE 10.2. Morphological properties of various algal sizes and shapes, plotted in a log/log matrix to distinguish the diminishing surface/volume (sv^{-1}) ratio of algae of increasing size, and the effect of attenuation (into markedly nonspherical shapes) in preserving high surface/volume ratios of filaments and needle-like cells, where m = the maximum linear dimension. The categories *C, S,* and *R* are based upon the ecology of the species concerned, which, it is suggested, is not independent of the specific morphology. The C-species are small-celled, fast-growing invasive colonists (*Chla, Chlamydomonas; Chlo, Chlorella; Mon, Monodus; Rho, Rhodomonas; Sth, Stephanodiscus hantzschi; Syn, Synechococcus*); S-species are slow-growing acquisitive self-regulating species, usually colonial or large-celled (*Ana, Anabaena; Aph, Aphanizomenon; Cer, Ceratium; Cry, Cryptomonas; Eud, Eudorina; Mic, Microcystis; Per, Peridinium; Vol, Volvox*); R-species are mostly acclimating species that will become well attuned to intercept light in vigorously mixed, turbid water columns offering severely truncated photoperiods (*Ank, Ankistrodesmus; Ast, Asterionella; Din, Dinobryon; Fra, Fragilaria; Mel, Melosira* [now mainly *Aulacoseira* spp.]; *Scq, Scenedesmus quadricauda; Tab, Tabellaria; Osca, Oscillatoria* [now *Limnothrix redekei*]). (Redrawn from Reynolds 1993b.)

tion of particular suites of adaptations associated with the three primary plant strategies he recognized. The same labels were adopted, albeit with slightly different definitions (Reynolds 1988b): C- (for colonist) species, primarily adapted for rapid transmission, fast growth, and main investment in cell replication but liable to heavy cropping by grazers; R- (for ruderal) species, tolerant of high frequency-disturbance episodes, especially where these impinge on underwater light availability; and S- (for stress-tolerant, biomass-conserving) species, which are typically "large" (sometimes through aggregation of cells into colonies) and sufficiently mobile to make controlled migrations through structured thermal gradients to avoid or to gain access to local properties of the environment. S-species grow relatively slowly, especially at low temperatures (<6°–10°C); generally they have higher light saturation requirements than algae of the other two primary categories.

Although in the first instance Grime's C-, S-, and R-categories were adopted to apply to functional groups already termed r, K, and w, respectively, (Reynolds et al. 1983), it is important to expunge the notion of interchangeability of r- and K-selection with C-, S- and R-strategies. r-Selection refers to investment of anabolic production in population replication rate; K-selection alludes to a proportionately larger investment in maintenance, including through competitive advantage near to the capacity limit of the critical carrying resource and survival of cells, K (MacArthur and Wilson 1967). It *is* true that C-strategists are primarily r-selected (e.g., *Chlorella, Ankyra*) and that S-strategists are likely to be strongly K-selected (as is *Microcystis* or *Ceratium*); R-strategists, however, include species that are no less K-selected (e.g., *Oscillatoria* species) than *Microcystis* while many diatoms (*Asterionella*, small *Stephanodiscus* spp.) rely heavily on a rapid growth rate (r) when appropriate conditions obtain.

I have been able to collect considerable quantitative evidence to show that the development of species-specific populations in situ follows well-established or well-developed theoretical models (see, for instance, diatoms in Reynolds 1973a; Reynolds and Wiseman 1982; cyanobacteria in Reynolds 1972, 1973b, 1975; Reynolds et al., 1981; dinoflagellates in Reynolds 1978 and in Lund & Reynolds 1982; *Ankyra* in Reynolds et al. 1982; *Volvox* in Reynolds 1983). The early increase of colonist populations in resource-replete environments is described by the exponential growth equation (Eq. 10.1):

$$n_t = n_o \cdot e^{rt} \tag{10.1}$$

where n_o is the initial mass (or cell number, or other analog) of the population and n_t is the population after time t. r is the exponent of net mass replication to the natural logarithmic base e. Rearranging Eq. 10.1, r is theoretically solved from the natural logarithm of the factor of increase in the population during a known period:

$$r = \ln(n_t/n_o)/t \tag{10.2}$$

It is often also convenient to solve the generation time, or mean time of population doubling, (t_G). Thus, if $(n_t/n_o) = 2$, then

$$t_G = \ln 2/r \tag{10.3}$$

In Figure 10.3, increases in two natural populations are illustrated, first on a scale of absolute concentration and then on a logarithmic scale. The latter would be expected to normalize (or render the curve as straight line) the data, while the plot is itself a series of curves, reflecting a change in exponential net replication rate. This itself prompts the question, which rate is the true one and what caused the acceleration/deceleration of rate? Unfortunately, the answer can be reliably determined only in laboratory cultures and under defined, controlled conditions. Some robust patterns have been discerned, however, that distinguish among the replication rates of different species under conditions of nutrient saturation and continuous saturating irradiance and at similar, constant temperatures. Similarly, the sensitivities of species-specific replication rates to altered temperatures and, at a given temperature, to truncated photoperiods are correlated to certain morphological properties of cells (Reynolds 1989a; see also Fig. 10.4). Indeed, these relationships have been assembled into algorithms predicting rates of specific population increase in defined waters that, in several instances, have fairly reconstructed actual increase phases.

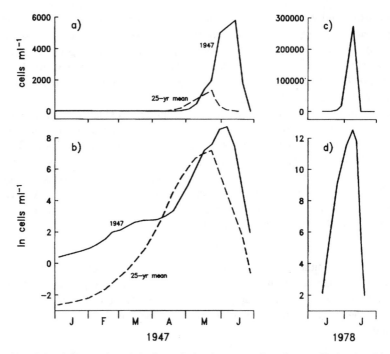

FIGURE 10.3a–d. Examples of algal population increase, plotted normally (a, c) and logarithmically (b, d) for (a, b) *Asterionella formosa* in the north basin of Windermere, 1947, and contrasted with the mean course for the 25 years, 1964–1988 (from Reynolds 1990), and for (c, d) a population of *Ankyra judayi* raised in the plankton of a large limnetic enclosure and studied by Reynolds et al. (1982).

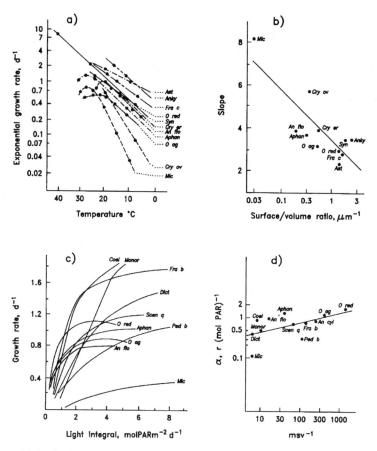

FIGURE 10.4a–d. Temperature- and light-sensitive growth rates of planktonic algae as a function of their morphology. (a) Arrhenius plots of exponential growth rates against temperature; in (b), the corresponding slopes of the normalized curves are plotted against the surface/volume ratio of the alga; reported growth rates are shown in (c) as a function of the daily integral irradiance, in mol photons PAR m^{-2} d^{-1}, and the slopes of the steepest rising portions of these curves are plotted (d) against the products of maximum linear dimension and surface/volume ratio (msv^{-1}) for the respective algae. Algal abbreviations: *An cyl, Anabaena cylindrica; An flo, Anabaena flos-aquae; Anky, Ankyra judayi; Aph, Aphanizomenon flos-aquae; Ast, Asterionella formosa; Coel, Coelastrum microporum; Cry er, Cryptomonas erosa; Cry ov, Cryptomonas ovata; Dict, Dictyosphaerium pulchellum; Fra b, Fragiliaria bidens; Fra C, Fragilaria crotonensis; Mic, Microcystis aeruginosa; Monor, Monoraphidium; O ag, Oscillatoria agardhii; O red, Oscillatoria redekei; Ped b, Pediastrum boryanum; Scen q, Scenedesmus quadricanda; Syn, Synechococcus.* (Redrawn from data assembled in Reynolds 1989a.)

It is interesting to note, however, that the maximum temperature-determined resource-saturated growth rate remains a cell-specific function and can be considerably lower than the potential of the resource- and energy-harvesting processes.

Reynolds (1990) argued that at their cell-specific maxima *Chlorella* cells could take up sufficient phosphorus to sustain the next cell generation in just ~7 min and to fix sufficient carbon in 4.4 h to sustain a doubling of the parental cell but still require more than 9 h in which to assimilate the carbon and to assemble the relevant carbon skeletons and cytological structures of the cells of the next generation. In this context, it becomes difficult to conceive of growth-rate limitation until it can be shown that there is a sufficient impairment of the photon flux, or of the carbon supply, or a sufficient exhaustion of phosphorus or silicon from the pool available, to ensure that the minimum period required for cell doubling is exceeded (Reynolds et al. 1985). Until that point is reached, the cell has the opportunity to store nutrients in excess of immediate requirements and to vent excessive photosynthate from the cell, through either accelerated respiration or photorespiration or by excretion of various photosynthetic intermediates (e.g., glycollate) and of high molecular weight polymers. Such homeostatic properties assist the deduction of the resource(s) most likely to "run out" first, i.e., the determination of carrying capacity. In this context, light energy (photon flux) as well as specific nutrient supplies can be treated as potentially limiting resources and quantified as K. Divided by the cell quota for each resource (eventually the minimum quota, q_o), the smaller carrying capacity (m_{pot}) can be solved:

$$m_{pot} = K/q_o \tag{10.4}$$

The formulation can be used to define the time scale of capacity-filling because the number of generations (G) that can be sustained is given by

$$G = \ln(K/q_o)/\ln 2 \tag{10.5}$$

which should be completed in not less than $G \times t_G$, i.e., in

$$\frac{\ln(K/q_o)}{\ln 2} \times \frac{\ln 2}{r} = \frac{\ln (K/q_o)}{r} \tag{10.6}$$

Finally, it is important to observe that the factors themselves alter through time, that not all do so in the same way, and that factor interaction varies in consequence. Cells are able to adapt to variation in the supply of some of those factors (e.g., photon flux) but others (such as the cell silicon requirement) are more fixed. In this way, the capacity ceiling for *Asterionella* in Windermere continues to be set by silicon availability (Lund 1950) (see also Fig. 10.5). However, the rate of capacity attainment is, at first, light determined, pending a historical switch to rate limitation by the depleted phosphorus supply. One consequence of phosphorus enrichment that has occurred since the mid-1960s (Talling and Heaney 1988) is an earlier attainment of the silicon-limited capacity. This is, at least, more readily explicable in terms of an altered capacity factor than a changed logistic growth response in respect of any of the factors (Fig. 10.6). This clearly anticipates an advance in the time at which the time of the silicon-limited maximum of *Asterionella* in Windermere may actually be reached.

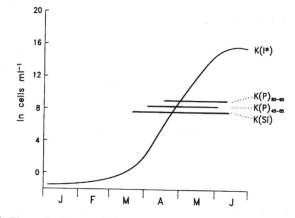

FIGURE 10.5. Theoretically extended curve of the seasonally changing light-determined carrying capacity, $K(I^*)$, of Windermere north basin for *Asterionella formosa*, and the positions of the ceiling capacities provided by the silicon available, $K(Si)$, by the phosphorus typically available before the recent enrichment began, $K(P)_{45-65}$, and by that available in the last decade or so, $K(P)_{80-90}$. The representation is offered as explaining the annual bloom of the alga in Windermere as being controlled by the increasing light income until the population exhausts the silicon. The eutrophication of the lake (phosphorus enrichment) has not directly influenced the size of the maximum, which continues to be determined by the silicon-determined carrying capacity.

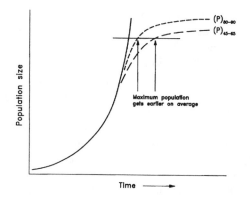

FIGURE 10.6. Logistic plots of *Asterionella* crop predicted by the availability of phosphorus in Windermere, 1945–1965 and 1980–1990, in relation to the curve of light-determined capacity and the threshold of silicon exhaustion. This representation is offered to support the argument that although the silicon ceiling continues to operate, the rate of its attainment may have been historically phosphorus limited. Thus, increasing the availability through eutrophication could explain an apparent advance in the timing of the maximum population.

The Establishment of Communities

In the plankton, as in other types of vegetation, it is rarely the case that only one species is represented at any one time. It is also recognized, rather more clearly in the case of plankton, that the second, the third, or the ith species may be similarly adapted and share similar predilections to the first, or they may not. If "not," they may well fall back in representation and become "rare," until such later time that one or another species is favored by an alteration to the environmental conditions; these changes could be gradual or abrupt, and they could be effected by the first species (internal change) or independently, from outside (external change). If the latter species are promoted by similar circumstances to the first, they may enrich the accreting structure of the assemblage. Moreover, if both or all species are actively increasing, and against the same spectrum of mutually beneficial resources, there is an increasing likelihood that they will eventually find themselves in direct interspecific competition to acquire the resources.

A substantial part of theoretical community ecology is founded on the perception of interspecific competition among the species present, and the community structure is itself an outcome of competitive processes (Kilham and Kilham 1980; Tilman et al. 1982). The mechanisms of exclusion, which have been compellingly demonstrated in batch-cultured planktonic algae, do not translate well to the natural environments of plankton (Reynolds 1992b). The competitive exclusion principle (Hardin 1960; also known as Gause's hypothesis) argues that two species sharing identical ecological "niches" cannot coexist and that the fitter will progressively exclude the latter. Given that open lakes and seas are randomized, isotropic environments, the bizarre array of several thousands of known species of planktonic algae is perplexing and paradoxical (Hutchinson 1961). Such diversity could be justified only if the pelagic environment is not uniform but infinitely and contemporaneously patchy (Richerson et al. 1970) or if the species are sufficiently different in their individual requirements to be other than in direct competition with each other (Petersen 1975).

The evidence from my observations and experiments suggests that the structure and dominance of assemblages is influenced primarily by the comparative dynamics of the (almost fortuitous) assembly of species present and *by the period over which they apply:* the species increasing fastest or for longest or, again, fielding the greatest inoculum at the outset, stands most likely to become dominant (Reynolds 1986). We have already considered the coherence between cellular growth rates, adaptability to low light doses, and cell morphology; the dynamics of population changes are additionally influenced by the instantaneous specific rates of population attrition (e.g., through cellular death, settlement loss, or grazing by zooplankton), which need not be simultaneously similar for all. In the broad sense, being a better competitor is, indeed, to fare better than the others, so that avoiding grazing or sedimentation to the extent that it makes up the disadvantage of a poorer rate of replication could be described as being competitive. One example of the establishment of dominant *Microcystis* population, illustrated in Figure 10.7, serves to show this dynamics-based *directionality* of community evolution.

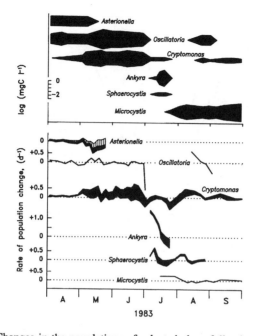

FIGURE 10.7. Changes in the populations of selected algae following the initiation of a new successional episode in a large limnetic enclosure, commencing at the end of June, through which first *Ankyra*, then *Sphaerocystis*, and then *Microcystis* dominated. The *lower plot* shows the net rates of population replication and the simultaneous rates of net population increase for each participant, and the *filled space* between them therefore represents the exponential rate of population loss (*solid*, grazing; *hatched*, sinking). The plot shows how *Ankyra* grows much faster than (e.g.) *Microcystis*, although it slows down, and the increasing rate of consumption eventually takes the population into net decline. *Microcystis*, however, is little effected by direct losses. (Simplified from Reynolds 1988b.)

The role of resource-based competition can be held to shape the outcome of between-species interaction provided that the resource is simultaneously limiting both species. This condition was set by Tilman (1982) but has rarely been fulfilled in the many instances in which competitive outcomes are claimed or persisted long enough for a community response to become manifest. The examples illustrated in Figure 10.8 show the course of events when several vernal diatoms were vying for prominence with *Oscillatoria* but with the ratios of phosphorus to silicon set differently and with the starting quantity of *Oscillatoria* also variable. The relative performances of the species do not show any behavior consistent with resource-competition theory; the overwhelming impression is that dominance is conditioned by the inertia of population replacement. Whether the dominance of the largest population is challenged by a smaller population of faster growing species may depend on whether the capacity-limiting resource level is attained before the dominance is surrendered. This latter point is crucial: for if it is not, or

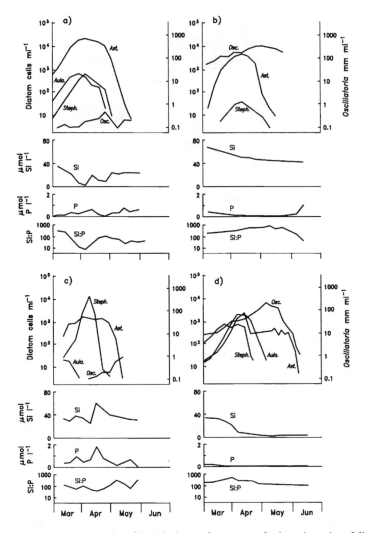

Figure 10.8a–d. Some examples of the relative performances of selected species of diatoms (*Ast, Asterionella formosa; Aulaco, Aulacoserira italica* ssp. *subarctica; Steph, Stephanodsicus astraea* var. *minutula*) and the cyanobacterium *Oscillatoria agardhii* var. *isothrix* during the vernal bloom periods in a large limnetic enclosure and under controlled chemical conditions in relation to the ambient concentrations of dissolved reactive silicon and phosphorus and to the ambient ratio between them. (Plotted from various sources reviewed in Reynolds 1986.)

is unlikely to do so, then the assumptions about competition, exclusion, and "equilibrium" processes are unlikely often to control community structure.

If this appreciation can be acknowledged in species assemblages that change rapidly, at least in human perception, it is tempting to look for it at other trophic levels or in other systems that change more slowly. If it is indeed true that many assemblages exist at far short of their equilibrated, competitively excluded outcome and that typically their composition owes as much to precedence and fortuity as to intense, resource-based competition, then the interpretations of the scale and the energetics of trophic-level interaction should be tempered accordingly.

It is inappropriate, in this chapter, to discuss the wide disparity of results or, more significantly, generalized interpretations about food web regulation arising from various field observations and manipulative experiments. It is acknowledged that there is a continuing debate about whether ecosystems are controlled by the resources available to the primary producer level of the food chain [as is implicit in Vollenweider's (1976) averaging; the so-called bottom-up control] or whether the cascading effects of the top, carnivorous feeders regulate the lower trophic levels (the "top-down" control promoted by Carpenter et al. 1985). It is not a question of which is correct but how often either is true: as I have argued (Reynolds 1991), the effects of either would, or do, collapse into one another as the steady state is struck. Before that, altering factors may see the abundances of feeders lurch between heavily depleting their foods to being unable to control them. This particular view was shaped by our own attempts (Reynolds 1986) to impose artificial cropping regimes on phytoplankton through lowering or enhancing the ambient intensity of *Daphnia* feeding in the Blelham Tarn limnetic mesocosms (Reynolds et al. 1982).

The mechanics of the manipulations, indeed, involve analogous relationships between zooplankton and the next trophic level. Removing the fish to allow *Daphnia* "free dynamic rein" was relatively easier to attain than was the maintenance of a population of planktivorous fish sufficient to suppress the development of *Daphnia* populations (Reynolds 1986). Ultimately, however, the size structure of the husbanded population was biased against larger sized *Daphnia* individuals; careful analysis (George and Reynolds; in manuscript) showed that even then it was still possible for *Daphnia* populations (as well as those of rotifers and ciliates) to increase or decrease more according to the quality of the food resource available than in tune with relatively constant planktivory. The dynamic responses of even the Daphnia populations, unconstrained by their fish predators, were too slow to prevent the initial stages of algal increase adding substantially to the plant component of the food web or, sometimes, to prevent the alga from eventually reaching a resource-based capacity. Nevertheless, responses often passed from resource-led opportunities to a consumer-driven demise (Fig. 10.9).

The evident success of imposed and ongoing grazer control reported elsewhere (Shapiro 1990, 1993; Moss 1992), to the extent that the technique can be advanced as a legitimate and effective means of regulating producer biomass, has led me to question our very different experiences. I concluded (Reynolds 1991, 1992b) that, at least for cladoceran zooplankton, the essential difference is an alternative refuge for food (detritus, bacteria in sufficient concentration) and, probably, predation-protected physical refuges (e.g., among submerged macrophytes) in which sufficient biomass can be maintained for an immediate and critical control on the rapid development of planktonic autotrophs to be applied. That such refuges are not available in the deep-water enclosures in Blelham Tarn but

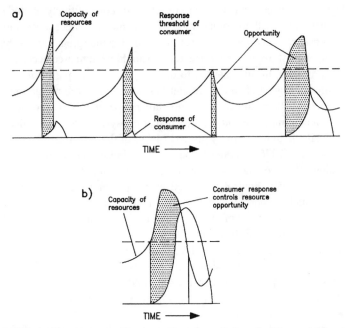

FIGURE 10.9a,b. The response of a consumer to a regularly pulsed but variable magnitude resource. (a) The response is scaled to the opportunity; (b) the opportunity is sufficiently extended for the consumer to deplete and regulate the resource directly. (Redrawn from Reynolds 1991.)

have been shown to have been critical to cases reviewed by Moss (1992), or are likely to have been so in just the shallow enriched ponds in which biomanipulative experiments have been successful (see especially McQueen 1990; De Melo et al. 1992), may explain the different experiences.

The conclusion may be helpful, too, in reminding field experimenters that altering the ambient relationship between the predator and the prey species, whose individual generation times differ considerably, introduces several simultaneous (and possibly significant) side effects. Changing the population of the predator may affect the population of the prey species' food (the desired effect in biomanipulation), the prey's competitors, the food species other consumers (where appropriate), or the populations of the prey species' other predators (Reynolds 1991; see also Pahl-Wostl 1990). This question of trophic-level scaling also gives rise to doubts about the alleged efficiency of trophic transfer and tight cycling of resources, save at demonstrably steady states (Fig. 10.10).

The Approach to a Climactic Steady State

This phase of (successional) "accumulation" is followed by one of "maturation" (terminology of Price 1984) wherein diagnostic developmental trends are said to

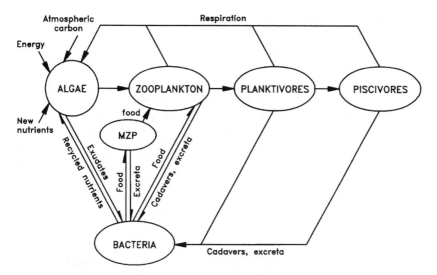

FIGURE 10.10. Scheme of material cycling through the pelagic ecosystem. *MZP*, micro-zooplankton (in reality, to include heterotrophic nanoflagellates and microzooplanktonic ciliates).

be evident (Odum 1969): increasing organic biomass and increasing gross production but decreasing productivity and net production; increasingly intrabiotic resources, greater diversity, increasing spatial organization and food web connections; larger and more specialized organisms with increasingly complex life cycles; decreased nutrient exchange and the increased dependence on efficient cycling and conservation; an increasing emphasis on K-selection; more elaborate symbiosis and greater resilience to external perturbations; and greater acquired information and decreasing entropy.

None of these hallmarks of ecological succession is necessarily less evident in the phytoplankton, provided (Reynolds 1980, 1988a) such sequences are (autogenically) driven from within the community, that is to say, alterations of the growth environment that owe primarily to the activities of the organisms already present. A habitat stripped of a critical nutrient by a fast-growing, pioneer C-species might be rendered untenable save by a migratory nutrient scavenger, like *Ceratium* or *Microcystis, S*-species in which energy investment is directed principally into biomass conservation. Note that, unlike major terrestrial ecosystems, direct grazing on the phytoplankton by filter-feeding microcrustacea, on *Chlorella* but not *Ceratium,* is ultimately likely to advance the successional trend rather than to retard it (as, say, large grazing mammals resist the transformation of grassland into scrub and forest). Alternatively, overall biomass increases without exhausting free nutrients but determines a premium on the underwater availability of light energy. This may favor the more light-efficient, R-species of filamentous cyanobacteria, like the *Oscillatoria* spp. that can ascend to almost total dominance of particular systems (Berger 1975; Faafeng and Nilssen 1981). These instances provide further examples of how the pioneer species, with their rapid rates of colonization, are progressively overtaken by species that, even though they are inferior in terms of their maximal rates of growth, are nevertheless better adapted to self-maintenance. In the case of *Microcystis,* for example, the coenobia are often too large or unpalatable to become direct food of planktonic crustacea and rotifer, while their ability to regulate buoyancy permits them to avoid permanent sedimentation and to adjust to frequent (two per day) changes in vertical mixing intensity (Reynolds 1989b). So long as the same ambient conditions persist (which might be, effectively, continuously in low-latitude shallow lakes, or until the end of the high-latitude summer), other algae have little chance of outcompeting the *Microcystis* and are, as predicted by equilibrium models, competitively excluded. "Succession rate" (Jassby and Goldman 1974; Lewis 1978) or the rate of community change (Reynolds 1980) falls to zero, as do the indices of diversity and of equitability (Reynolds 1988a; see Fig. 10.11).

This pattern is arguably fairly typical for commonly observed phytoplankton successions, at least insofar as their strong directionality is concerned. Nevertheless the eventual competitively excluded outcome may favor any one of a number of conspicuously different species, usually the one best suited to the characteristics of the water body. Moreover, particular lakes at particular times frequently achieve a seasonal steady state of overwhelming dominance, typically by such genera as *Oscillatoria, Microcystis, Ceratium,* or *Peridinium.* Many other "extreme" systems seem to arrive at a long, low-diversity phase, determined by the environmental conditions imposed and the predilection of the algae concerned. Thus, where nitrogen supplies do not match freely available phosphorus, it is not uncommon to observe dominant populations of nitrogen-fixing cyanobacteria such as *Anabaena, Aphanizomenon,* or *Cylindrospermopsis.* In clear, oligotrophic, and nitrogen-deficient Crater Lake (Oregon), at the time it was investigated by Utterback et al. (1942), the dominant *Anabaena* was stratified at depths offering optimal combinations of sufficient phosphorus, invasive nitrogen, and light to sustain nitrogen fixation as well as photosynthesis. In the turbid, phosphorus-rich epilimnion of the explosion crater of Rotongaio (North Island, New Zealand), the advantage might be expected to fall to a nitrogen-fixing *Oscillatoria,* were such an organism known to science; in fact, the dominant alga is a solitary, straight-filamented *Anabaena* which almost fulfils that description (Walsby et al. 1989). In much the same way, extremes of acidity, alkalinity, and dystrophy encourage particular dominants (references in Havens 1992); frequent or continuous flushing or continuous flow, as in rivers, also select for particular morphological (small size) or physiological (rapid growth) features (Reynolds 1992c, 1993b; Descy 1993) and, in turn, for low-diversity communities.

It is also worth observing that diminution in the range of primary producers often limits the range of consumption strategies. Certainly, with many instances of cyanobacterial dominance, the food chains become shortened and simplified, often to dependent bacteria and their protozoan grazers, to the detriment of ecological interest and practicality of human exploitation. These are not the only outcomes, of course. I have drawn attention to the similarity of community structure

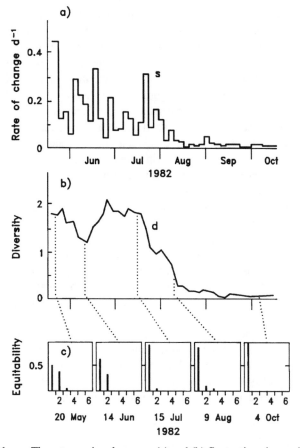

FIGURE 10.11a–c. The rate species change, *s*, (a) and (b) fluctuations in species diversity, *d*, in a limnetic enclosure during the establishment of steady-state, near-monocultural dominance of a *Microcystis* population. (c) The changes in equitability are shown through a series of dated "snapshots," giving the proportions of the total phytoplankton mass (= 1) invested in the first, second,... most abundant species. (Figure composites from Reynolds 1988a.)

in very stably stratified waters, with pronounced and evolved vertical gradients of physical and chemical properties. These are usually accompanied by organismic layers, or "plates," each analogously dominated by particular species and in recurrent sequences (Reynolds 1992a). There is thus a situation in which several species can be simultaneously present but are niche differentiated and physically separated. Each may well have been the winner of competitive exclusion *within the layer* where it is best adapted but, elsewhere in the gradient, to have been excluded by another and locally better adapted competitor. Constancy of conditions over long time spans is essential for niche-separated diversity to develop and stabilize (Reynolds 1992a).

The unresolved arguments concerning the outcome of terrestrial successions may well be answered by the observations on planktonic communities! These examples of aquatic successional climax and biological self-organization are not well recognized among ecologists generally, despite having been vigorously advocated by Margalef (1961, 1968). Moreover, the available data withstand scrutiny, even against the most stringent tests of thermodynamic theory (Boltzmann 1896; Prigogine 1955). For instance, the second law of thermodynamics indicates that all energy in a system, itself supposedly emanating from the cosmic "big bang" said to have initiated the universe, must eventually be dissipated as heat. There should be an accompanying increase in disorder, stagnation, and entropy until, ultimately, the system comes to a complete rest and to complete thermodynamic equilibrium; i.e., it "dies."

At first sight, living ecosystems seem to counter this second law by acquiring energy and order directed to the achievement of a competitively excluded steady state. In fact, they are dissipative structures that continue to consume energy in their anabolism, only a proportion of which is used to increase structure. Nevertheless, living systems are characterized by a dominating ability to reduce the abiotic rate of entropic dissipation (Lovelock 1979). This may be recognized at several levels of biological organization; ultrastructurally, cytologically, biochemically and physiologically, organelles, cells, and individual organisms maintain themselves against the gradients of dissipation. In entirely analogous ways, the successional self-organization of communities and ecosystems not only resists disorder but reduces the rate of their entropic breakdown. In short, it tends to bring a system to a steady-state condition (Odum's "ecological climax") wherein the organization (biomass) maintained is the maximum that the incoming energy flux will support.

Energetic States of Planktonic Communities

By extension of this idea, the organizational state of an ecosystem and its progress toward its steady-state climactic condition should be measurable by the balance of its entropic fluxes, or its "exergy" (Mejer and Jørgensen 1979) or, roughly, its "negative entropy." The concept has been explored and developed in relation to aquatic ecosystems during the subsequent decade by Jørgensen (see his 1992 review) and his various associates (e.g., Nielsen 1992). In essence, they have taken Prigogine's (1955) appreciation of an open system allowing fluxes of energy and matter to determine a probabilistic expression for its entropy. This approach recognizes that while the biological processes represent unidirectional dissipative structures, the surroundings were open to radiative fluxes in and heat fluxes out (Fig. 10.12). The rate of change in the entropic state of the whole system under consideration, dS, might be expressed: (10.7)

$$dS = dS_i \pm dS_e$$

where dS_i is the entropic flux of the biological system and dS_e represents the net

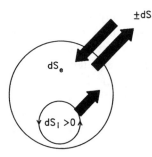

FIGURE 10.12. "Entropy" fluxes in ecosystem (as seen by Nielsen 1992). The abiotic, external exchanges receive direct radiant energy and export long-wavelength radiation as heat, providing an instantaneous entropy flux (dS_e) that may be positive or negative but which over time tends to become zero. The imposition of a biological system consumes part of the energy in elaborating structure and then loses a part of that energy in its maintenance (dS_i): it is always dissipative, in the sense of "spilling" an entropic quantity. Thus, the ability of the ecosystem to build is gauged by whether the negative external flux exceeds the quantity of the internal flux, so $dS_e + dS_i < 0$.

flux exchanging in the surroundings. For the biological system to survive, it is necessary, on balance, for the influx of low-entropy radiative energy, minus its direct rate of dissipation as heat, still to exceed the maintenance loss from the biological component; i.e., that

$$dS_e + dS_i < 0 \qquad (10.8)$$

or, that entropy flux is negative.

In their ascendant stages, biological systems self-organize because and for so long as the condition is maintained. It may be supposed that, while the incoming radiation is steady the magnitude of dS_i and $-dS_e$ increase and that, for a time, the latter will do so absolutely more rapidly. As the system moves further from thermodynamic equilibrium, however, the system expends relatively more of its energy on maintenance until at its climactic steady state the biological entropy balances its energy consumption; i.e., when

$$dS_e + dS_i = 0$$

The remaining contingency (when $dS_e + dS_i > 0$) covers those occasions that represent a loss of structural organization, because it is dissipating more rapidly than it is acquiring its low entropy: this condition is explored in a later section.

The simple terms of negative entropy belie what is a rather sophisticated concept. However, it is one that suffers from a lack of empirical evidence. Following is a series of calculations, based on our data for pelagic systems but transposed into energetic terms, to show the utility of the concept.

Let us consider a lake in a temperate latitude and consider the maximum probable radiation income to be that obtaining around the time of the summer solstice, some 26.7 MJ m^{-2} d^{-1}. Given a 16-h day, this is equivalent to an average flux of 464 W m^{-2} ($\equiv$ 464 kg s^{-3}) during the daylight period. If we suppose that 47% of this flux is within the photosynthetically active waveband (400–750 nm) and that 1 mol photon $\equiv$ 218 kJ, the visible light is close to 218 W m^{-2} and 1 mmol photon m^{-2} s^{-1}, and equivalent to a daily flux of "usable" energy of 57.6 mol photon m^{-2} (12.6 MJ m^{-2} d^{-1}). At the energetic equilibrium, this 12.6 MJ m^{-2}, together with the nonvisible balance of the 26.7 MJ m^{-2} d^{-1} flux, is transmitted as high-entropy heat. Intercepted by photosynthetic organisms, however, a fraction is "siphoned off" to contribute the energetic cost of elaboration of a biological community.

Let us now take the photosynthetic attributes of a *Chlorella* population as being representative of planktonic primary producers. Of particular relevance here is how much energy a *Chlorella* cell can harvest and how much is required to augment (say, double) the standing biomass. Reynolds (1990) has calculated that a single *Chlorella* cell, 4 μm in diameter, can receive light, at flux retaining, on the basis of the area of light fluid it projects, equivalent to $\pi(4/2)^2 \approx 12.6 \ \mu m^2$. Taking the carbon content of the cell to be 7.3 pg, or 0.61×10^{-12} mol C (cell)$^{-1}$ (Reynolds 1984b), the cell-carbon-specific project is found to be

$$(12.6 \times 10^{-12} \ m^2) \div (0.61 \times 10^{-12} \ \text{mol cell C}) = 20.6 \ m^2 \ (\text{mol cell C})^{-1}$$

The interception of photosynthetically active energy at an average photon flux density of 1 mmol m^{-2} is then

$$20.6 \ m^2 \ (\text{mol cell C})^{-1} \times 1 \times 10^{-3} \ \text{mol photons m}^{-2} \ s^{-1}$$
$$= 20.6 \times 10^{-3} \ \text{mol photons (mol cell C)}^{-1} s^{-1}$$

The photosynthetic pigments absorb only in certain wavebands, said to represent approximately 13.7% of the visible spectrum. The light energy which could be used by the cell is then

$$0.137 \times 20.6 \times 10^{-3} \ \text{mol photon (mol cell C)}^{-1} \ s^{-1}$$
$$= 2.82 \times 10^{-3} \ \text{mol photon "active" radiation (mol Cell C)}^{-1} \ s^{-1}$$

We should next consider the effectiveness of the photon absorbance and energy transfer reaction in the photosynthetic units. From relationships worked out by Emerson and Arnold (1933; see also Harris 1978), the theoretical minimum requirement for the fixation of 1 mol of reduced carbon is 8 mol photons and 2400 molecules of chlorophyll per reaction; to absorb 1 mol photon of active radiation requires 0.268×10^6 g chlorophyll a (268 kg!) Given that the reaction takes some $10^{-2.6}$/ s, however, the same amount of chlorophyll can absorb ~400 mol photons s^{-1}, or about 1.49×10^{-3} mol photon (g chl a)$^{-1}$ s^{-1}. Substituting the supposed maximum yield of fixed carbon (0.125 mol C per mol photon), the fixation capacity is

$$0.125 \times 1.49 \times 10^{-3} = 0.186 \times 10^{-3} \ \text{mol C fixed (g chl } a)^{-1} \ s^{-1}$$

and the visible photon flux to sustain it is equivalent to

$$1.49 \times 10^{-3} \div 0.137 = 10.9 \times 10^{-3} \ \text{mol photon (g chl } a)^{-1} \ s^{-1}$$

The amount of chlorophyll per cell is not constant, but 0.5% of the ash-free dry weight (itself almost 50% carbon) is a reasonable estimate for cells grown in high light. The chlorophyll content of the cell may then be expressed as 10 mg chl a (g cell C)$^{-1}$, or 0.12 g chl a (mol cell C)$^{-1}$, or 8.33 mol cell C (g chl a)$^{-1}$. We may now also deduce the cell-specific absorption and fixation equivalents. The sustaining flux is equivalent to 10.9×10^{-3} mol photon (g chl a) $s^{-1} \div 8.33$ mol cell C (g chl a)$^{-1}$ = 1.31×10^{-3} mol photon (mol cell C)$^{-1}$ s^{-1} and the fixation capacity is 0.186×10^{-3} mol C fixed (g chl a)$^{-1}$ $s^{-1} \div 8.33$ mol cell C (g chl a)$^{-1}$ = 22.4×10^{-6} mol C fixed (mol cell C)$^{-1}$ s^{-1}.

The rate of increase in mass may be subject to other constraints but there is always a finite maintenance cost. This we will equate to the dark respiration rate of *Chlorella* cells, as being about 5% of the maximal photosynthesis rate or 1.1×10^{-6} mol C (mol cell C) s^{-1}. Thus the growth rate that we can suggest is sustained by the deduced rate of net photosynthesis is $(22.4 - 1.1) \times 10^{-6} = 21.3 \times 10^{-6}$ mol C (mol cell C) s^{-1}; the time taken to double the cell mass, deduced from (ln 2/ln 1.0000213), = 32.5×10^{5} s or ~9.04 h.

If we now place our *Chlorella* in the envisaged light field of the lake, we may suppose that this sort of growth rate could be sustained through each light hour (i.e., ~1.2 d^{-1}), until such time as *Chlorella* cells near the surface begin to shade out those beneath them. Even if the population is gently mixed, the probability is that individual cells will experience a light field that is on average weaker and will pass part of the daylight period in the dark. This critical point is reached when the light field is occluded by a canopy of *Chlorella* cells, i.e., at an areal concentration of $1/(20.6$ m^2 mol cell C^{-1}), or 0.0485 mol cell C m^{-2}. (This is equivalent to 0.586 g C or ~6 mg chlorophyll m^{-2}). Beneath this threshold, maximum growth by all the population can be assumed. Above it, the cell-specific carbon yield decreases, while the cell-specific maintenance remains constant. Thus, for increasingly larger populations, maintenance costs increase absolutely and directly; fixation increases absolutely but, while the radiation flux is constant, with an exponentially diminishing efficiency (Fig. 10.13).

At Odum's (1969) energetic steady state, we expect the production and maintenance to balance. The maximum standing crop that could be theoretically maintained is the one which disperses 12.6 MJ m^{-2} d^{-1}, mainly as heat energy, this balancing the visible radiant energy absorbed in generating the replacement biomass (12.6 MJ m^{-2} d^{-1} = 57.8 mol photon m^{-2} d^{-1}). At this steady state, there is no net change in cell carbon despite maintenance consumption of 1.1×10^{-6} mol C (mol cell C)$^{-1}$ $s^{-1} \times 86,400$ s d^{-1}, or 0.095 mol C respired d^{-1}. The photon cost is 0.095 mol C fixed (mol cell C)$^{-1}$ d^{-1} $\times$ (8 ÷ 0.137) mol photon (mol C fixed) = 5.55 mol photon (mol cell C)$^{-1}$ d^{-1}. Thus, the maximum sustainable crop is given from the daily photon flux as 57.8 mol photon m^{-2} d^{-1} ÷ 5.55 mol photon (mol cell C)$^{-1}$ d^{-1} = 10.5 mol cell C m^{-2} (or about 1300 mg chlorophyll m^{-2}).

The construction of Figure 10.13 recognizes the exponential decline in productivity above a biomass of 0.0485 mol cell C m^{-2} to its balance with the visible energy flux at 10.5 mol cell C m^{-2}. The curves, however, are described in terms of their equivalent negative entropies. The balance of the external energy is based upon a constant radiant flux 26.7 MJ m^{-2} d^{-1}, of which a constant 12.6 MJ m^{-2} d^{-1} comprises visible wavelengths. A fraction of the daylight flux is used to develop the biological structure, which, at 0.0485 mol cell C, reduces the external exchange by ~0.8 MJ m^{-2} d^{-1}. The maintenance energy (over 24 h) is equivalent to

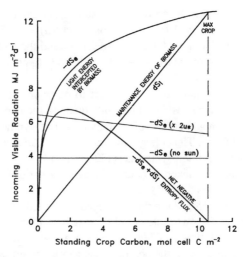

FIGURE 10.13. Biomass and energetics of a hypothetical pelagic vegetation, based on characters of *Chlorella* under conditions supposing the potential radiative fluxes to a temperate lake at the summer solstice. The energy exploited by the vegetation ($-dS_e$) is increasingly consumed in maintenance (dS_i) at increasingly larger standing biomass, up to a maximum sustainable level. The difference between the curves (the net negative entropy flux) is also inserted. The effect of increasing wind mixing ($\times 2u^*$) or occluding the sun by cloud (*no sun*) have profound impacts on the carrying capacity and the energetic steady state.

(0.0485/10.5) $\times$ 12.6 MJ m^{-2} d^{-1}, or ~0.06 MJ m^{-2} d^{-1}. At this point, $dS_e + dS_i = -0.8 + 0.06 \ll 0$.

The accretion of negative entropy flux reaches a hypothetical maximum (i.e., dS achieves its greatest negative value) when the biomass is about 2 mol cell C m^{-2}. Comparing the rate of attainment of biomass doubling, the growth rate becomes increasingly energy limited and so slows down thereafter (Fig. 10.14). As biomass builds higher and moves further away from energetic equilibrium, so the net entropy flux begins to weaken back to balance. The sequences in Figures 10.13 and 10.14 assist us to define the scales at which structure is elaborated, information is assembled, and entropy is diminished.

So far as the application of the concept of negative entropy is concerned, this simple example serves to reinforce the view that a successfully self-organizing ecosystem will be distinguished first by traits that exploit radiation flux for the rapid accumulation of structure (analogous to *r*-selection). Later in the sequence, it will be beneficial to move toward a minimization of dissipation (keeping dS_i small, for instance, by forcing *K*-selection). The perpetuation of high biomass, relative to resources, may be seen to become increasingly unstable. Attainment and survival of steady-state communities are evidently dependent on high environmental constancy. In the next section, we may consider how often or, rather, how rarely, this is encountered.

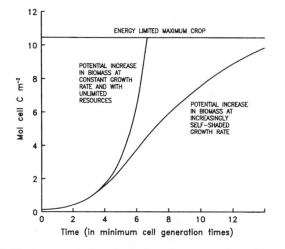

FIGURE 10.14. The increase of a hypothetical steady-state maximum biomass commences exponentially but, as a consequence of increasing energy limitation of the individual participants, the rate of increase slows toward zero at maximum.

Heterogeneity and the Maintenance of Diversity

To those of us brought up on the understanding of density dependence, competition, and the survival of the fittest, as well as the implicit pre-Darwinian concept of plenitude (which has it that the world is so full of organisms that they are in constant contention for the resources available and thus these conditions lead inexorably toward fitter individuals with improved characters), the recognition that the real world is only rarely like this is scarcely intuitive. Nevertheless, it is difficult in the extreme to nominate any of the world's great ecosystems that yet approaches its climactic, energetically balanced, and competitively excluded steady state in compliance with Odum's or Nielsen's criteria.

The most likely candidates might include the bathypelagic communities of the oceans, although these are incomplete because their basal primary products must be imported from elsewhere. I have suggested (Reynolds 1992a) that certain kinds of physically stabilized and largely isolated water columns (e.g., those under permanent ice or reinforced by salinity gradients) maintain stable, niche-differentiated planktonic assemblages of algae, bacteria, and protozoa, probably at an ecologically near-steady state (i.e., close to the extreme right-hand side of a curve similar to that shown in Figure 10.13 but asymptotic to a much lower ceiling of photon flux).

Certainly, the steady-state condition appears not to embrace the long-established and intensively biodiverse ecosystems of coral reefs and low-latitude rain forests. The fact that these evolved and complex systems support *so many* species occupying apparently similar niches (why so many canopy-forming tree species, for instance?) has perplexed ecological theorists for some time. Some of the same ecologists have offered solutions to the paradox, but the best known and most widely accepted view is properly attributed to Connell (1978). Before the succession completes its progress toward its climactic steady state, the system is disturbed (or "attacked") by forces external to the system, such as fire or storm, to the extent that existing structure is damaged and, perhaps, abruptly set back to a more primitive condition; the unidirectional self-organization is, at best, slowed down or is halted or spontaneously returned to a less developed, "more minimal" (Pickett et al. 1989) condition. Connell (1978) reasoned that the same could be true on a number of occasions: were there some degree of recovery of biological structure in the intervening periods and renewed progress toward an ecological climax, the overall effect would be to preserve a vigorous but nonsteady state whose character might be determined by the frequency and severity of these intermediate disturbances. The biodiverse reef or forest just has not had long enough to achieve its climactic state since its last major disturbance. Accordingly, most other terrestrial and aquatic systems that are overtly less organized or carry less biomass must represent accumulation-phase stages in succession that are still more primitive and which are probably not subject to species selection through severe competition.

This is not a view of the world that is necessarily supported by compelling quantitative evidence—ecologists do not live long enough. However, the concept translates well to the planktonic vegetation, where the large numbers of algal species inhabiting the supposedly monotonous, isotropic environment are analogously "paradoxical" (Hutchinson 1961). Fortunately, the generation times of the key components are in human terms quite brief, while the successional trends have been characterized (Reynolds 1988a, 1992a): the 16 or so generations required to establish steady-state dominance can be accommodated within a few weeks to months, making processes indeed amenable to meaningful study and searching experiment. The intervention of more sustained physical forcing on a sequence running from a colonist nanoplankton to depth-segregated niche dominance could impose a diatom-, desmid-, or, ultimately, an *Oscillatoria*-dominated community. Alternation between stronger and weaker forcing could see brief episodes of incomplete dominance by several species simultaneously (some in increase, others in decline). The frequency of disturbances will determine whether dominance or a steady state will ever be achieved or whether it will be prevented, and the system is perpetuated in a permanently immature, transitional condition (Reynolds 1993a). The proviso to this statement is that the forcing does reduce structure (i.e., it is a disturbance according to the definition of Pickett et al. 1989) and that it does so to the extent that the species composition of the community is moved significantly away from its competitively excluded potential.

This scheme will have many attractions to the ecologist. Its most helpful feature is an explanative mechanism for the manifestly nonsteady state for most pelagic ecosystems. Case studies assembled by Padisák et al. (1993) conform to Connell's (1978) hypothesis: the picture which emerges is that an appropriate frequency of preclimactic disturbances not only renews the opportunities to a wider selection of species but also enables them to survive in between. In fact, at any given moment, the probability is that some species will be decreasing while others are increasing: the "snapshot" sample shows a broad mix of species in apparent coexistence.

When the frequency of alternation is increased, to less than a generation in time, the conditions are perceived by the individual as being a "constant" feature and the system thus appropriately "selects" for fewer species: Reynolds (1993a) cited low-diversity, near-monocultures of *Oscillatoria* and *Microcystis* as examples. If the frequency is decreased to the scale of several generations, the succession advances further toward a competitively excluded single-species dominance. Either way, the diversity is related to the periodicity of disturbance (Sommer 1993), with a maximum falling at the scale of two to four generation times, which is only about 5–15 days in the plankton. Note, however, that diversity and disturbance frequency are not precisely interdependent variables but separate expressions of the same structural complexity.

Significant as these deductions are claimed to be and as worthy of further scrutiny and investigation they may be, they do introduce several impediments to a ready acceptance of the scheme (Juhasz-Nagy 1993): it has theoretical and practical drawbacks that need to be addressed if its attractiveness is not to be lost but transformed into an acceptable, potentially falsifiable theory.

Of fundamental importance is the fact that the same given level of physical forcing invokes no unique or even standard community response. Partly this is frequency related, in that the impact of a storm will be less dramatic in the case of a community regrouping after a similar storm only a few days previously than of one last disturbed several weeks beforehand. Indeed, the succession may have advanced so far that the opposite is true: the evolved structure acquires resilience and survives the forcing event, at least in terms of species dominance and numbers of individuals. This apparent paradox is illustrated by Jacobsen and Simonsen's (1993) demonstration that a violent storm and heavy rainfall increased the loss from an *Aphanizomenon* population in Lake Godstrup but failed to break its dominance. Other cases of late-successional complexity governing the community response to external forcing are described by Eloranta (1993) and Moustaka-Gouni (1993). Further, the simultaneous forcing event can induce different results in adjacent water bodies, owing to differences in their morphometry (Sommer 1993) or their water chemistry (Holzmann 1993). The real question is "has the forcing significantly altered the status quo?" Enhanced wind-mixing of a lake 2 m deep may not greatly change the growth conditions, but deepening a 2-m epilimnion to 5 or 6 m in a 10-m lake *should* have profound effects on the community. It will not do so if that lake is typically very acid or chronically short of nutrient: the previously limiting condition is not overcome. Simply, the forcing has *not* disturbed the status quo.

The problem may be seen as one of semantics rather than of ecology, although clearly the definition of disturbance (Pickett et al. 1989) is framed in terms of the biotic response. No storm-forced shift from the low-diversity *Microcystis*-dominated plankton means "no disturbance," not that "the disturbance failed to dislodge the *Microcystis*." In contrast, the rapid increase of diatoms into a *Sphaerocystis*-dominated plankton represents a strong disturbance, despite having been initiated only by convectional weakening of near-surface stratification of Grasmere (data of Reynolds and Lund 1988).

Such counterintuitive statements do not make for ready assimilation into ecological theory. This is a pity because the intermediate disturbance hypothesis is
too useful to reject (Reynolds et al. 1993). I would take this opportunity to venture
that forcing and responses be accorded quantities and that the currency of entropy
flux lends itself to an explanation in terms of thermodynamics. In this way, it is
fairly simple to recognize that while dissipative external energy is raised spontaneously, the critical component remains the magnitude of the sign on the external
exchanges: so long as $(dS_e + dS_i)$ remains a negative quantity, the system survives
relatively undisturbed.

The plots in Figure 10.13 help to explain this proposal. The addition of a disruptive, mechanical-energy influx should be seen to reduce the negative entropy
flux. Curiously, the level of the disruptive energy makes only a very minor direct
impact on $-dS_e$. For instance, we can derive the mechanical energy required to disperse 26.7 MJ m^{-2} uniformly through 1 m depth by rearranging the
Monin–Obukhov equation, which counterposes the kinetic energy flux $\rho_w(u*)^3$,
and the buoyancy flux to a layer of known thickness, h_b (= 1 m in this instance).

$$\frac{g\gamma h_b \cdot Q*}{2\sigma} = \rho_w(u*)^3 \tag{10.9}$$

where g is gravitational acceleration and ρ_w is the density of water; γ is the coefficient of thermal expansion at the appropriate temperature and σ the specific heat
of the water (4186 J kg^{-1}K^{-1}). $Q*$ is the heat flux and $u*$ is the friction velocity of
the turbulence down to h_b = 1 m. Rearranging and solving for 20°, γ = 2.1 ×
10^{-4} K^{-1}; ρ_w = 998.2 kg m^{-3}, and $Q*$ = 26.7 MJ m^{-2}d^{-1}:

$$u* = \left(\frac{g\gamma h_b \cdot Q*}{2\sigma\rho_w} \right)^{\frac{1}{3}}$$

$$= \left(\frac{9.81 \text{ ms}^{-2} \cdot 2.1 \times 10^{-4} \text{ K}^{-1} \cdot 1 \text{ m} \cdot 26.7 \text{ MJ m}^{-2} (86400)^{-1}}{2 \times 416 \text{ J kg K}^{-1} \times 998.2 \text{ kg m}^{-3}} \right)^{\frac{1}{3}}$$

$\approx$ 0.0042 m s^{-1} (which would be generated by a wind of 3.4 m s^{-2}). The kinetic
energy flux is equivalent to $\rho(0.0042)^3 \approx 7.6 \times 10^{-5}$ W m$^{-2} \approx$ 65 J m^{-2} d^{-1}.

If windspeed is increased by a factor of 2 (to 6.8 m s^{-1}), $u*$ increases linearly to
0.0085 m s^{-1}. The kinetic flux, $\rho_w(u*^3)$, increases by the cube, to 6.1 × 10^{-4} W m^{-2}
(53 J m^{-2} d^{-1}), as does the new equilibrium depth of the mixed layer:

$$h_b = \frac{2\sigma\rho_w(u*)^3}{g\gamma Q*} = 8 \text{ m}$$

Even if windspeed were increased 10 fold (to 34 m s^{-1}), the kinetic energy flux
(6.5 KJ m^{-2} d^{-1}) still has a very small effect in terms of the negative entropy.

In contrast, it is the extension of the mixed layer, even with the same phytoplankton diluted through the layer, which has the profound, indirect effect on the

average insolation experienced by individual entrained cells (Reynolds 1989a). From the starting point of the Beer–Lambert equation, which relates I_z, the light remaining at a given depth, to a given incident radiation, I_o, we can derive a mean insolation ($I*$) for cells mixed through the column h, from 0 to z beneath the surface:

$$I* = (I_o \cdot I_z)^{1/2} \tag{10.10}$$

$$\text{where } I_z = I_o \cdot e^{-\varepsilon h} \tag{10.11}$$

and where e is the base of natural logarithms and ε, the vertical extinction coefficient (units, m^{-1}), comprises components due to the water ($\varepsilon_w = 0.2$ m^{-1}, is an arbitrarily chosen value for an uncolored water with no suspended particles, apart from algae) and the light-absorbing phytoplankton, on the basis $\varepsilon_a = n \cdot k_a$ where k_a is the area of light field subtended by a unit of chlorophyll a (as a first approximation, $k_a = 0.01$ m^2 mg^{-1}); and n is the algal population in terms of its chlorophyll concentration (n mg m^{-3}). As the bottom of the 1-m layer ($h = 1$), we can deduce

$$I_z = I_o \cdot \exp \{-(nk_a + 0.2)1\}$$

and, at the bottom of the 8-m mixed layer, with the *same* population now dispersed throughout its depth, that

$$I_z = I_o \cdot \exp \{-(0.125 \, nk_a + 0.2)8\}$$

For a variety of values of n to 100 mg chl m^{-2}, mixing n down from $h = 1$ to $h = 8$ m decreases $I*$ by a factor close to 0.5 at low concentration and to 0.47 at 100 mg chl m^{-2} (just under 4 mol C m^{-2}) or to a potential reduction in the negative entropy flux of more than 6 MJ m^{-2} d^{-1}. If it is also considered that when the sky is overcast the visible radiation may diminish to as little as about 0.3 of the clear-sky value for the same day of the year, then the opportunities for spontaneous reductions in the energetic carrying capacity of the system become strongly apparent (see Fig. 10.13).

It may now be seen just how day-to-day weather fluctuations can affect the progress and eventual outcome of plankton successions. At initiation, the biological organization is, in any case, weak. As the succession proceeds, the organization first acquires more resistance to external change (Fig. 10.13) but later becomes increasingly at risk from outside forcing. A hypothetical sequence through a series of consecutive days or groups of days is shown in Figure 10.15; the carrying capacity, or potential steady-state "target," is updated by day-to-day change in the permitting weather conditions. These ultimately determine whether the community might maintain progress toward an unchanged steady state, or whether the target condition is itself perhaps revised, or whether the community is abruptly destabilized by physical conditions with which it is unable to contend. The attainment of a true climactic steady-state vegetation would seem a remote prospect from Figure 10.15 without the imposition of physically very constant conditions. In the majority of physically variable systems (with the frequency of variations coming at the scale of several generation times), the failure to achieve an evolved steady state is recognizably persistent.

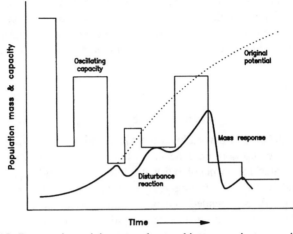

FIGURE 10.15. Progress in attaining a steady-state biomass against an oscillating energy-based capacity. The original potential is resisted and delayed, perhaps indefinitely, simply by the periodic inability of the energy flux to support the maintenance of the existing structure.

Indeed, such external regulation of community organization is suggested to apply to the majority of pelagic communities and to explain why so few of them display for long the characteristics of a successional steady state (Reynolds 1992a, 1993a). In Figure 10.16, the directionality of internal processes is shown by the sigmoid curve of maintained biomass drawn from Figure 10.14, only now labeled with the terms of self-assembly against a scale of elaboration (initiation, accumulation, maturation; terms of Price 1984). The impacts of external forcing events of sufficient intensity to decrease biomass and organization are represented by the series of abrupt returns. Also represented is a sequence of low-amplitude disturbances imposed at increased frequencies: the system can acquire and support an increase in the mass of specialist (R) organisms but with little overall structural organization of the community.

The extremes of these representations are finally encapsulated in a triangle, which has the apices of Grime's (1979) three primary strategies among plants, C, S, and R, although as defined by Reynolds (1988b) these are called colonists, stress-tolerators, and ruderals and are characterized (Reynolds 1993b) according to their invasivenes, acquisitiveness, and attunement capacity. Superimposition of the triangle attempts to unify the concepts of ecosystem development variously proffered in terms of successional markers (Odum 1969), of structural complexity (Margalef 1961, 1968), of thermodynamics (Prigogine 1955; Nielsen 1992), and of the strategic adaptations of species (Grime 1979) within a single matrix. The dominance of communities by species selected through time in processes represented by the insertion of the two courses of biomass accretion (C → S, C → R), as well as of the routes of return toward the origin and the reestablishment of colonist species (→ C), is adequately accommodated. It also emphasizes the

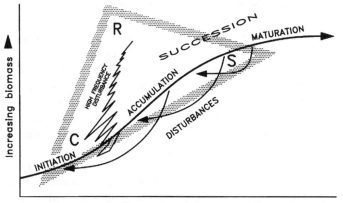

FIGURE 10.16. The states of community succession and organization, between initiation and maturation, in relation either to a constant energy-determined potential or to a fluctuating potential, and the organismic strategies best suited to the selective pressures operating.

effects of differing frequencies of external disturbances in preventing the self-organizing attributes of communities from attaining any ecologically steady-state outcome. Most of all, it demonstrates how heterogeneous fluctuations, at the appropriate scale of alternation, assist in the maintenance of species diversity.

Other Ecosystems and the Pelagic "Model": The Importance of Scaling

The final segment of this essay is directed toward the applicability of the foregoing analysis of community processes in pelagic environments to the organization and function of other ecosystems. All ultimately share the same fundamental structural components featuring photoautotropic production driven by solar radiation and cropped by herbivores and successive levels of carnivores and its reprocessing by heterotrophic microorganisms. The principle of selection of the organisms available, those best adapted to the environmental conditions with or without niche differentiation, is also recognizably a common behavior, while the number of species making up the bulk of the community biomass (generally five to eight are needed to make 99%) also seems to be frequently encountered (see individual contributions in Gee and Giller 1987). Many are biologically productive, with the production exceeding respiration and with a fraction therefore being exported, if not always accreting as new biomass, in ways which the present consideration would associate with subclimactic communities in nonsteady state and not subject to "equilibrium dynamics " (Connell 1978; Miles 1985). The successional replacement of r-selected by K-selected dominants is also clearly common to the reported development of most well-studied ecosystems. There should be no surprise, then, about the striking similarities between the independently constructed figures of (say) Reynolds et al. (1983) and those of Miles (1988); quite as fascinating is the ready assimilability of Grime's (1979) concept of primary strategies (C–S–R) to the separation by Reynolds et al. (1983) of phytoplankton population behaviors, with analogously distinctive ecologies and adaptive morphologies (see Reynolds 1988b). "Ecology" does not alter simply because its factual basis is assembled from observations on different systems.

It has to be admitted, however, that this level of interchangeability of unifying concepts (van Dobben and Lowe-McConnell 1975) between plankton biologists and more "traditional" (terrestrial) ecologists has remained, until recently at least, underdeveloped. There may be several contributory factors to this hiatus but the main one is surely that of scale. So long as the supposition prevails that "the spring bloom" is analogous to the annual appearance of flowers or fruits on a shrub, rather than, as countered by Reynolds (1993a), to the equivalent number of generations of *Betula* or *Pinus* that was required before the first forest canopies were established over the lowlands exposed at the end of the most recent glaciation, this difficulty will persist. Moreover, if the later planktonic successional series occurring in summer and autumn months are similarly equated to the postglacial periods of *Ulmus-Alnus* and *Quercus-Fagus* dominance of the terrestrial vegetation, then the 50–70 generations of succeeding dominant species accommodated in the subsequent interstadial occupy just about a calendar year in the plankton. In much the same way, both will have experienced ameliorative and recessive climatic fluctuations, with concomitant changes in the species best adapted and thus in the potential steady-state dominant. The proximal events during the calendar year (such as altered daylength, heat fluxes, and temperature; stratification and mixing; and hydraulic and nutrient exchanges) assume comparable significance in the planktonic vegetation processes.

This cross-scaling comparison is represented in Figure 10.17: with very little adjustment to the labels, Figure 1.1 of Miles (1987) and Figure 14 of Reynolds (1987) can be mutually superimposed but for the difference in the finite scale of time. Interpolating the classes of plant responses on to the added scale, the logarithmic cycles are not equal: at the vegetation scale, the pelagic operates at some 10^5 times faster; at the scale of populations the ratio is about 10^3; at the scale of phenology it is nearer 10^2. The scales coincide, quite properly within the realms of cell physiology and biochemical reaction, somewhere to the left of Miles' (1987) figure, at about 10^{-5} years (or ~300 s). This not only lends authenticity to the comparison but also makes more exciting those at the large-scale end. To me, the great attraction of working with phytoplankton is that the very questions that tax ecologists' minds—how diversity is maintained and whether it increases or decreases during succession and the extent to which competition and predation determine the structure of communities—are not only regularly answered in the vegetation processes of the pelagic but have been extensively tested in ecological experiments (Lund 1975; Reynolds 1986). Such possibilities of scaling equivalence deserve to be explored and developed by ecologists more deeply and extensively than hitherto.

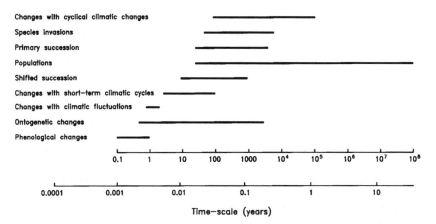

FIGURE 10.17. Various changes in terrestrial vegetation and the time scales they occupy, according to Miles (1987) (0.1–10⁸ years). With little change in terms, the corresponding processes in planktonic communities can be fitted to the lower time scale used by Reynolds (1987). Note, however, the time divisions are not equivalent and the scales coincide somewhere to the left, in the realm of cell physiology, in fact. (Compilation from the cited sources.)

Summary

The model of ecological succession in the phytoplankton and the ability of developing communties to withstand externally imposed disruptive forcing is argued to be applicable to other, nonplanktonic successions. The recognition of a logical, steady-state outcome involves the eventual formation of spatially segregated niches and deep-located chlorophyll maxima, although most communities are arrested in a subclimactic or plagioclimactic condition. The anabolism of community structure can be measured in terms of its negative entropy, referring to the capacity of the energy resources and the acquisitiveness and net productivity of the biomass. External events increase the entropy flux, requiring the biotic system to restructure at a lower level. The principle of intermediate disturbance is thus argued to regulate community development, species dominance, and internal recycling, with abrupt reversals of direction, the maintenance of a state of incomplete accumulation, and the prolongation of high species diversity.

Acknowledgments. I am most grateful to Kirsty Ross and Trevor Furnass for technical assistance in the preparation of this paper.

Literature Cited

Atkinson, K.M. 1980. Experiments in dispersal of phytoplankton by ducks. British Phycological Journal 15:49–58.

Berger, C. 1975. Occurrence of *Oscillatoria agardhii* Gomont in some shallow eutrophic lakes. Verhandlungen der Internationale Vereinigung für Theoretische und Angewandte Limnologie 19:2687–2697.

Boltzmann, L. 1896. Vorlesungen über Gastheorie. Akademische Druck und Verlagsanstalt, Graz.

Carpenter, S.R., J.F. Kitchell, and J.R. Hodgson. 1985. Cascading trophic interactions and lake productivity. Bioscience 35:634–639.

Connell, J.H. 1978. Diversity in tropical rainforests and coral reefs. Science 199:1304–1310.

De Melo, R., R. France, and D.J. McQueen. 1992. Biomanipulation: hit or myth? Limnology and Oceanography 37:192–207.

Descy, J.-P. 1993. Ecology of the phytoplankton of the River Moselle: effects of disturbances on community structure and diversity. Hydrobiologia 249:111–116.

Eloranta, P. 1993. Diversity and succession of the phytoplankton in a small lake over a two-year period. Hydrobiologia 249:25–32..

Emerson, R. and W. Arnold. 1933. The photochemical reaction in photosynthesis. Journal of General Physiology 16:191–205.

Faafeng, B.A. and J.P. Nilssen. 1981. A twenty-year study of eutrophication in a deep soft-water lake. Verhandlungen der Internationale Vereinigung für Theoretische und Angewandte Limnologie 21:412–424.

Gee, J.H.R. and P.S. Giller. 1987. Organization of Communities, Past and Present. Blackwell Scientific, Oxford.

Grime, J.P. 1979. Plant Strategies and Vegetation Processes. Wiley-Interscience, Chichester.

Hardin, G. 1960. The competitive exclusion principle. Science 131:1292–1297.

Harris, G.P. 1978. Photosynthesis, productivity and growth: the physiological ecology of phytoplankton. Ergebnisse der Limnologie 10:1–171.

Havens, K.E. 1992. Acidification effects on the plankton size spectrum: an in situ mesocosm experiment. Journal of Plankton Research 14:1687–1696.

Holzmann, R. 1993. Seasonal fluctuations in the diversity and compositional stability of phytoplankton communities in small lakes in upper Bavaria. Hydrobiologia 249:101–109.

Hutchinson, G.E. 1961. The paradox of the plankton. American Naturalist 95:137–147.

Jacobsen, B.A. and P. Simonsen. 1993. Disturbance events affecting phytoplankton biomass, composition and species diversity in a shallow, eutrophic, temperate lake. Hydrobiologia 249:9–14.

Jassby, A.D. and C.R. Goldman. 1974. A quantitative measure of succession rate and its application to the phytoplankton of lakes. American Naturalist 108:688–693.

Jørgensen, S.E. 1992. Structural dynamic eutrophication models. In: D.W. Sutcliffe and J.G. Jones, eds. Eutrophication: Research and Application to Water Supply, pp. 59–72. Freshwater Biological Association, Ambleside.

Juhász-Nagy, P. 1993. Notes on compositional diversity. Hydrobiologia 249:173–182.

Kilham, P. and S.S. Kilham. 1980. The evolutionary ecology of phytoplankton. In: I. Morris, ed. The Physiological Ecology of Phytoplankton, pp. 571–597. Blackwell Scientific, Oxford.

Lewis, W.M. 1978. Analysis of succession in a tropical phytoplankton community and a new measure of succession rate. American Naturalist 112:401–414.

Lovelock, J.E. 1979. Gaia: A New Look at Life on Earth. Oxford University Press, Oxford.

Lund, J.W.G. 1950. Studies on *Asterionella formosa* Hass. II. Nutrient depletion and the spring maximum. Journal of Ecology 38:1–35.

Lund, J.W.G. 1975. The use of large experimental tubes in lakes. In: R.E. Youngman, ed. The Effect of Storage on Water Quality, pp. 291–312. Water Research Centre, Medmenham.

Lund, J.W.G. and C.S. Reynolds. 1982. The development and operation of large limnetic enclosures in Blelham Tarn, English Lake District, and their contribution to phytoplankton ecology. In: F.E. Round and D.J. Chapman, eds. Progress in Phycological Research, Vol. 1, pp. 1–65. Elsevier, Amsterdam.

MacArthur, R.H. and E.O. Wilson. 1967. The Theory of Island Biogeography. Princeton University Press, Princeton.

Margalef, R. 1961. Communication of structure in planktonic populations. Limnology and Oceanography 6:124–128.

Margalef, R. 1968. Perspectives in Ecological Theory. University of Chicago Press, Chicago.

McQueen, D.J. 1990. Manipulating lake community structure: where do we go from here? Freshwater Biology 23:613–620.

Mejer, H. and S.E. Jørgensen. 1979. Energy and ecological buffer capacity. In: S.E. Jørgensen, ed. State-of-the-Art of Ecological Modelling, Environmental Sciences and Applications. Proceedings of the 7th Conference, pp. 829–846. International Society for Ecological Modelling, København.

Miles, J. 1985. The pedogenic effects of different species and vegetation types and the implications of succession. Journal of Soil Science 36:571–584.

Miles, J. 1987. Vegetation succession: past and present perceptions. In: A.J. Gray, M.J. Crawley, and P.J. Edwards, eds. Colonization, Succession and Stability, pp. 1–29. Blackwell Scientific, Oxford.

Miles, J. 1988. Vegetation and soil change in the uplands. In: M.B. Usher and D.B.A. Thompson, eds. Ecological Change in the Uplands, pp. 57–70. Blackwell Scientific, Oxford.

Moss, B. 1992. The scope of biomanipulation for improving water quality. In: D.W. Sutcliffe and J.G. Jones, eds. Eutrophication: Research and Application to Water Supply, pp. 73–81. Freshwater Biological Association, Ambleside.

Moustaka-Gouni, M. 1993. Phytoplankton succession and diversity in a warm monomictic relatively shallow lake: Lake Volvi, Macedonia, Greece. Hydrobiologia 249:33–42.

Nielsen, S.N. 1992. Application of Maximum Energy in Structural Dynamic Models. Ministry of the Environment, København.

Odum, E.P. 1969. The strategy of ecosystem development. Science 164:262–270.

Padisák, J. 1992. Seasonal succession of phytoplankton in a large shallow lake (Balaton, Hungary)—a dynamic approach to ecological memory, its possible role and mechanisms. Journal of Ecology 80:217–230.

Padisák, J., C.S. Reynolds, and U. Sommer. 1993. Intermediate Disturbance Hypothesis in Phytoplankton Ecology. Developments in Hydrobiology Series, DH 81. (Reprinted from Hydrobiologia 249:1–199.) Kluwer Academic, Dordrecht.

Pahl-Wostl, C. 1990. Temporal organisation: a new perspective on the ecological network. Oikos 58:293–305.

Petersen, R. 1975. The paradox of the plankton: an equilibrium hypothesis. American Naturalist 109:35–49.

Pickett, S.T.A., I. Kolasa, I.I. Armesto, and S.L. Collins. 1989. The ecological concept of disturbance and its expression at various hierarchical levels. Oikos 54:129–136.

Price, P.W. 1984. Alternative paradigms in community ecology. In: P.W. Price, C.N. Slobodchikoff, and W.S. Gaud, eds. A New Ecology: Novel Approaches to Interactive Systems, pp. 353–383. Wiley-Interscience, New York.

Prigogine, I. 1955. Thermodynamics of Irreversible Processes. Interscience, New York.

Reynolds, C.S. 1972. Growth, gas-vacuolation and buoyancy in a natural population of a blue-green alga. Freshwater Biology 2:87–106.

Reynolds, C.S. 1973a. The seasonal periodicity of planktonic diatoms in a shallow eutrophic lake. Freshwater Biology 3:89–110.

Reynolds, C.S. 1973b. Growth and buoyancy of *Microcystis aeruginosa* Kütz. emend. Elenkin in a shallow eutrophic lake. Proceedings of the Royal Society London, Series B 184:29–50.

Reynolds, C.S. 1975. Interrelations of photosynthetic behaviour and buoyancy regulation in a natural population of a blue-green alga. Freshwater Biology 5:323–338.

Reynolds, C.S. 1978. The plankton of the North-West Midland meres. Occasional Papers of the Caradoc and Severn Valley Field Club, No. 2. Shrewsbury.

Reynolds, C.S. 1980. Phytoplankton assemblages and their periodicity in stratifying lake systems. Holarctic Ecology 3:141–159.

Reynolds, C.S. 1983. Growth-rate responses of *Volvox aureus* Ehrenb. (Chlorophyta, Volvocales) to variability in the physical environment. British Phycological Journal 18:433–442.

Reynolds, C.S. 1984a. Phytoplankton periodicity: interactions among form, function and environmental variability. Freshwater Biology 14:111–142.

Reynolds, C.S. 1984b. The Ecology of Freshwater Phytoplankton. Cambridge University Press, Cambridge.

Reynolds, C.S. 1986. Experimental manipulations of the phytoplankton periodicity in large limnetic enclosures in Blelham Tarn, English Lake District. Hydrobiologia 138:43–64.

Reynolds, C.S. 1987. Community organization in the freshwater plankton. In: J.H.R. Gee and P.S. Giller, eds. Organization of Communities, Past and Present, pp. 297–325. Blackwell Scientific, Oxford.

Reynolds, C.S. 1988a. The concept of ecological succession applied to seasonal periodicity of freshwater phytoplankton. Verhandlungen der Internationale Vereinigung für Theoretische und Angewandte Limnologie 23:683–691.

Reynolds, C.S. 1988b. Functional morphology and the adaptive strategies of freshwater phytoplankton. In: C.D. Sandgren, ed. Growth and Reproductive Strategies of Freshwater Phytoplankton, pp. 388–437. Cambridge University Press, New York.

Reynolds, C.S. 1989a. Physical determinants of phytoplankton succession. In: U. Sommer, ed. Plankton Ecology, pp. 9–56. Springer-Verlag, New York.

Reynolds, C.S. 1989b. Relationships among the biological properties, distribution and regulation of production by planktonic Cyanobacteria. Toxicity Assessment 4:229–255.

Reynolds, C.S. 1990. Temporal scales of variability in pelagic environments and the responses of phytoplankton. Freshwater Biology 23:25–53.

Reynolds, C.S. 1991. Lake communities: an approach to their management for conservation. In: I.F. Spellerberg, F.B. Goldsmith, and M.G. Morris, eds. The Scientific Management of Temperate Communities for Conservation, pp. 199–225. Blackwell Scientific, Oxford.

Reynolds, C.S. 1992a. Dynamics, selection and composition of phytoplankton in relation to vertical studies in lakes. Ergebnisse der Limnologie 35:13–31.

Reynolds, C.S. 1992b. Eutrophication and the management of planktonic algae: what Vollenweider couldn't tell us. In: D.W. Sutcliffe and J.G. Jones, eds. Eutrophication: Research and Application to Water Supply, pp. 4–29. Freshwater Biological Association, Ambleside.

Reynolds, C.S. 1992c. Algae. In: P. Calow and G.E. Petts, eds. The Rivers Handbook, Vol. I, pp. 195–215. Blackwell Scientific, Oxford.

Reynolds, C.S. 1993a. Scales of disturbance and their role in plankton ecology. Hydrobiologia 249:157–171.

Reynolds, C.S. 1993b. Swings and roundabouts: engineering the environment of algal growth. In: K.N. White, E.G. Bellinger, A.J. Saul, M. Symes, and K. Hendry, eds. Urban Waterside Regeneration: Problems and Prospects, pp. 330–349. Ellis Horwood, Chichester.

Reynolds, C.S. and J.W.G. Lund. 1988. The phytoplankton of an enriched, soft-water lake subject to intermittent hydraulic flushing (Grasmere, English Lake District). Freshwater Biology 19:379–404.

Reynolds, C.S. and S.W. Wiseman. 1982. Sinking losses of phytoplankton maintained in closed limnetic systems. Journal of Plankton Research 4:489–522.

Reynolds, C.S., G.P. Harris, and D.N. Gouldney. 1985. Comparison of carbon-specific growth rates and rates of cellular increase of phytoplankton in large limnetic enclosures. Journal of Plankton Research 7:791–820.

Reynolds, C.S., J. Padisák, and U. Sommer. 1993. Intermediate disturbance in the ecology of phytoplankton and the maintenance of species diversity: a synthesis. Hydrobiologia 249:183–188.

Reynolds, C.S., G.H.M. Jaworski, H.M. Cmiech, and G.F. Leedale. 1981. On the annual cycle of the blue-green alga *Microcystis aeruginosa* Kütz. emend. Elenkin. Philosophical Transactions of the Royal Society of London, Series B 293:419–477.

Reynolds, C.S., J.F. Thompson, A.J.D. Ferguson, and S.W. Wiseman. 1982. Loss processes in the population dynamics of phytoplankton maintained in closed systems. Journal of Plankton Research 4:561–600.

Reynolds, C.S., S.W. Wiseman, B.M. Godfrey, and C. Butterwick. 1983. Some effects of artificial mixing on the dynamics of phytoplankton populations in large limnetic enclosures. Journal of Plankton Research 5:203–234.

Richerson, P., R. Armstrong, and C.R. Goldman. 1970. Contemporaneous disequilibrium, a new hypothesis to explain the paradox of the plankton. Proceedings of the National Academy of Sciences of the United States of America 67:1710–1714.

Saxby, K.J. 1990. The Physiological Ecology of Freshwater Chrysophytes, with Special Reference to *Synura petersenii*. Ph.D. thesis, University of Birmingham.

Shapiro, J. 1990. Current beliefs regarding dominance by blue-greens: the case for the importance of CO_2 and pH. Verhandlungen der Internationale Vereinigung für Theoretische und Angewandte Limnologie 24:38–54.

Shapiro, J. 1993. Theory and practice in the control of algal blooms. In: K.N. White, E.G. Bellinger, A.J. Saul, M. Symes, and K. Hendrey, eds. Urban Waterside Regeneration—Problems and Prospects, pp. 350–357. Ellis Horwood, Chichester.

Sommer, U. 1993. Disturbance–diversity relationships in two lakes of similar nutrient chemistry but contrasting disturbance regimes. Hydrobiologia 249:59–65.

Talling, J.F. and S.I. Heaney. 1988. Long-term changes in some English (Cumbrian) lakes subjected to increased nutrient inputs. In: F.E. Round, ed. Algae and the Aquatic Environment, pp. 1–29. Biopress, Bristol.

Tilman, D. 1982. Resource Competition and Community Structure. Princeton University Press, Princeton.

Tilman, D., S.S. Kilham, and P. Kilham. 1982. Phytoplankton community ecology: the role of limiting nutrients. Annual Reviews in Ecology and Systematics 13:349–372.

Utterback, C.L., L.D. Phifer, and R.J. Robinson. 1942. Some chemical, planktonic and optical characteristics of Crater Lake. Ecology 23:97–103.

van Dobben, W.H. and R.H. Lowe-McConnell. 1975. Unifying Concepts in Ecology. Junk, Den Haag.

Vollenweider, R.A. 1976. Advances in defining critical load levels for phosphorus in lake eutrophication. Memorie dell'Istituto Italiano di Idrobiologia 33:53–83.

Walsby, A.E., C.S. Reynolds, R.L. Oliver, and J. Kromkamp. 1989. The role of gas vacuoles and carbohydrate content in the buoyancy and vertical distribution of *Anabaena minutissima* in Lake Rotongaio, New Zealand. Ergebnisse der Limnologie 32:1–25.

11

Food Web Structure and Biodiversity in Lake Ecosystems

KEIICHI KAWABATA AND MASAMI NAKANISHI

Introduction

Aquatic ecosystems are different from terrestrial ones in many aspects, and their existence increases the biodiversity of the earth. Littoral ecosystems are structured around solid substrata, particularly for aquatic plants and algae, and consequently resemble terrestrial ones. In contrast, pelagic ecosystems are unique in being suspended in the water. In addition, aquatic ecosystems involve complex interactions (Carpenter 1988) that may promote biodiversity within the systems.

With a single water sampler, we can collect all components of pelagic ecosystems except fishes, at least qualitatively. It is thus possible to estimate the biodiversity of whole plankton communities, not just that within one taxonomic group. This merit of pelagic ecosystems, however, involves the methodological difficulties of working with such diverse organisms at one time. In this study, we mainly used the size of organisms for describing biodiversity in lake ecosystems and thereby analyzed the food web structure.

Description of Plankton Diversity

Size, taxon, functional group, and trophic level have been used in several combinations to describe plankton diversity (Porter et al. 1988). Although none of these criteria works by itself, size has been the main criterion for classifying plankton communities; taxon and ecological traits then constitute subordinate parameters. To collect and separate plankters, we use filters of certain pore sizes. Size is, however, relevant not only to methodology but also to biology (Sieburth et al. 1978). In a general way, size correlates with taxon and ecological features in aquatic ecosystems.

In this study, organisms were first divided into size classes, and then each class or each group of classes was labeled with an appropriate combination of taxonomic and ecological characteristics. Taxonomic classification is difficult for microbes and insufficient for metazoa. Molecular biological techniques recently have enabled us to study bacterial diversity in situ (Pedrós-Alió 1993). On the other hand, crustaceans often show ontogenetic niche shifts (Werner and Gilliam 1984), thereby necessitating categorization by developmental stage as well as by species. Thus, taxa at different levels should be used appropriately for different size classes. As for ecological classification, only autotrophs and nonautotrophs, here called heterotrophs, were distinguished. Other ecological characteristics were described for each constituent.

Size Classes of Plankton

In planktology, the following size fractions have been recognized: 0.2–2.0 µm, picoplankton; 2.0–20 µm, nanoplankton; 20–200 µm, microplankton; and 0.2–20 mm, mesoplankton (Sieburth et al. 1978). Fortunately, the prefixes (pico, nano, and micro) correspond to the approximate live weights (pg, ng, and µg) of the organisms at the upper end of their range.

In lake ecosystems, the picoplankton is composed mainly of prokaryotes, both autotrophs and heterotrophs; nanoplankton, of autotrophic and heterotrophic flagellates and small algae; microplankton, of large algae, ciliates, and rotifers; and mesoplankton, of crustaceans (Table 11.1). The complex interactions within a microbial food web include predation, competition, and commensalism (see Fig. 5, in Azam et al. 1983).

Autotrophs fix CO_2 by photosynthesis and release organic matter at the same time. Although the physiological mechanism behind algal exudation is still uncertain, active release and passive permeation are two major explanations (Bjørnsen 1988). Bacteria recover the dissolved organic carbon (DOC) exuded from healthy cells in addition to that released from senescent and dead cells. Thus, bacteria utilize the excretions of autotrophs on the one hand and compete with autotrophs for nutrients such as P and N on the other. Another source of organic substrate for bacteria is zooplankton activity, that is, release from fecal material, excretory release, and fragmentation of prey during handling (Peduzzi and Herndl 1992).

TABLE 11.1. Simplified size classes of a freshwater plankton community

Size class	Autotrophs	Heterotrophs
Picoplankton:		
0.2–2.0 µm	Cyanobacteria	Bacteria
Nanoplankton:		
2.0–20 µm	Flagellates Unicellular diatoms	Flagellates
Microplankton:		
20–200 µm	Green algae Colonial diatoms Dinoflagellates	Ciliates Rotifers
Mesoplankton:		
0.2–20 mm		Crustaceans

Bacteria and picoalgae are consumed by heterotrophic flagellates, ciliates, and rotifers. In addition, ciliates and rotifers utilize flagellates and nanoalgae. Crustaceans prey on nanoalgae, protozoa, and rotifers. Mesozooplankton thus consume the production of autotrophic microbes directly and indirectly via heterotrophic microbes (Sherr and Sherr 1991). Microbes repack small organisms into larger parcels edible for crustacea, thereby raising the overall trophic efficiency of planktonic food webs (Porter et al. 1985).

Biomass Size Spectrum

One promising tool providing insight into various ecological processes in complex pelagic food webs is the analysis of biomass size spectra (Gaedke 1993). The biomass size spectrum comprises a scale of body mass on the horizontal axis and one of biomass concentration on the vertical axis (Fig. 11.1). Several theoretical analyses of biomass size spectrum have been developed (e.g., Thiebaux and Dickie 1993). However, Gaedke (1992a) criticized the current theoretical concepts for ignoring the important roles played by bacteria as well as for considering only a biomass flux exclusively from small to large organisms. In fact, carbon is fixed not only in the smallest size classes but also in the medium-size classes, including algae. Carbon is then processed within the microbial food web, and a part is transferred to larger consumers.

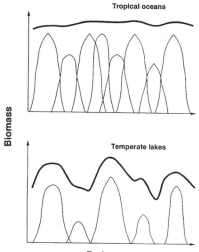

FIGURE 11.1. Diagrammatic biomass size spectra showing component distributions *(fine lines)* and resultant general shape *(heavy lines)* for tropical oceans and temperate lakes (after Sprules 1988, used by permission of E. Schweizerbart'sche Verlagsbuchhandlung). Common log of biomass concentration is plotted against binary log of body mass.

Plankton size spectra have been reported to be smooth for tropical oceans and irregular for temperate lakes (Sprules 1988). A size spectrum usually consists of a series of overlapping size distributions of the component organisms (see Fig. 11.1). The pattern of a size spectrum is accordingly determined by the species richness and distribution overlaps of the components (Sprules 1988). These parameters are larger in tropical oceans than in temperate lakes, thereby resulting in a different shape of the size spectrum.

Gaedke (1992a) also showed that plankton biomass accumulates in certain size classes. The mode in the size spectrum may be formed by inedible algae or by daphniids, which feed on a broad range of prey organisms. The biomass accumulation thus expresses the specific adaptations by constituent organisms to the constraints imposed by nature (Gaedke 1992a).

Lake Biwa

The Lake Biwa Plankton Research Group studied the plankton community of Lake Biwa, a large (675 km^2) and deep (maximum, 104 m) lake located near 35°N 136°E in June, July, August, September, and November of 1992 (Urabe et al. 1995, unpublished data). The densities and sizes of organisms were determined by microscopy, and their body masses were then estimated using conversion factors; finally, their biomass concentrations were calculated.

Figure 11.2 shows the biomass size spectrum averaged for the growing season of 1992. The highest biomass concentrations exceeding 2 g C/m^2 were found between 1 and 16 ng C body mass. These size classes consisted of large phytoplankters, such as the desmids *Closterium aciculare* and *Staurastrum dorsidentiferum,* the colonial diatom *Fragilaria crotonensis,* and the dinoflagellate *Ceratium hirundinella* (Fig. 11.3). The next peaks, of 1–2 g C/m^2, were observed in 0.01–0.03 pg C and 0.1–0.3 ng C classes, corresponding to bacteria and unicellular diatoms, respectively. In addition, nanoflagellates of 4–16 pg C body mass formed moderate modes. On the contrary, the biomass concentrations of crustacea of 0.5–4 µg C body mass were much lower than those of other modes, not exceeding 0.3 g C/m^2. Thus, irregularity and a low crustacean mode were the characteristics of the biomass size spectrum in Lake Biwa. The results suggest that carbon is processed within the microbial food web of Lake Biwa, but that a bottleneck in carbon flow exists between algae and crustacea (see following).

In Lake Biwa, crustacea were scarce not only in biomass but also in species number. Dodson (1991) has shown a significant species–area relationship for pelagic crustacea in 32 European lakes, including Lake Constance. His regression equation predicts 14 species from a lake as large as Lake Biwa with a lower 95% confidence limit of 6 species. In fact, only 7 species have been found by extensive plankton survey since 1983 (Table 11.2). Dodson (1991) also listed 7 lakes larger than 100 km^2 and 4 lakes deeper than 100 m; the minimal species number was 10 in these large lakes and 13 in these deep lakes. Thus, the crustacean fauna of Lake Biwa was particularly poor.

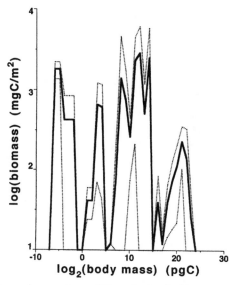

FIGURE 11.2. Biomass size spectrum of the pelagic plankton community in Lake Biwa (Lake Biwa Plankton Research Group, unpublished). *Solid line* shows the mean from June to November of 1992; *dotted lines* are SDs.

Of four species of large algae that became abundant in Lake Biwa, *Ceratium hirundinella* and *S. dorsidentiferum* are inedible to crustacea (see Fig. 11.3). Besides, *F. crotonensis* is vulnerable only to the late developmental stages of the copepod *Eodiaptomus japonicus,* and *Closterium aciculare* only to large individuals of the branchiopod *Daphnia galeata* (Kawabata 1987a). Thus, only a small proportion of the herbivorous crustaceans were predators of large algae, consuming less than 1% of the biomass of large algae in a day (Kawabata 1987a). Although omnivorous crustaceans, the branchiopod *Leptodora kindtii* and the late copepodid stages of cyclopoid copepods in Lake Biwa, are also able to ingest large algae, their biomass occupied only 12% ± 9% of all crustacean biomass.

The plankton community of Lake Biwa was thus characterized by the dominance of large algae and a poor crustacean fauna. The fact that large algae are inedible to most crustacea suggests that poor food conditions depressed the fauna. There is evidence that planktonic crustacea in Lake Biwa are under poor food conditions (Okamaoto 1984; Kawabata 1989). This may be the direct consequence of the dominance of inedible large algae. Also plausible is the following indirect effect; with sufficient food, zooplankters are expected to perform diel vertical migration to avoid visual predators (Gliwicz and Pijanowska 1988). In fact, planktonic crustacea in Lake Biwa do not migrate vertically, probably because of food deficiency (Kawabata 1987b). Dominance of the inedible large algae thus raises the possibility that predation impacts upon zooplankters indirectly.

While omnivores are adapted to handling large particles, stoichiometry predicts that they should prefer more nutritious animal prey to large algae (e.g., Stoecker

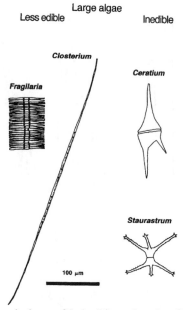

FIGURE 11.3. Large phytoplankters of Lake Biwa: *Ceratium hirundinella, Staurastrum dorsidentiferum, Fragilaria crotonensis,* and *Closterium aciculare.* Scale, 100 μm.

and Capuzzo 1990). Thus, large algae seldom suffer significant predatory loss, irrespective of crustacean fauna. Large algae thus show size escape from predation. On the other hand, large size carries a cost to the alga; for instance, the maximum growth rate is a negative function of algal size (Porter 1977). The diverse algal size hence reflects strategic differences (Reynolds 1988). Miyajima et al. (1994) listed temperature, insolation, wind, and nutrients as abiotic determinants of algal population dynamics in Lake Biwa.

Other Lakes

Gaedke (1992b) reported the plankton biomass size spectrum averaged for the growing season from Lake Constance, a large (476 km²) and deep (maximum, 252 m) lake located near 47°N 9°E. It had three distinct peaks of similar heights of 0.5–1.0 g C/m² at about 0.02 pg C, 0.06–0.26 ng C, and 4–34 μg C body mass; these were formed by bacteria, algae, and crustacea, respectively. In Lake Constance, carbon is effectively processed within the microbial food web and then transferred to crustacea.

The differences in the biomass size spectrum between Lake Biwa and Lake Constance were most conspicuous for the horizontal position of both phyto- and zooplankton modes and for the vertical height of zooplankton mode. In Lake

TABLE 11.2. Species list of pelagic crustacea in Lake Biwa

Subclass Copepoda
    Order Cyclopoida
        Family Cyclopidae
            Cyclops vicinus Uljanin 1875
            Mesocyclops dissimilis Defaye et Kawabata 1993
    Order Calanoida
        Family Diaptomidae
            Eodiaptomus japonicus (Burckhardt 1913)

Class Branchiopoda
    Order Anomopoda
        Family Daphnidae
            Daphnia galeata Sars 1864
        Family Bosminidae
            Bosmina longirostris (O. F. Muller 1785)
    Order Ctenopoda
        Family Sididae
            Diaphanosoma brachyurum (Lieven 1848)
    Order Haplopoda
        Family Leptodoridae
            Leptodora kindtii (Focke 1844)

Biwa, phytoplankton were distributed in larger size classes than in Lake Constance while zooplankton were smaller and less abundant.

Figure 11.4 shows the difference in biomass composition between Lake Constance (Geller et al. 1991) and Lake Biwa. Autotrophs, particularly large algae, were much more abundant in Lake Biwa. The carbon fixation rate of large algae in Lake Biwa was 54% ± 15% (mean ± SD of 5 months) of that of all autotrophs (Urabe et al. 1995). Nevertheless, the difference in the biomass of heterotrophs between the lakes was small, even though it was dominated by crustacea in Lake Constance and by bacteria in Lake Biwa. Of all heterotrophs, the annual mean proportion of bacteria and crustacea was 23% and 61% in Lake Constance, and 81% ± 7% (mean ± SD) and 14% ± 7% in Lake Biwa, respectively. Furthermore, the daphniids, which transfer pico- and nanoplankton production to larger organisms with high efficiency, were 58% of all crustaceans in Lake Constance and 20% ± 26% in Lake Biwa. Thus, the crustacea of Lake Biwa were not only low in biomass but also inefficient in material transfer.

Less complete data sets have been published for North American lakes. Sprules et al. (1983) showed bimodal biomass size spectra with peaks centered at phytoplankton and zooplankton for 37 lakes in Ontario. Studying size spectra from 15 temperate lake sites in southern Quebec, Ahrens and Peters (1991) reported that more oligotrophic systems have a more uniform biomass distribution. Because of lack of data, only the ratio in volumetric biomass concentration of total phytoplankton to total zooplankton for annual average can be compared quantitatively; the ratio was 4.7 in Lake Biwa, 1.6 in Lake Michigan (Sprules et al. 1991), and 0.14 in Lake Constance (Geller et al. 1991). In addition, the median ratio from

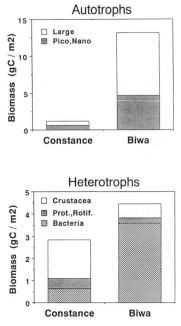

FIGURE 11.4. Biomass of autotrophs *(upper)* and heterotrophs *(lower)* averaged over the growing season for Lake Constance in 1987 (Geller et al. 1991) and Lake Biwa in 1992.

Ontario lakes was 1.64; 1st and 3rd quantiles were 0.77 and 3.52 (Sprules et al. 1983). Again, the zooplankton biomass of Lake Biwa was particularly low relative to phytoplankton biomass.

Summary

The trophic interactions involving the microbial food web in Lake Biwa are shown in Figure 11.5. Biomass was large in the fractions of large algae and bacteria. Large algae fixed carbon dioxide more than the smaller algae do, but these algae are not utilized directly. Crustacea mainly consumed nanoalgae and flagellates. Also, omnivores that can prey on ciliates and rotifers were scarce.

Large algae release carbon when alive and after death. Bacteria utilize the released carbon. The biomass of bacteria and small autotrophs is then processed within the microbial food web. It is true that longer food chains reduce transfer efficiency from autotrophs to mesozooplankton. Nevertheless, microbes recover photosynthetic production otherwise not utilized at all. Thus, microbes play an active part in biomass transfer.

Dominance of large algae in Lake Biwa has two opposing relationships with biodiversity. Large algae promote microbial diversity by exuding organic matter

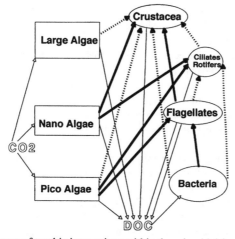

FIGURE 11.5. Diagram of trophic interactions within the microbial food web of Lake Biwa based on carbon flow. Autotrophs are shown in *rectangles* and heterotrophs in *ovals*. *Fine lines*, excretion and uptake; *heavy solid lines*, major predation; *heavy dotted lines*, minor predation.

and depress crustacean diversity by preventing direct trophic interactions. Regrettably, our knowledge is limited to less diverse organisms, large algae, and crustacea; we have just begun studying the diversity and activity of microbes. There are difficulties in separation and manipulation of these tiny organisms, and breakthroughs in methodology are needed.

Acknowledgments. We are grateful to C.S. Reynolds, S. Nakano, and anonymous reviewers for comments on the manuscript. Thanks are also due to the members of Lake Biwa Plankton Research Group, H. Haga, Y. Kusuoka, T. Nagata, S. Nakano, K. Ninomiya, K. Takai, and J. Urabe for permitting us access to their unpublished data. This study was partly supported by a fund from Lake Biwa Museum Project Office, Shiga Prefecture, Japan.

Literature Cited

Ahrens, M.A. and R.H. Peters. 1991. Patterns and limitations in limnoplankton size spectra. Canadian Journal of Fisheries and Aquatic Sciences 48:1967–1978.

Azam, F., T. Fenchel, J.G. Field, J.S. Gray, L.A. Meyer-Reil, and F. Thingstad. 1983. The ecological role of water-column microbes in the sea. Marine Ecology Progress Series 10:257–263.

Bjørnsen, P.K. 1988. Phytoplankton exudation of organic matter: why do healthy cells do it? Limnology and Oceanography 33:151–154.

Carpenter, S.R. 1988. Complex Interactions in Lake Communities. Springer-Verlag, New York.

Dodson, S. 1991. Species richness of crustacean zooplankton in European lakes of different sizes. Verhandlungen Internationale Vereinigung für Theoretische und Angewandte Limnologie 24:1223–1229

Gaedke, U. 1992a. Identifying ecosystem properties: a case study using plankton biomass size distributions. Ecological Modelling 63:277–298.

Gaedke, U. 1992b. The size distribution of plankton biomass in a large lake and its seasonal variability. Limnology and Oceanography 37:1202–1220.

Gaedke, U. 1993. Ecosystem analysis based on biomass size distributions: a case study of a plankton community in a large lake. Limnology and Oceanography 38:112–127.

Geller, W., R. Berberovic, U. Gaedke, H. Muller, H.-R. Pauli, M.M. Tilzer, and T. Weisse. 1991. Relations among the components of autotrophic and heterotrophic plankton during the seasonal cycle 1987 in Lake Constance. Verhandlungen Internationale Vereinigung für Theoretische und Angewandte Limnologie 24:831–836.

Gliwicz, M.Z. and J. Pijanowska. 1988. Effect of predation and resource depth distribution on vertical migration of zooplankton. Bulletin of Marine Science 43:695–709.

Kawabata, K. 1987a. Ecology of large phytoplankters in Lake Biwa: population dynamics and food relations with zooplankters. Bulletin of Plankton Society of Japan 34:165–172.

Kawabata, K. 1987b. Abundance and distribution of *Eodiaptomus japonicus* (Copepoda: Calanoida) in Lake Biwa. Bulletin of Plankton Society of Japan 34:173–183.

Kawabata, K. 1989. Natural development time of *Eodiaptomus japonicus* (Copepoda: Calanoida) in Lake Biwa. Journal of Plankton Research 11:1261–1272.

Miyajima, T., M. Nakanishi, S. Nakano, and Y. Tezuka. 1994. An autumnal bloom of the diatom *Melosira granulata* in a shallow eutrophic lake: physical and chemical constraints on its population dynamics. Archiv für Hydrobiologie 130:143–162.

Okamoto, K. 1984. Diurnal changing patterns of the *in situ* size selective feeding activities of *Daphnia longispina hyalina* and *Eodiaptomus japonicus* in a pelagic area of Lake Biwa. Memoirs of the Faculty of Science, Kyoto University, Series of Biology 9:107–132.

Pedrés-Alió, C. 1993. Diversity of bacterioplankton. Trends in Ecology & Evolution 8:86–90.

Peduzzi, P. and G.J. Herndl. 1992. Zooplankton activity fueling the microbial loop: differential growth response of bacteria from oligotrophic and eutrophic waters. Limnology and Oceanography 37:1087–1092.

Porter, K.G. 1977. The plant-animal interface in freshwater ecosystems. American Scientist 65:159–170.

Porter, K.G.., E.B. Sherr, B.F. Sherr, M. Pace, and R.W. Sanders. 1985. Protozoa in planktonic food webs. Journal of Protozoology 32:409–415.

Porter, K G., H. Paerl, R. Hodson, M. Pace, J. Priscu, B. Riemann, D. Scavia, and J. Stockner. 1988. Microbial interactions in lake food webs. In: S.R. Carpenter, ed. Complex Interactions in Lake Communities, pp. 209–227. Springer-Verlag, New York.

Reynolds, C.S. 1988. Functional morphology and the adaptive strategies of freshwater phytoplankton. In: C.D. Sandgren, ed. Growth and Survival Strategies of Freshwater Phytoplankton, pp. 388–433. Cambridge University Press, New York.

Sherr, E.B. and B.F. Sherr. 1991. Planktonic microbes: tiny cells at the base of the ocean's food webs. Trends in Ecology & Evolution 6:50–54.

Sieburth, J. McN., V. Smetacek, and J. Lenz. 1978. Pelagic ecosystem structure: heterotrophic compartments of the plankton and their relationship to plankton size fractions. Limnology and Oceanography 23:1256–1263.

Sprules, W.G. 1988. Effects of trophic interactions on the shape of pelagic size spectra.

Verhandlungen Internationale Vereinigung für Theoretische und Angewandte Limnologie 23:234–240.

Sprules, W.G., J.M. Casselman, and B.J. Shuter. 1983. Size distribution of pelagic particles in lakes. Canadian Journal of Fisheries and Aquatic Sciences 40:1761–1769.

Sprules, W.G., S.B. Brandt, D.J. Stewart, M. Munawar, E.H. Jin, and J. Love. 1991. Biomass size spectrum of the Lake Michigan pelagic food web. Canadian Journal of Fisheries and Aquatic Sciences 48:105–115.

Stoecker, D.K. and J.M. Capuzzo. 1990. Predation on protozoa: its importance to zooplankton. Journal of Plankton Research 12:891–908.

Thiebaux, M. L. and L.M. Dickie. 1993. Structure of the body-size spectrum of the biomass in aquatic ecosystems: a consequence of allometry in predator–prey interactions. Canadian Journal of Fisheries and Aquatic Sciences 50:1308–1317.

Urabe, J., M. Nakanishi, and K. Kawabata. 1995. Contribution of metazoan plankton to the cycling of nitrogen and phosphorus in Lake Biwa. Limnology and Oceanography 40:232–241.

Wemer, E.E. and J. F. Gilliam. 1984. The ontogenetic niche and species interactions in size-structured populations. Annual Review of Ecology and Systematics 15:393–425.

12

The Role of Species in Ecosystems: Aspects of Ecological Complexity and Biological Diversity

JOHN H. LAWTON

Introduction

The international cooperative research program known as the "SymBiosphere" project (Kawanabe et al. 1993) has as its organizing principles that ecological complexity plays a key role in promoting biodiversity, and that it is fundamentally important, for the conservation and recovery of biodiversity, to understand how this diversity is created and sustained. In this chapter, I want to explore three aspects of the role of species in ecosystems that illuminate different facets of "ecological complexity," how it promotes biodiversity, and what the consequences of loss of biodiversity may be for the maintenance of that complexity. The essay was originally published in *Oikos* (Lawton 1994), and appears here in a modified form, with the kind permission of that journal.

The three, interrelated themes are as follows. I start with a theoretical question, and ask what are the possible relationships between ecosystem processes [Likens' (1992) "transformation and flux of energy and matter"; see following] and the species richness of communities? How might loss of species impair ecosystem function? Crudely, does it matter to ecosystem functioning if biomass is divided among few or many species (Walker 1992; Lawton and Brown 1993; Vitousek and Hooper 1993)?

The second theme involves a recent attempt to answer these questions experimentally with artificial terrestrial ecosystems in a controlled environment facility known as the *Ecotron* at Silwood Park (Lawton et al. 1993; Naeem et al. 1994, 1995), and examines one aspect of the feedbacks postulated by the SymBiosphere program between biodiversity and the maintenance of ecosystem function and hence, ultimately, ecological complexity.

The third theme arises from a preliminary experiment in the Ecotron (Thompson et al. 1993) that draws attention to the role of earthworms as "ecosystem engineers" (Jones et al. 1994; Lawton 1994). Ecosystem engineering by many taxa (not just earthworms) is a good example of a complex interaction central to SymBiosphere research, because it is an important way in which particular species promote and maintain biodiversity, albeit rather poorly studied.

SymBiosphere emphasizes the interdependence of organisms, their environment, and entire ecosystems. Despite this obvious interdependence, however, for almost three decades ecosystem and population ecology have ploughed their own independent furrows and developed their own paradigms, approaches, and questions (e.g., Cherrett 1989). Yet populations and communities live in, and are part of, ecosystems. As Gene Likens recently pointed out (Likens 1992), ecology is "the scientific study of the processes influencing the distribution and abundance of organisms, the interaction among organisms, and the interactions between organisms and the transformation and flux of energy and matter." If we wish ecology to progress and to develop the more integrated, holistic understanding of nature envisioned by SymBiosphere, we have no choice but to break down the barriers that currently divide population and ecosystem science (Lawton and Jones 1993; Jones and Lawton 1995). This chapter lies at the interface of these two disciplines, and attempts modest integration between them.

Theories About Ecosystem Function and Species Richness

Four hypotheses summarize the possible general responses of ecosystem processes to reductions in species richness (Naeem et al. 1995) (Fig. 12.1). (Obviously, if we were to remove all the species in an entire trophic level—all the plants, or all the decomposers, for example—ecosystem processes would be drastically impaired; *reductio ad absurdum,* the hypotheses are not interesting.) The first, the redundant species hypothesis, suggests that there is a minimal diversity necessary for proper ecosystem functioning, but beyond that, most species are redundant in their roles (Walker 1992; Lawton and Brown 1993).

A second, contrasting view is the rivet hypothesis, which suggests that all species make a contribution to ecosystem performance (Ehrlich and Ehrlich 1981). This hypothesis likens species to rivets holding together a complex machine and postulates that functioning will be impaired as its rivets (species) fall out. A range of possible responses, indicated on Figure 12.1, are compatible with this hypothesis. A third view, the idiosyncratic response hypothesis, suggests that ecosystem function changes when diversity changes, but the magnitude and direction of change is unpredictable because the roles of individual species are complex and varied. Finally, the null hypothesis (not illustrated) is that ecosystem function is insensitive to species deletions or additions.

Vitousek and Hooper (1993) have provided a broadly similar, but not identical, set of possibilities. Their type 1 response (a linear decline in function with declining species richness) roughly corresponds with an extreme version of the rivet hypothesis. A type 2 response has an asymptotically curvilinear relationship between function and diversity; species are functionally redundant on the asymptote (see also Swift and Anderson 1993). Vitousek and Hooper's type 3 response postulates no effect of diversity on function at any level of diversity—the null hypothesis. Vitousek and Hooper have stated nothing equivalent to the idiosyncratic hypothesis.

These hypotheses serve primarily as heuristic devices to facilitate and sharpen the debate about the role of species in ecosystems. A colleague who refereed this

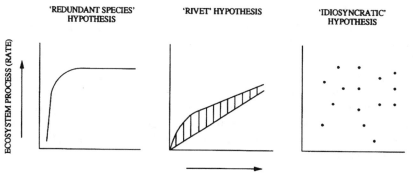

FIGURE 12.1. Three hypothetical relationships between the rate of an ecosystem process (e.g., primary production, rate of decomposition) and ecosystem species richness (see text).

chapter, for example, objected to the simple (not to say simplistic) models displayed in Figure 12.1 on two grounds. First, they believed that the redundant species hypothesis should predict an increase in ecosystem processes up to a threshold number of species, followed by a sharp break to a plateau (rather than the gentle curve shown). Perhaps, but ecological processes rarely show razor-sharp breaks. Second, they suggested that the idiosyncratic hypothesis is simply a "high-variance variant of the null hypothesis," which is indeed one way to think about the problem. But I believe that it is useful to distinguish between average trends and the variance around those trends and to focus on the variance when it becomes very large (which is what the idiosyncratic hypothesis does). What constitutes "very large" is, of course, subjective, and highlights some of the subtle, hidden assumptions and problems in Figure 12.1. For instance, there are no scales on the axes. If species richness is relatively low, a rescaled and redrawn version of the redundant species hypothesis would be hard to distinguish from the upper-bound curve of the rivet hypothesis, and so on. Nevertheless, the main point of Figure 12.1 is to suggest that we do experiments in which we vary species richness and then measure ecosystem processes. Which model(s) best describe the results?

Another implicit feature of the models displayed in Figure 12.1 is that they make different assumptions about whether species deletion is a fixed or random factor in the experiment. The redundant, rivet, and null hypotheses treat loss of species as a random factor. Which species, and the order in which they are removed, are not important; on average, species loss will produce the trends shown in the first two panels of Figure 12.1, or the horizontal line postulated by the null hypothesis. The sequence in which species are removed does, however, matter for the idiosyncratic hypothesis, as do their identities. The idiosyncratic hypothesis proposes that species have particular (fixed) effects on ecosystem processes and that deleting species in the sequence A,B,C will produce a graph very different from the sequence C,B,A, which may be different again from deleting species D, E, and F. The average of several such experiments, with species selected in random order, will yield a graph with very wide confidence intervals around an average trend that is hard to distinguish from any of the three alternatives.

These conceptual issues aside, the experiments are also hard to do, particularly in the field. One such attempt is currently being made by Tilman and his colleagues in a Minnesota grassland (D. Tilman, personal communication). The first, direct experimental tests were recently carried out, not in the field, but in the Ecotron facility at Silwood Park (Naeem et al. 1994, 1995). It is to this facility and to these experiments that I now turn.

The Ecotron

The Ecotron consists of a series of 16 controlled environment chambers, in 2 banks of 8, capable of simulating natural environments in diurnal light:dark cycle, temperature, humidity, and rainfall. Each chamber has a floor area of 2 x 2 m, accommodating small terrestrial ecosystems 1 m^2 in area and about 40 cm deep. A full description of the facility is given by Lawton et al. (1993). Each bank of 8 chambers has a separate air-handling unit that collects, reconditions, and recycles the air within the chambers. However, the ecosystems are not totally isolated. Scientists enter the chambers to take measurements, and approximately 5% of the 130 kg of air in each bank is exchanged with filtered outside air on each cycle (every 50 s).

Ecosystems are established in 0.3 m^3 of relatively infertile soil (pH 6.4, 60% loam and 40% sand, laid on top of a 0.1-m^3 washed gravel base) sterilized initially with methyl bromide. Undefined microbial communities are reintroduced by the addition of a filtered soil wash. Plants, insect herbivores, parasitoids, and soil fauna can then be added to the system, as required. Experiments are run long enough for the plants to have more than one generation and for most of the animals to have several generations.

Many ecologists are suspicious of laboratory experiments in artificial ecosystems. My own view is that they act as a valuable halfway house between the simplicity of most plant pot and greenhouse experiments, or mathematical models involving one or a few species, and the full complexity of the field (Kareiva 1994; Lawton 1996). In particular, they allow precise, experimental control of the species, species interactions, and ecological complexity that are central to Sym-Biosphere.

The Effect of Earthworms in a Simple Plant Community

The first experiment in the Ecotron established a simple detritivore and plant community, lacking insect herbivores. The animal assemblage consisted of earthworms (*Lumbricus terrestris* and *Aporrectodea* spp., mostly *A. longa*), snails

(*Helix aspersa,* which also acted as herbivores), and Collembola. Plants were all small, self-compatible annuals: a nitrogen fixer, *Trifolium dubium,* a grass, *Poa annua,* and a composite, *Senecio vulgaris.* Treatments consisted of manipulating the detritivore assemblage (i) with snails, (ii) with worms, (iii) with both snails and worms, and (iv) with neither. All treatments contained Collembola. A full description of the experiment is in Thompson et al. (1993). Here, for brevity, I concentrate on the role of the earthworms.

Despite the simplicity of the communities, some of the results were unexpected. For example, worms increased soil phosphates, and affected the establishment, growth, and cover of *Trifolium* in the middle phase of the experiment, via several mechanisms. Worms increased root nodulation by *Trifolium,* increased germination of *Trifolium* seeds by creating "safe sites" in worm casts, and decreased competition from *Poa* and *Senecio,* by differentially burying the seeds of these two species in burrows.

In the present context this experiment makes two main points. First, the animal and plant assemblages were very simple; for instance, there were just two abundant species of earthworms. Would it have made any difference if we had used more species of earthworms, or alternatively just one species? Second, the earthworms had their main effects on *Trifolium* dynamics by none of the processes ecological textbooks say are important in structuring ecological communities (predominantly trophic and competitive effects). It is unclear why earthworms increased *Trifolium* nodulation, but they may have done so by concentrating rhizobia in the rhizosphere. Their other main effects are certainly "mechanical" in nature—depositing worm casts that act as safe *Trifolium* germination sites and burying seeds of competitors too deep for them to germinate successfully. In these roles, earthworms are, in fact, acting as ecological engineers, to which I will return in due course. Before doing so, however, we need to look in more detail at the effects of species richness on ecosystem processes.

Species Richness and Ecosystem Processes

The second Ecotron experiment used just a single species of earthworm (not two as in the first experiment), but it was not designed to investigate effects of varying earthworm species richness. Rather, we chose to explore more fundamental changes in species richness by creating ecosystems with three levels of higher plant and animal diversity (Naeem et al. 1994, 1995); no attempt was made to manipulate or characterize the soil microbial community. There were four replicates of the low-diversity system each with 9 species in total (2 species of plants, 3 species of herbivores, 1 parasitoid, and 3 detritivores), four replicates with intermediate diversity (15 species), and six high-diversity replicates (31 species, made up of 16 plants, 5 herbivores, 2 parasitoids, and 8 detritivores). Species were assembled so that lower diversity systems contained a subset of the higher diversity assemblages; in other words, the lower diversity communities resembled depauperate descendants of higher diversity communities that had lost species uniformly from all trophic categories (plants, herbivores, decomposers, and parasitoids).

We measured five ecosystem processes: (1) community respiration (CO_2 fluxes); (2) decomposition, both short term (surface litter bags) and long term (buried wood); (3) soil nutrient retention (available N, P, and K); (4) plant net primary productivity (inverse of % 400-700nm light transmittance); and (5) water retention (rate of outflow). [Full details were published by Naeem et al. (1994, 1995).] Most of the ecosystem processes varied significantly with species richness; only long-term decomposition of buried wood showed no response. Short-term decomposition, N, P, and K soil nutrient levels, and the rate of water outflow all varied significantly with species richness, but not in any systematic way. Uptake of CO_2 and plant productivity both declined as species richness declined.

This experiment is the first time that the hypotheses illustrated in Figure 12.1 have been tested experimentally under controlled conditions, involving the simultaneous manipulation of four trophic levels. The results support three of the four hypothetical possibilities (or more precisely, particular hypotheses can be rejected for some processes). Only long-term decomposition conformed to the null hypothesis (no effects), and this may be because the experiment was not run long enough to detect changes or because microorganisms capable of decomposing wood were lacking from the ecosystems (J. Anderson, personal communication). Uptake of CO_2 and plant productivity conformed to the predictions of the rivet hypothesis. The remaining ecosystem processes varied idiosyncratically as diversity declined; in other words, species richness did make a difference to the performance of ecosystems, but not in a consistent manner.

The mechanisms underpinning these responses are not well understood, except those for CO_2 and plant productivity. In the experiment, the mechanism appears to be very simple. Higher diversity ecosystems intercept more light because they have better three-dimensional, space-filling canopies, i.e., as emphasized by SymBiosphere, they are more structurally complex. Most natural historians could probably have predicted this on common sense grounds, at least post facto. In science, however, it is sometimes necessary to test common sense experimentally.

Caveats and Elaborations

Four further comments are in order. Although only plant productivity and CO_2 uptake declined with declining diversity, the mechanism implies that the response may be quite general. Other things being equal (for example, no addition of artificial fertilizers), replacing multilayered, species-rich plant communities by monocultures should generally impair CO_2 uptake and reduce productivity. This prediction requires testing but has some support from the intercropping literature (Vandermeer 1988; Swift and Anderson 1993). If the mechanism is indeed a general one, it suggests (Naeem et al. 1994) that major simplification of plant communities in the real world will lead to a reduction in the ability of terrestrial ecosystems to absorb anthropogenic CO_2 (Körner and Arnone 1992; Diaz et al. 1993; Wolfsy et al. 1993; Dixon et al. 1994), with potentially far-reaching implications.

Second, although we varied total species richness roughly fourfold and plant richness eightfold, the richest systems were still extremely species poor (and very small) compared with real terrestrial ecosystems. It is not at all clear how applicable these results are to systems containing hundreds of plant species and their associated fauna—that is, to the full complexity and richness of real ecosystems embraced by SymBiosphere. Species redundancy in ecosystem processes is much more likely in very diverse systems. Indeed, the plateau in the redundant species hypothesis (the Vitousek and Hooper type 2 response) may be reached at quite small numbers of species. Reviewing the intercropping literature, Swift and Anderson (1993) argued that "the relationship between plant species number and efficiency and stability of ecosystem functions may...be a hyperbolic one...in which the plateau is reached at quite low species number."

In a related experiment, Tilman and Downing (1994) showed that primary productivity in more diverse grassland communities was more resilient to, and recovered more quickly from, a major drought. But the response was curvilinear, with drought resistance becoming essentially independent of plant species richness in communities with more than 8 species (see the legend to Fig. 1 in Tilman and Downing 1994). In other words, more than halving plant diversity from 25 to about 10 (and possibly even fewer) species in Tilman and Downing's communities had no detectable effect on the way in which productivity resisted and recovered from drought. Similar caveats apply to the Ecotron experiment with a maximum of 16 plant species. We have no experimental evidence that addresses the role of species richness in assemblages with more than about 20 species of plants.

Third, it would be foolish to read too much into the Ecotron experiment, even for relatively simple natural assemblages. It is one experiment, done under controlled and artificial conditions. It would be wise to await further studies, on a range of natural, seminatural, and artificial ecosystems, for a variety of ecosystem processes, before generalizing about the role of species richness in ecosystem processes.

Finally, we need to be very clear what we are trying to understand, predict, and make generalizations about. Figure 12.1 refers to changes in ecosystem processes under reasonably constant or average environmental conditions. The environment in the Ecotron is carefully controlled to avoid extreme events (no droughts, no killing frosts, no pollution incidents). Our experiment says nothing about the role of species richness in buffering ecosystems against rare or extreme events; this is the focus of Tilman and Downing's (1994) experiment. Numerous rare species that are apparently redundant under average, benign environmental conditions may maintain ecosystem functions during extreme events (Schulze and Mooney 1993). Anthropomorphically, they may provide the "backup systems" for ecosystem processes. We understand very little about what rare species do in ecosystems and need, urgently, to know more.

Organisms As Ecosystem Engineers

It is a measure of how little ecologists have thought about the role of species in promoting and maintaining biodiversity and ecosystem processes that an entire, fundamentally important role has been essentially ignored. Return, for example, to the earthworms in the Ecotron. As already pointed out, the earthworms had their main effects on *Trifolium* dynamics by none of the processes ecological textbooks say are important in structuring ecological communities. Rather, their main role was as "engineers"—moving soil, making casts, and burying seeds. The concept of organisms as ecosystem engineers has been developed at length elsewhere (Jones et al. 1994; Jones and Lawton 1995). I therefore only provide a brief résumé.

A Definition of Engineering

Ecosystem engineers are organisms that directly or indirectly modulate the availability of resources (other than themselves) to other species, by causing physical-state changes in biotic or abiotic materials. In so doing they modify, maintain, or create habitats. Autogenic engineers change the environment via their own physical structures, i.e., their living and dead tissues. Allogenic engineers change the environment by transforming living or nonliving materials from one physical state to another via mechanical or other means.

Examples

The concept is best illustrated by an example. Beaver (*Castor canadensis*) are classical allogenic engineers, taking materials in the environment (in this case trees, but in the more general case it can be any living or nonliving material), and turning them (engineering them) from physical state 1 (living trees) into physical state 2 (dead trees in a beaver dam). This act of engineering then creates a pond, and it is the pond that has profound effects on a whole series of resource flows used by other organisms. "These activities retain sediments and organic matter in the channel,...modify nutrient cycling and decomposition dynamics, modify the structure and dynamics of the riparian zone, influence the character of water and materials transported downstream, and ultimately influence plant and animal community composition and diversity" (Naiman et al. 1988). A critical characteristic of ecosystem engineering is that it must change the availability (quality, quantity, distribution) of resources utilized by other taxa, excluding the biomass provided directly by the population of allogenic engineers. Engineering is not the direct provision of resources in the form of meat, fruits, leaves, or corpses. Beaver are not the direct providers of water, in the way that prey are a direct resource for predators or leaves are food for caterpillars. Earthworms are also good examples of allogenic engineers (Lal 1991). Their burrowing, mixing, and casting activities change the mineral and organic composition of soils, provide safe sites for seed germination, affect nutrient cycling, and alter hydrology and drainage, with (as we saw in the first Ecotron experiment) impacts on plant population dynamics and community development.

Now consider autogenic engineers. Simple examples are the growth of a forest or a coral reef. Trees and corals are direct sources of food and living space for numerous organisms, but the production of branches, leaves, or living coral tissue does not, in itself, constitute engineering; rather, it is the direct provision of resources. However, the development of the forest or the reef results in physical structures that do change the environment and modulate the distribution and abundance of many other resources. This modulation constitutes autogenic engineering. Trees alter hydrology, nutrient cycles, and soil stability, as well as humidity, temperature, windspeed, and light levels (see also Holling 1992); corals modulate current speeds, siltation rates, and so on. In the Ecotron, development of the plant canopy markedly changes the microclimate and hydrology within these miniature ecosystems. It is obvious, but surprisingly rarely explicitly stated, that numerous inhabitants of the habitats so created are dependent on the physical conditions modulated by the autogenic engineers and on resource flows which they influence but do not directly provide; without the engineers, most of these other organisms would disappear.

There are probably no ecosystems on earth that are not engineered in some way by one or more species. To take but one unexpected example, microalgae growing in and under Antarctic sea ice are important autogenic engineers in this most inhospitable of ecosystems. By scattering and absorbing light, they enhance the melting and breakup of the ice (Buynitskiy 1986; Arrigo et al. 1991).

However, species clearly differ markedly in their capacity to act as ecosystem engineers. Most probably never perform this function; a few have major impacts. Six factors scale the impact of engineers. They are (i) the lifetime per capita activity of individual organisms; (ii) the population density; (iii) the spatial distribution, both locally and regionally, of the population; (iv) the length of time the population has been present at a site; (v) the durability of constructs, artifacts, and impacts in the absence of the original engineer; and (vi) the number and types of resource flows that are modulated by the constructs and artifacts and the number of other species dependent on these flows.

The most obvious allogenic engineering is attributable to species with large per capita effects, living at high densities over large areas for a long time, and giving rise to structures that persist for millennia and which affect many resource flows; for instance, the mima mounds created by fossorial rodents (Naiman et al. 1988). Autogenic engineers may also have massive effects; as Holling (1992) succinctly states: "To a degree,...the boreal forest 'makes its own weather' and the animals living therein are exposed to more moderate and slower variation in temperature and moisture than they would otherwise be." Boreal forest trees have large per capita effects on hydrology and climatic regimes, occur at high densities over large areas, and live for decades, but their impacts as autogenic engineers rapidly disappear if the forest is logged. Organisms with small individual impacts can also have huge ecological effects, providing that they occur at sufficiently high densities over large areas for sufficient periods of time. Burrowing marine meofauna (Reichelt 1991) and bog-forming *Sphagnum* mosses (Tansley 1949) are good examples.

Engineers, Keystone Species, and Ecosystem Processes

We need, now, to return to Figure 12.1. Although a majority of species may have no role to play as engineers, for certain species this prediction is clearly not true. Eliminating beavers has a dramatic, cascading effect on many other taxa and ecosystem processes; there is no redundancy built into the role beavers play in ecosystems. There may sometimes be redundancy among engineers—the unresolved question of whether one species of deep-burrowing, casting earthworm would be as effective as a guild of several similar species comes to mind—but in general we know too little about the impacts of engineers to predict how reducing the diversity of engineers might change ecosystem function. For example, I know of very few field manipulation experiments designed to quantify the impact of ecosystem engineers by removing or adding species, although the few studies that have been done are revealing. Bertness (1984a,b, 1985) provided some excellent examples of manipulative experiments on both allogenic and autogenic engineers. Removing burrowing fiddler crabs (*Uca pugnax*) or ribbed mussels (*Geukensia demissa*) from salt marshes on the eastern seaboard of the United States, for instance, changed rates of drainage, decomposition, primary production, sedimentation, and erosion.

What this serves to remind us is that all species are not equal in their impacts on ecosystems. Although the term "keystone species" is overworked and has been used in ways that extend far beyond its original meaning (Mills et al. 1993), it is nonetheless useful shorthand for species whose removal has big effects on many other taxa and ecosystem functions (Paine 1969; Krebs 1985; Daily et al. 1993)—species with a key role to play in promoting and maintaining biodiversity. However, we need to carefully distinguish the mechanisms involved (Mills et al. 1993).

Most ecologists probably regard trophic links as the main mechanism by which keystone species act; but engineering is arguably more important (recall all those autogenic engineers called trees). SymBiosphere will need to study engineering as a key "complex and intricate interaction among various forms of life" (Kawanabe et al. 1993, p. 5). In the frequently cited example of sea otters (*Enhydra lutris*), removal of otters leads to an increase in sea urchins (*Strongylocentrotus* sp.) and hence to the disappearance of kelp beds, which in turn changes wave action and siltation rates, with profound consequences for other inshore flora and fauna (Estes and Palmisano 1974). This cascade is not just a trophic effect of otters. Kelp are autogenic engineers and removal of kelp by urchins is, among other things, allogenic engineering. In other words, in this familiar example the species traditionally regarded as the keystone (sea otter) has major effects because it changes the impact of one engineer (urchin) on another (kelp), with knock-on effects on other species in the web of interactions.

Whatever the mechanism(s) by which keystone species (broadly defined) impose their effects, their existence means that ecosystem processes may respond dramatically and suddenly to the loss of just one species. A thought experiment may be helpful. Suppose we blindly and ignorantly remove species at random from a system about which we know little and measure ecosystem processes as we proceed. Occasionally we will hit on a keystone species, and by definition, the effects will be dramatic. Some ecosystem processes will undoubtedly be impaired, but the rates of others may increase. In other words, the existence of species with strong effects, be they trophic or engineering, means that the overall response of ecosystems to species removal will more often than not conform to the idiosyncratic hypothesis. We expect to see effects when we haphazardly remove species, but the magnitude and direction of these effects are often unpredictable. This is, of course, primarily what we observed for most of the processes measured in the second Ecotron experiment; they responded idiosyncratically to species deletion.

Symbiosphere: Ecological Complexity for Promoting Biodiversity

In this chapter I have done no more than explore a tiny sample of the complex interactions between species and what the loss of species may mean for the maintenance of ecosystem function and hence, ultimately, for the maintenance of biodiversity. SymBiosphere postulates that not only do complex and intricate interactions promote, enhance, and maintain biodiversity, but also that biodiversity feeds back into, and influences, complex interactions and habitat heterogeneity (Kawanabe et. al. 1993).

First and foremost, it is becoming increasingly obvious that loss of plant species may change and impair ecosystem processes under average environmental conditions and reduce the ability of ecosystems to withstand, and recover from, extreme events, with clear implications for maintenance of biodiversity in the longer term. However, it is still unclear when the law of diminishing returns starts to apply, and whether there is significant functional redundancy in species-rich plant communities. No experiments have yet been performed to explore the effects of changing animal species richness independently of plant richness. We know that removing single species of animals, usually top predators, can have major population and ecosystem consequences, but do species-rich pollinator, decomposer, or herbivore guilds work better than simpler guilds? We do not know.

Second, the very existence of some ecosystems depends on particular species, the major autogenic and allogenic engineers that create the system, and all ecosystems are probably modulated and modified to a significant extent by at least one species of engineer. Engineering is a fundamental but poorly studied form of complex and intricate interaction.

If nothing else, the holistic approach to ecology espoused by SymBiosphere forces increased integration of ecosystem and population ecology, I believe to the benefit of the subject as a whole.

Acknowledgments. A version of this paper was first delivered in Lund in October 1993 as the Per Brink Lecture, and subsequently in a modified form in Kyoto in December 1993 at an International Symposium on "Ecological Perspectives of Biodiversity." I am extremely grateful to colleagues in the Centre for Ecological Research, Kyoto University, for making my visit to Japan possible, and for the invitation to prepare a written version for publication. Colleagues in the Centre for Population Biology, particularly Phil Heads, Ros Jones, Shahid Naeem, Lindsey Thompson, and Richard Woodfin, made the Ecotron experiments possible. Clive Jones and Moshe Shachak forced me to think about ecosystem engineering. Clive Jones, Hefin Jones, and Lindsey Thompson made valuable comments on the manuscript.

Literature Cited

Arrigo, K.R., C.W. Sullivan, and J.N. Kremer. 1991. A bio-optical model of Antarctic sea ice. Journal of Geophysical Research 96(C6):10581–10592.

Bertness, M.D. 1984a. Habitat and community modification by an introduced herbivorous snail. Ecology 65:370–381.

Bertness, M.D. 1984b. Ribbed mussels and *Spartina alterniflora* production in a New England salt marsh. Ecology 65:1794–1807.

Bertness, M.D. 1985. Fiddler crab regulation of *Spartina alterniflora* production on a New England salt marsh. Ecology 66:1042–1055.

Buynitskiy, V.K. 1968. The influence of microalgae on the structure and strength of Antarctic sea ice. Oceanology 8:771–776.

Cherrett, J.M., ed. 1989. Ecological Concepts. The Contribution of Ecology to an Understanding of the Natural World. Blackwell Scientific, Oxford.

Daily, G.C., P.R. Ehrlich, and N.M. Haddad. 1993. Double keystone bird in a keystone species complex. Proceedings of the National Academy of Sciences of the United States of America 90:592–594.

Diaz, S., J.P. Grime, J. Harris, and E. McPherson. 1993. Evidence of a feedback mechanism limiting plant response to elevated carbon dioxide. Nature 364:616–617.

Dixon, R.K., S. Brown, R.A. Houghton, A.M. Solomon, M.C. Trexler, and J. Wisniewski. 1994. Carbon pools and flux of global forest ecosystems. Science 263:185–190.

Ehrlich, P.R. and A.H. Ehrlich. 1981. Extinction. The Causes and Consequences of the Disappearance of Species. Random House, New York.

Estes, J.A. and J.F. Palmisano. 1974. Sea-otters: their role in structuring nearshore communities. Science 185:1058–1060.

Holling, C.S. 1992. Cross-scale morphology, geometry, and dynamics of ecosystems. Ecological Monographs 62:447–502.

Jones, C.G. and J.H. Lawton, eds. 1995. Linking Species and Ecosystems. Chapman & Hall, New York.

Jones, C.G., J.H. Lawton, and M. Shachak. 1994. Organisms as ecosystem engineers. Oikos 69:373–386.

Kareiva, P. 1994. Diversity begets productivity. Nature 368:686–687.

Kawanabe, H., T. Ohgushi, and M. Higashi, eds. 1993. Symbiosphere. Ecological Complexity for Promoting Biodiversity. Biology International Special Issue 29, International Union of Biological Sciences, Paris.

Körner, C. and J.A. Arnone III. 1992. Response to elevated carbon dioxide in artificial tropical ecosystems. Science 257:1672–1675.

Krebs, C.J. 1985. Ecology. The Experimental Analysis of Distribution and Abundance, 3rd Ed. Harper & Row, New York.

Lal, R. 1991. Soil conservation and biodiversity. In: D.L. Hawksworth, ed. The Biodiversity of Microorganisms and Invertebrates: Its Role in Sustainable Agriculture, pp. 89–103. CAB International, Wallingford.

Lawton, J.H. 1994. What do species do in ecosystems? Oikos 71:367–374.

Lawton, J.H. 1996. The Ecotron facility at Silwood Park: the value of 'big bottle' experiments. Ecology 77:665–669.

Lawton, J.H. and V.K. Brown. 1993. Redundancy in ecosystems. In: E.-D. Schulze and H.A. Mooney, eds. Biodiversity and Ecosystem Function, pp. 255–270. Springer-Verlag, Berlin.

Lawton, J.H. and C. G. Jones. 1993. Linking species and ecosystem perspectives. Trends in Ecology & Evolution 8:311–313.

Lawton, J.H., S. Naeem, R.M. Woodfin, V.K. Brown, A. Gange, H.J.C. Godfray, P.A. Heads, S. Lawler, D. Magda, C.D. Thomas, L.J. Thompson, and S. Young. 1993. The Ecotron: a controlled environmental facility for the investigation of population and ecosystem processes. Philosophical Transactions of the Royal Society of London B, Biological Sciences 341:181–194.

Likens, G.E. 1992. The Ecosystem Approach: Its Use and Abuse. Ecology Institute, Oldendorf/Luhe.

Mills, L.S.M., M.E. Soulé, and D.F. Doak. 1993. The keystone-species concept in ecology and conservation. Bioscience 43:219–224.

Naeem, S., L.J. Thompson, S.P. Lawler, J.H. Lawton, and R.M. Woodfin. 1994. Declining biodiversity can alter the performance of ecosystems. Nature 368:734–737.

Naeem, S., L.J. Thompson, S.P. Lawler, J.H. Lawton, and R.M. Woodfin. 1995. Empirical evidence that declining biodiversity may alter the performance of terrestrial ecosystems. Philosophical Transactions of the Royal Society of London B, Biological Sciences 347:249–262.

Naiman, R.J., C.A. Johnston, and J.C. Kelley. 1988. Alteration of North American streams by beaver. Bioscience 38:753–762.

Paine, R.T. 1969. A note on trophic complexity and community stability. American Naturalist 103:91–93.

Reichelt, A.C. 1991. Environmental effects of meiofaunal burrowing. Symposia of the Zoological Society of London 63:33–52.

Schulze, E.-D. and H.A. Mooney, eds. 1993. Biodiversity and Ecosystem Function. Springer-Verlag, Berlin.

Swift, M.J. and J.M. Anderson. 1993. Biodiversity and ecosystem function in agricultural systems. In: E.-D. Schulze and H.A. Mooney, eds. Biodiversity and Ecosystem Function, pp. 15–41. Springer-Verlag, Berlin.

Tansley, A.G. 1949. Britain's Green Mantle. Allen & Unwin, London.

Thompson, L., C.D. Thomas, J.M.A. Radley, S. Williamson, and J.H. Lawton. 1993. The effect of earthworms and snails in a simple plant community. Oecologia 95:171–178.

Tilman, D. and J.A. Downing. 1994. Biodiversity and stability in grasslands. Nature 367:363–365.

Vandermeer, J.H. 1988. The Ecology of Intercropping. Cambridge University Press, Cambridge.

Vitousek, P.M. and D.U. Hooper. 1993. Biological diversity and terrestrial ecosystem biogeochemistry. In: E.-D. Schulze and H.A. Mooney, eds. Biodiversity and Ecosystem Function, pp. 3–14. Springer-Verlag, Berlin.

Walker, B.H. 1992. Biodiversity and ecological redundancy. Biological Conservation 6:18–23.

Wofsy, S.C., M.L. Goulden, J.W. Munger, S.-M. Fan, P.S. Bakwin, B.C. Daube, S.L. Bassow, and F.A. Bazzaz. 1993. Net exchange of CO_2 in a mid-latitude forest. Science 260:1314–1317.

V

Management for Biodiversity Conservation

13

Sources and Management of Biodiversity in the Russian Far East

Yuri N. Zhuravlev

Introduction

As a part of Siberia, the Russian Far East (RFE) currently represents the world's primary resource of conifers. This biospheric significance of the RFE will have a tendency to increase while tropical forests diminish. A second RFE contribution to global biotic resources is its biological diversity, which saturates the RFE as a whole, especially its southern part, including the Primorye region. The numerous floral and faunal elements—indigenous, endemic, and newcomers from adjoining areas—create a biodiversity here at different levels: forest, meadow, and tundra formations, with a wide variety of ecosystems and communities. These structures display interconnecting and variable cooperation, some of them undergoing succession, others rapidly evolving; some are very close to extinction following intense anthropogenic stress.

Such complicated patterns call for a more complete understanding of the origin, structure, and function of all these communities. On the other hand, the growing conflict between industrial use of resources in the region and the goals of conservation of these resources (in part for the purpose of long-term industrial use) demands a program of sustainable regional economic development. The major principles oriented to maintaining the functional state of ecosystems and communities have to be elaborated with regard to local conditions.

Thus, the aims of this chapter are to discuss the origins of biodiversity formation in the RFE and more specifically in Primorye, and to design, on this basis, a more suitable approach to conserving its biodiversity in the near future.

The Origin of Biodiversity in the RFE

The RFE, by itself, is a somewhat narrow territory stretching lengthwise along the Pacific Ocean (Fig. 13.1). The physicogeographical map of the USSR divides the land part of the RFE into five landscape ecozones (Parmusin 1964) (see Fig. 13.1). Three ecozones (Amur-Primorye, Far North-East, and Kamchatka-Kuril) belong completely to the RFE, while the other two (East-Siberian and Verkho-

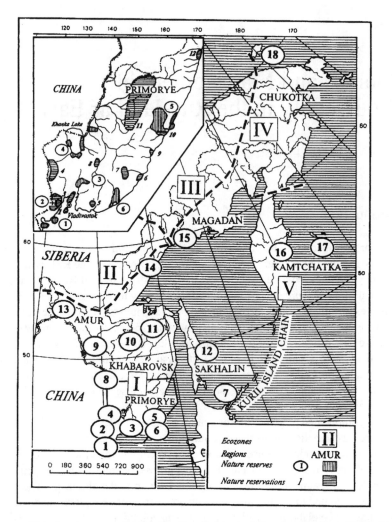

FIGURE 13.1. Landscape and administrative divisions of Russian Far East (RFE). *Eco-zones* (by Parmusin 1964): *I*, Amur-Primorye; *II*, East-Siberian; *III*, Verkhojansk-Kolyma; *IV*, Far North-East; *V*, Kamchatka-Kuril. *Nature Reserves* (in part by Sokolov and Syroechkovsky 1985): *1*, "Kedrovaya Pad"; *2*, Far East Sea; *3*, Ussuri; *4*, Khanka; *5*, Sikhote-Alin; *6*, Lazo; *7*, Kuril; *8*, Bolsche-Khekhtsir; *9*, Khingan; *10*, Buryea; *11*, Komso-mol; *12*, Poronaisk; *13*, Zeya; *14*, Jugjur; *15*, Magadan; *16*, Kronotsk; *17*, Komandor; *18*, Wrangel Island. *Nature Reservations* (by Long-Term Program 1993): *1*, group of islands in Great Peter's Bay; *2*, national park on the Murajev-Amurskii's peninsula; *3*, "Barsovy" reservation for Amur subspecies of the leopard; *4*, "Poltavskii" reservation for complex of oak forests and Khanka prairies; *5*, Partisank medicinal plants reservation; *6*, Vasilkova game refuge; *7*, "Beresovy" game refuge; *8*, "Tikhii" game refuge; *9*, "Black Rocks" reser-vation for *Nemorhaedus caudatus* Milne-Edwards; *10*, "Goral" reservation for *Nemorhae-dus caudatus* Milne-Edwards; *11*, nut-industrial zone; *12*, "Losiny" reservation for elk. *Scale*, kilometers. (See also symbol key on figure.)

jansk-Kolyma) belong mostly to Siberia. Some details of description of the first three ecozones are given in Table 13.1.

Concerning the remaining two ecozones, it is extremely important to emphasize that they have formed under the influence of the stable land of the East Siberian platform which first arose in the (Pre)Archean. The east part of the platform, and primarily its Aldan Upland, invade into the middle RFE territory, leaving north- and southward two stretched areas of Mesozoic plication (Fig. 13.2). The north-west part of the Mesozoic plication is lying on a rigid base of Paleozoic origin (the Kolyma massif) and includes a number of structures of more ancient origin (Chemekov 1975; Markov and Pushcharovsky 1980). In the south, the influence of the Siberian platform was restricted by the independent formation of a circum-oval superstructure called Amuriya (Solotov 1976).

Tectonic processes are not presently completed in Chukotka and Kamchatka, as well as in the Kuril Archipelago. All those elements, together with Sakhalin Island, are represented now by a Cenozoic plication of a different age. The primary chain of the Kuril Islands is composed of late Tertiary formations with many younger volcanic mountains, while a group called the Shikotan Islands, lying outside the main Kuril volcanic chain, is mostly composed of Cretaceous formations with flat-topped terraces (Hara 1959; Arkhipov and Nikolaev 1972).

Table 13.1. Specific characters of the Russian Far East (RFE) ecozones

Zones	Tectonics, Relief	Climate	Vegetation	Soils	Landscape
Far North-East	Main Cenozoic plication. Mountain-plain relief. Numerous signs of quite recent volcanism and mountain-valley glaciation. Present-day valley-corrie glaciation.	Marine arctic and sub-arctic climate. Pacific influence is restricted to the 50- to 250-km stretch westward. Monsoon climate is weakly pronounced.	Combined Yakut, Okhotsk, and Beringian floras. Poor species diversity because of northern location.	Primitive, gley soils of low fertility.	Herbaceous and bush tundra landscapes with patches of hardwood forests in flood plains. Larch open woodlands in the south. High bogs and sphagnum bogs. Wide distribution of mountain arctic deserts.
Amur-Primorye zone	Main Mesozoic plication. Rare signs of ancient volcanism and glaciation. Recent glaciation is absent.	Typical monsoon climate, affecting a stretch about 700 km from the Pacific coast. Relatively thin snow cover, dry spring, rainy summer.	Combined Manchurian, Okhotsk, Daurian, and Yakut floras. Species richness and diversity.	Podzolic, soddy podzolic, forest-brown, meadow-dark soils of high fertility.	Various landscapes of coniferous, broadleaved, and mixed forests. Far-East prairies, maris, and meadows. No mountain arctic deserts.
Kamchatka-Kuril zone	Recent late Cenozoic plication is still going on. Intensive present-day volcanism. Seismic activity. Recent glaciation of specific Kamchatka in northern areas.	Typical oceanic climate. Thick snow cover. Wet summer.	Combined American, Japanese, Kamchatka, and Okhotsk floras. Contrast of poor and rich species diversity resulting from volcanism and island location.	Soils composed of hummus and volcanogenic layers.	Landscapes of forest-tundras, open-stone birch forests in the north, and coniferous-broad-leaved forest in the south. Tall herbaceous vegetation with bamboo. Specific altitudinal zonality.

Modified from Parmusin (1964).

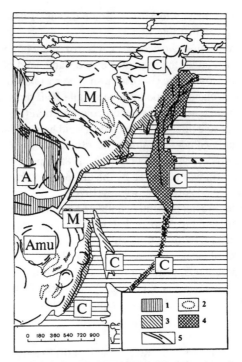

FIGURE 13.2. Geostructural elements of RFE. Simplified map based on data of Arkhipov and Nikolaev (1972); Chemekov (1975); Solotov (1976); Markhov and Pushcharovsky (1980). A, Siberia plate with Archean core; Amu, circumoval structure "Amuriya"; M, Mesozoic plication; C, Cenozoic plication; *1*, stable land; *2*, local ancient display; *3*, recent tectonic activity and late volcanism; *4*, present-day volcanism; *5*, main mountain ranges. *Scale*, kilometers. (See also symbol key on figure.)

Areas of Cenozoic plication and volcanogenic activity are located close to the Pacific Rim. Volcano remnants are found in Primorye, the Khabarovsk region, and Chukotka. More than 20 volcanoes of different activity grades were registered in Kamchatka and 35 in the Kuril Islands. Mud volcanoes are found on Sakhalin (Arkhipov and Nikolaev 1972; Markov and Pushcharovsky 1980).

The origin of the high tectonic and volcanic activity observed in the RFE is usually connected with the specific location of the territory at the junction of Pacific and continental geomorphological structures that have shown the greatest contact activity in mid-Paleozoic times (Krassilov 1977). Particularly large-scale tectonic activity has taken place in the southern part of the contact where the processes, starting in the Paleozoic, led in Mesozoic times to the formation of a gigantic, roughly oval structure, called Amuriya (Solotov 1976). The present Amuriya contour coincides with the outline of the Amur River Basin, i.e., embracing the southern RFE (Amur-Primorye ecozone) and a part of northern China.

Volcanism, magmatism, emergence and submergence, and other geological processes have changed the face of the initial Amuriya by means of the formation of new structures of lower categories: spheres, calders, ovals of depression, and various circular structures (Solotov 1976). These secondary processes, which are especially intensive in contact zones, have produced the freakish combinations of mountain chains in recent RFE mountain areas, particularly in Sikhote-Alin, the main mountain chain in Primorye. The combinations have resulted in a complicated pattern of slopes, valleys, and various secluded places widely varying in altitude, degree of protection from winds, rate and accumulation of precipitation, temperature, moisture, and so on. Such areas of differing isolation have served as the ecological basis for the appearance of various kinds of biological formations and different types of ecosystems and communities.

The RFE was divided into three floristic zones: arctic, boreal, and East-Asian (Komarov 1964). The arctic zone includes Wrangel Island and northern Chukotka; the boreal zone consists of the rest of Chukotka, Kamchatka, north Kuril, Magadan, north Sakhalin, north Khabarovsk, and the northwest regions of Amur, thus including 12 flora regions. The seven flora regions of the East-Asian zone correspond to south Kuril, south Sakhalin, main Amur, Khabarovsk, and Primorye administrative regions. These 20 flora regions join about 6 different forest formation complexes, about 40 forest ecosystem types, more than 200 meadow ecosystem types, and about 400 types of tundra and marsh ecosystems that can be distinguished today in the RFE (Elyakov 1993).

Some authors argue that there is a coincidence between the geological formation steps and those of organic evolution (Krassilov 1977). Indeed, such geological events as volcanic eruptions and oceanic invasions have had usually dramatic consequences for life in affected places, sometimes expressed by complete changes of vegetation and of animal and microbial populations (see examples in Wilson 1992). Primorye, particularly, underwent several extensive oceanic invasions in the late Tertiary, when the sea level was elevated by more than 150 m (Lindsberg 1965). Geological changes were of great importance for the formation of the junction between RFE and North America floras and fauna, those of Kamchatka and Kuril Island, and (not so trivially) between Arctic and East-Asian elements; the latter occurred when the narrow stretch of land near the Jugjur chain (currently located in the territory of the Jugjur Reserve; Fig. 13.1) was submerged. The junction of Sakhalin and some other present-day islands with the continent was periodically destroyed and restored (Arkhipov and Nikolaev 1972; Chemekov 1975; Markov and Pushcharovsky 1980).

As well as changes in the coastline, global climatic fluctuations have accompanied these transgressions (Krassilov 1977). The average temperature and humidity deviations in that time were quite distinct in various regions of the RFE. Dramatic climatic changes over the Pleistocene were accompanied by repeated shifts of northern range of vegetation zones as much as 8°–10° to the south and 700–800 m vertically. However, the forest type of vegetation persisted here for the whole period; in addition, the southern part of the RFE was never covered by solid ice (Arkhipov and Nikolaev 1972). Then, as a result of the ecological complexity developed there, many refuges were provided for local representatives to survive the harsh period. Many species of the Tertiary flora common in Primorye, the Caucuses, and North America (but not in Siberia, which was subjected entirely to glaciation) were retained in the southern part of the RFE: yew (*Taxus cuspidata*), hornbeam (*Carpinus cordata*), some maples (*Acer* spp.), ferns such as *Phyllitis japonica* and *Adiantum pedatum,* and various other elements of North America and RFE genera such as *Panax, Oplopanax,* and *Onoclea* (Krishtofovich 1958).

The well-known latitudinal gradients of species diversity (for references see Stevens 1992) can be readily observed among the RFE regions: there is a clear decline in species richness toward the higher latitudes. Thus, the flora and fauna of the northern Magadan region have only half the number of species of equivalent areas in Primorye; the number of vascular plant species in the southern Sakhalin area is 1167, while in Kamchatka it is only 828 (Komarov 1927–1930 and Sugawara 1937–1940, in Tolmachev 1959). More recent data of species richness in different RFE regions are presented in Tables 13.2 and 13. 3.

The idea of species richness in RFE is generally based on the widespread conviction that this territory is saturated by rare species. This conviction can be illustrated through the presentation of RFE species in the Russia Red List, where 25 species of RFE mammals (total number in Red List, 94), 39 species of birds (total, 80), and 131 species of vascular plants (total, 465) are enumerated, although the RFE occupies only about one-seventh of the total Russia territory (Sokolov and Syroechkovsky 1985). It is, however, rather difficult to confirm this idea more directly. The species numbers in different territories have not been studied in equal detail. Data for the nature reserves are more complete; however, the areas of some reserves differ from each other on the order of 10–100 fold. If the data are transformed according to the formula of MacArthur and Wilson (1967) ($S = c\,A^z$),

TABLE 13.2. Number of plant and animal species in some RFE nature reserves of different biogeographic affinities

Nature reserve	Area, km^2	Vascular plants	Mammals	Birds
Wrangel Island	7,670	310/33*	19/2*	40/4*
Kronotsk	19,990	800/68	33/3	179/15
Zeya	825	621/117	50/9	160/30
Khingan	822	782/148	44/8	304/57
Bolsche-Khekhtsir	449	755/164	35/8	191/42
Sikhote-Alin	3,470	>1000/130	61/8	340/44
Lazo	1,171	1212/209	48/8	286/49
Kuril	653	776/155	21/4	227/45
Far East Sea, land part	10	706/388	9/5	306/170
Ussuri	404	868/193	49/11	99/22
"Kedrovaya Pad"	179	862/233	57/15	250/68

*In denominator, * is number of species per (area)$^{1/4}$ according to MacArthur and Wilson (1967).
Sources: Bromley et al. (1977); Gorovoj (1972); Kharkevich and Buch (1994); Kostenko et al. (1989); Sokolov and Syroechkovsky (1985); Voronov (1963); V. Barkalov, personal communication, Kuril Reserve flora.

Table 13.3. Rare and endangered species distribution of RFE

Group of endangered species	Primorye	Sakhalin	Khabarovsk	Amur	Magadan	Kamtschatka
Mammals[a]	23	13	16	7	12	15
Fishes[a]	6	2	7	3	7	4
Reptiles[a]	3	4	1	0	0	0
Amphibians[a]	1	0	0	0	0	0
Birds[a]	92	62	56	43	41	34
Total animal species	125	81	80	53	60	53
Vascular plants[b]	149	115	75	27	36	29

[a]From Kostenko et al. (1989).
[b]From Kharkevich and Katchura (1981).

at z power of 0.25, the resulting figures clearly satisfy the latitudinal gradient concept (see Table 13.2). The extreme value of plant species richness in the Far East Sea Reserve (388 spp.; Table 13.2) reflects the habitat variability in the many islands and coast elements of the reserve.

The unique, specific feature of the RFE biota is more fully illustrated by the richness of rare species in different RFE regions, also revealing the latitudinal gradient and species abundance in regions located immediately on the Pacific coast (see Table 13.3). These data are much less variable because of the comparability of the administrative regions.

Hypotheses advanced to explain the biotic richness in the tropics (references in Stone 1993) apply here as well: precipitation levels, temperature, time of evolution, spatial heterogeneity of habitats, competition and predation, climatic stability, and local productivity all decline to the north. These factors must also have played an important or decisive role in forming the biodiversity of the RFE, particularly the three historical aspects of the problem: geological, biological, and anthropogenic. Each is important in the formation of biodiversity as such, as well as in the development of ecological complexity promoting local biodiversity (Walker 1992; Symbiosphere 1993). Investigation of the problem in local conditions, however, reveals many features specific for RFE regions, and the general northward decline of plant species richness will show, at closer examination, some disordering in species richness of selected plant genera (Table 13.4). For instance, the *Papaver* genus shows the greatest species abundance in Wrangel Island; *Potamogeton* species distribution reflects the altitudinal position of the territory and availability of still waters, and so on. Ecological preferences and biogeographic membership impact the species partitioning shown in Table 13.4.

Thus, geological processes are primarily responsible for the peculiar ecological complexity of the RFE, while the following geological and biological components also have made important contributions. For instance, niche differentiation increases the capacity of forests to serve as shelter and to accumulate litter and sediments proportional to the density of the stands; the thickening soil sheath gradually becomes more effective in sediment accumulation and moisture control;

TABLE 13.4. Species richness for principal plant genera in the different RFE regions

Genus	KPRe	Vl	UsRe	USM	SAM	RFE	SkKU	KnIS	KuM	Km	KrRe	ThRi	Wrls
Carex	35	47	47	24	33	261	121	60	45	50	57	36	23
Taraxacum	2	6	3	1	1	91	25	4	7	28	12	3	17
Polygonum	20	28	18	1	0	84	33	19	5	16	7	28	2
Saxifraga	1	2	1	—	9	68	15	4	9	27	15	15	17
Artemesia	11	18	15	4	4	67	26	9	6	18	7	5	7
Potentilla	9	18	7	2	5	66	14	10	6	17	8	9	21
Salix	11	13	12	7	10	60	56	8	11	17	18	15	14
Oxytropis	0	1	0	0	2	55	13	2	5	12	8	7	9
Poa	8	9	11	5	8	49	40	7	11	27	18	8	14
Viola	15	17	12	6	6	45	18	13	8	7	7	2	—
Pedicularis	2	2	1	0	7	37	16	3	9	14	10	7	10
Aconitum	7	7	6	3	4	37	12	1	1	4	12	1	1
Calamagrotis	3	9	4	2	9	35	19	7	5	10	11	4	5
Ranunculus	6	7	5	0	2	34	20	7	5	21	5	7	11
Juncus	3	8	6	5	3	34	25	16	7	16	9	5	4
Senecio	8	8	3	1	3	31	5	4	3	12	9	5	6
Draba	1	1	1	0	3	29	5	1	2	16	7	12	24
Stellaria	6	6	4	2	3	28	11	7	3	15	9	4	10
Potomogeton	0	3	0	0	0	26	11	9	0	10	11	0	0
Papaver	0	2	0	0	2	25	4	0	1	8	3	3	15
Luzula	2	3	2	2	7	19	12	4	10	2	13	6	3
Galium	5	5	9	6	2	18	5	8	3	6	4	1	0
Allium	4	10	5	0	6	17	7	3	3	3	3	—	1
Sedum	4	3	4	3	3	15	12	6	4	4	2	—	0
Agrostis	2	5	4	2	3	14	25	7	5	12	11	—	0
Epilobium	2	4	4	3	2	13	11	7	5	10	5	—	1
Geranium	8	9	6	0	2	13	4	4	1	1	1	—	0
Platanthera	2	2	0	0	0	11	10	8	4	4	4	—	0
Pyrola	3	3	5	1	1	8	8	8	3	5	4	—	1

Km, Kamchatka; KrRe, Kronostk Reserve; KPRe, Reserve "Kedrovaya Pad"; KnIs, Kunashir Island; KuM, Mountains of Kuril Islands Chain; RFE, total RFE; SAM, Mountains of Sikhote-Alin; SkKu, total Sakhalin and Kuril Islands; ThRi, 25-km^2 area of the middle Tchegitun River in Chukotka; UsM, mountains adjacent to the Ussuri River source; UsRe, Ussuri Reserve; Vl, Vladavostok (city, neighborhood, and adjacent national park); Wrls, Wrangel Islands; dash (—), data unavailable.
Data based in part on Alexeeva (1983); Azbukina and Kharkevich (1984); Barkalov (1988); Kharkevich (1978); Kharkevich and Buch (1994); Kharkevich and Cherepanov (1981); Petrovsky (1988); Tolmachev (1974); Vorobjev (1982); Voroshilov (1982); Vyshin (1990); V. Yakubov, personal communication, Kronotsk Reserve flora.

and progressive litter accumulation then produces the food and shelter required by different insects and microbes. The fitness of these niches for the introduction of new, especially rare, species can be postulated from the fact that most endemic plants and animals, including various epiphytes, lianes, and relict butterflies, now inhabit only the old and virgin forests (Kurentsov 1961; Kurentsova 1968).

Exchanges between adjacent floras and faunas also have played (and continue to play) a significant role in the creation of local species richness. In the RFE, especially in its southern part, the East-Asian flora meets those of the Arctic, Siberia, and Beringia (Kharkevich 1985). In addition, the northeast outline of Amuriya serves as a natural barrier for the distribution of the East-Asian and, in part, Manchurian floras. Many species of these floras find here the northern limits of their range. The so-called Mayabe line (Tatewaki 1933; cited by Hara 1959) and its prolongation toward the continent limit the southern movement of most boreal elements and vice versa. A significant increase in species richness is usually observed in the areas of overlapping floras.

The biota of the RFE suffers, however, from the periodic seasonal changes that produce significant ecological consequences. Plants and most animals must survive the cold period of the year in a state of dormancy, hibernation, or anabiosis; only a few species can migrate to warmer areas. Here, then, must be adaptations that are not present in the equatorial latitudes. Narrow niche differentiation can be observed here only in cases in which a species is provided with a stable, although minimal, resource that ultimately limits it under these conditions. Examples include *Gynostemma pentaphyllum, Dioscorea batatum,* and other plants that inhabit the environs of hot springs and extend, in such a way, to the northern border of their ranges. Among the animals that show similar kinds of specialization can be mentioned monophagous butterflies such as *Sericinus talamon,* because its only food plants are species of the endemic genus *Aristolochia* (Kurentsov 1961). A large group of mollusks that are restricted to pools with specific water conditions also can be cited here (Satravkin and Bogatov 1983; Bogatov, personal communication).

Most local species show neither narrow food specialization nor any other specialization or narrow habitat attachment. Thus, wild boars (*Sus scrofa*) are omnivorous and change habitats seasonally; the same is found for two bear species (*Ursus arctos* and *U. thibetanus*) and the badger (*Meles meles*); a large array of prey items characterizes insect-eating birds, owls, and so on (Bromley et al. 1977). Many plants that utilize a wide range of habitat diversity also demonstrate a high degree of flexibility; for instance, one of the main species that compose the RFE forests, Korean pine (*Pinus koraiensis*), conforms easily to a variety of humidity and temperature regimes, producing dry, wet, and damp kinds of stands (Kolesnikov 1956). Another valuable and wide-ranging tree species, larch (*Larix dahurica*), shows a high plasticity in the Sikhote-Alin and Sakhalin forests, forming several subspecies (Kolesnikov 1947; Tolmachev 1959). These adaptations result in a decline in the total ecosystem saturation by species; they may, however, produce great stability in the populations and species themselves.

The other consequence of seasonal periodicity is the involvement of migrating species in faunal composition and the impact of these species on the life of local communities as well as on the dynamics of widespread local species.

Present State of RFE Biodiversity

Many ecobiological processes, varying in their type of influence, and human activities have contributed to the present-day biotic richness in the RFE. Thus, the total biodiversity dynamics results from these factors:

- Natural and anthropogenically induced succession
- Natural and anthropogenically induced immigration of species from the adjoining flora and fauna

- Natural and anthropogenically induced movements and contacts of local indigenous species
- Speciation and extinction

Various expressions of succession have taken place in the most heavily forested territories of the RFE. In Primorye, for instance, less than 50% of the area remains in its undisturbed state (Elyakov 1993). In local conditions, the types of succession are accompanied only by transient or relative increase of species in the composition of communities. The pioneer species usually appear in areas of the greatest disturbance, in open or cleared areas (Komarova and Prochorenko 1993).

Biodiversity at all levels has been reduced drastically in the basins of the Amur River and Khanka Lake because of agricultural improvements such as swamp drainage, tillage of steppes, grazing intensity, and use of pesticides and herbicides. Such unique natural complexes as relict steppe, with *Ephedra monosperma,* pine–oak psammophyte complex, and others, have been destroyed (Kurentsova 1968). Some increase in species numbers has taken place because of the cultivation of new crops, establishment of weeds, and mass reproduction of pests and their predators and parasites. In some cases, as for example in Sakhalin (Tolmachev 1959), weeds and other exotics may compose up to 10% of the indigenous flora. More recent data of this topic were summarized by Kharkevich and Buch (1994) (Table 13.5).

Patterns of historical migration also continue to function today. The Kamchatka Peninsula and the Kuril Islands volcanic chain, particularly, compose one route that promoted in the remote past the Beringian migration of North American species such as dantonian gramineous plants (Probatova 1993). Other late and recent migrations of boreal and subarctic elements from Kamchatka, as well as East-Asian elements from Hokkaido, continue to add to the flora of these islands (Hara 1959; Barkalov 1980), while the enrichment of Kamchatka with new elements from the North is evidently complete (Kharkevich and Cherepanov 1981).

At least two distinct patterns of plant speciation in the RFE may be differentiated today. The first is represented by hybridization in zones of contact between different floras, resulting from natural migration, anthropogenic transfer, or elimination of physical or geographical barriers between species. Such was the surge of hybridization among the *Sasa* species induced by elimination of species isolation caused by forest disturbance in Sakhalin (Tolmachev 1959). A similar situation with gramineous plants in the continental RFE has been followed by repeated hybridization, the rise of ploidy level, and the composition of new complex genomes. These changes then allowed intraspecies and even intrasectional hybridization (Probatova 1993). The aggregates of species subjected to repetitive hybridization, for example, *Hierochloe aggr. glabra* and *H. aggr. odorata* in this case, are now very difficult to distinguish in their combined habitats.

The second pattern of speciation is linked to narrow adaptation, habitat differentiation, and endemism. There are many examples, including the numerous endemics of the *Oxytropis* genus, distributed among the islands of the RFE (Kharkevich and Katshura 1981), and the Pacific-related species richness of conifers (Tolmachev 1959; Japan Forest Technical Association 1964) and of *Poa*

TABLE 13.5. Indices of species abundance, taxonomic diversity, and flora adventization regarding some territories in the northern hemisphere

Territory	Area km^2	Number of families	Number of general	Number of species Total	Adventitious	Adventization, %
Latitudinal direction:						
Experimental						
station "Vostok"	4	101	344	581	60	10
Far East						
Sea Reserve	10	88	348	706	25	3.5
Experimental station						
"Ryazanovka"	400	114	388	794	69	9
Vladivostok	615	125	469	184	246	21
Lazo Reserve	1200	131	533	222	152	12.5
Kronotsk Reserve	9640	82	266	702	38	5.5
Verkhoturov and						
Karaginsky						
Islands	2000	66	215	525	13	2.5
Middle Tchegitun	25	39	105	230	—	0.0
Wrangel Island	8000	34	113	387	—	0.0
Longitudinal direction:						
Calbefleish						
Experimental						
station	37	82	232	380	127	33.0
Karadag Reserve	24	92	430	1023	22	0.2

From Kharkevich and Buch (1994), with permission.

(Probatova 1993) and others. The variability of speciation and adaptation also requires differentiation of approaches in wildlife management and monitoring.

Initial Wildlife Conservation Activity

The first nature reserve, Kedrovaya Pad, was set up in Primorye in 1913; however, exploitation of the surrounding taiga and other natural wealth was then based on the idea of excess resources in the region. Human activity was considered minor compared to natural events, and wildlife protection was seen simply as an act of aesthetics or humanism. Five more nature reserves, created in Primorye at different times, have had as goals the protection of a specific unique territory. The role of nature reserves in supporting local ecosystem function was not considered, and restoration activity was nothing more than natural forest recruitment embracing a limited territory and a limited number of valuable tree species, e.g., larch, spruce (*Picea*), and Korean pine.

Limits to this approach were revealed in time. Some nature reserves were shown to be too small to maintain resident predators such as the Siberian tiger (*Panthera tigris*), leopard (*Panthera pardus*), and other large migrating animals (Bromley et al. 1977; Pikunov and Korkishko 1992). The result of forest restoration is thus now difficult to estimate; however, it is known that such work, if devoid of appropriate genetic monitoring, can lead to a loss of genetic variability (Rabinowitz et al. 1986). Such genetic loss has been documented in Sakhalin populations of artificially reproduced salmon, where the absence of some DNA repetitive fragments peculiar to natural populations has been shown (Ginatulina 1992; Ginatulina and Ginatulin 1992). Similar outcomes may take place during attempts to restore rare plants in natural habitats. For instance, cultivated populations of *Panax ginseng* were subjected to prolonged selection for yield traits; thus their return to natural habitats failed to produce populations of necessary heterogeneity (Zhuravlev et al. 1996a, 1996b). On the other hand, outbreeding depression can lead a population to extinction by its hybridization with a form that is not closely related (Templeton 1986).

The examples cited here demonstrate clearly the need for and inevitability of careful wildlife management. Similarly, it is evident that rational use of natural resources is closely dependent on biodiversity conservation. Questions that still remain: what, how, and to what extent to conserve? The idea that all wildlife should be conserved is attractive but unattainable. Conserving only rare species also raises numerous objections. Conserving selected territories in natural reserves is insufficient, as just mentioned. Then, what kind of program of biodiversity management should be designed?

Toward a Functional Management Strategy

To be accepted, a management program has to serve regional economic interests while providing for conservation and restoration of local biodiversity (Salwasser 1990; Aplet et al. 1992; Daily and Ehrlich 1992; Redford 1992). Redford called for integration of exploitation and conservation activities through an increase in the amount of land allocated to multiple-use areas (in contrast to timbering only). On the other hand, Aplet et al. (1992) suggested "we need practitioners of traditional resource management to become conservation biologists." Younes (1993), as well, emphasized the importance of education oriented to resource managers.

It is difficult to achieve this kind of integration, especially in the current conditions of Russia, even though some joint scientific management initiatives have been taken; "Tentative regulations on the territory of traditional nature use..." and a "Long-term program of nature conservation..." were adoped. In Primorye, the management strategy needs to be differentiated more clearly so as to fit its ecology better. Still, the main structures for such differentiation already exist in Primorye in the form of many refuges that vary in size and degree of protection. Here we have six nature reserves, five terrestrial and one marine, composing together about 4.5% of the protected territory of Primorye (see Fig. 13.1). In addition, 1 republic-wide and 13 regional reservations have been established and about 250 areas have been declared to be nature memorials; only limited activity is allowed in urban and suburban areas, in water-protection areas, and in the forest nut-industrial zones (forest areas selected for Korean pine fruit collecting as the only kind of commodity-producing activity) (Elyakov 1993). A select territory of traditional nature use was identified in north Primorye, which is inhabited by Udegeis and Nanaians (Regional Act 1993).

These refuges now are in need of full integration. In contrast to well-known, large-scale wildlife protection systems, for example, the U.S. National Wildlife Refuge System (Curtin 1993), this integration with local use must be detailed and functional. This latter is absolutely essential because, in the course of degradation, species and ecosystems lose functions before they actually disappear. As one example, Korean pine forests would lose their ability to provide food and shelter for numerous vertebrate species much sooner than they will show the loss of the pine itself. However, the loss of function by Korean pine forests would lead inevitably to reduced biodiversity of species and systems associated with those forests.

Furthermore, as has been mentioned, many Far Eastern animals cannot be limited to one or a few ecosystems. For instance, the Himalayan bear (*Ursus thibetanus*), listed in the Russia Red List, spends the early spring on the warm slopes of the mountains, forages in summer on agricultural crops, scavenges for food in oak forests and later in Korean pine forests during the autumn, and, finally, migrates to a place of winter hibernation, often moving over many tens of kilometers. Regular seasonal migrations connected with the search for food and with escape from mosquitoes, deep snow, and frozen snowcrusts are characteristic of the elk (*Alces alces*), red deer (*Cervus elaphus*), and wild boar (Bromley et al. 1977; and personal observations of author).

Because complete protection in a way that provides for all these habitats would require protection of the greatest part of the entire region, it seems reasonable to limit the integration to connecting present and future refuges by protected belts, similar in part to so-called corridors as proposed by some authors (Harris 1984; Harrison 1992; Ehrenfeld 1993). This approach calls for additional research in big animal ecology, for corridor structure elaboration, and, inevitably, additional areas for protection of newly discovered sites of episodic concentration of migrants. Concentration of other species in places of proposed corridors also could be considered, because these large vertebrates are not the only animals to be protected. The other and possibly more important goal of these interconnecting belts must be to conserve in their territory the local plant genotypes and to produce a migration route for plant species (via the gene flow in cross-pollinated species, for instance). The conjunction of conservation goals for plants and animals would allow the integrated refuge system to become a throughway, a framework supporting the local biota functional state.

New information concerning local ecosystem functions is still needed. Among the most urgent questions, the following may be cited:

- Which formations and which ecosystems in the refuges can be utilized for harvest of some commodity or for recreation?
- How much and what kinds of commodity production are compatible with the main function of every specific system that is regarded as a refuge?

- What statistical approach can be used for evaluating needs for system recruitment and for commodity production [experience with such statistics was developed for long-term rotation in collection of herbal medicines in tropical rain forests by Balick and Mendelson (1992)]?
- How does one determine the degradation threshold below which systems would irreversibly lose their functions?

One can suppose that the necessary levels of biodiversity conservation might be reached as a dynamic equilibrium between resource exploitation and conservation/restoration activities. To reach this equilibrium before the point of irreversible destruction in the main functions of natural systems, we must proceed from ideas about species diversity conservation (or ecosystem diversity conservation) and the protection of selected areas from selected human activity to the concept of integrated management of natural resources, including nature conservation as a field of management. If biodiversity conservation is necessary for ecosystem stability, it follows that stable functioning of the primary local natural systems underlies the conservation of local biodiversity.

Such a shift in biodiversity conservation toward the conservation of system functions may produce the impression that we are moving toward the old, well-known ideas of nature conservation. It is not so, however, because here biodiversity remains a goal and a tool of conservation.

Conclusion

The three main aspects considered here—historical, ecological, and manmade influences—have contributed to the biodiversity now observed in the RFE. The establishing of species under changing ecological complexity and the two basic types of species formation continue to this day to create a unique picture. To preserve the RFE natural systems, a program for managing natural resources in the region should include distinct criteria reflecting the capacity (functions) of major formations in the region to retain the processes for maintaining regional biodiversity. Existing refuges with different protective status should be integrated into a single nature-protecting system by creating passages that connect the still-isolated units into a future integrated system. A close partnership among representatives of conservation science and those who actually are involved in implementing regional policies is essential for this problem to be successfully resolved.

Acknowledgments. I thank Professor Hiroya Kawanabe for sponsoring my visit to Kyoto, funded by the Japan Ministry of Science and Education. My thanks also to Professor Takuya Abe for kindly participating in the official arrangement of my visit. I thank Professor Theodore Pietsch and anonymous reviewers for their valuable contributions to improve the English in my manuscript.

Literature Cited

Alexeeva, L.M. 1983. Flora of Kunashir Island (in Russian). Far Eastern Branch of [Russian] Academy of Science Press, Vladivostok.

Aplet, G.H., R.D. Laven, and P.L. Fiedler. 1992. The relevance of conservation biology to natural resources management. Conservation Biology 6:289–300.

Arkhipov, S.A. and V.A. Nikolaev, eds. 1972. South of Far East. Special issue of "History of Relief Genesis in Siberia and Far East" (in Russian). Nauka Press, Moscow.

Azbukina, Z.M. and S.S. Kharkevich, eds. 1984. Flora of Ussuri Mountain Station (in Russian). Far Eastern Branch of [Russian] Academy of Science Press, Vladivostok.

Balick, M.J. and R. Mendelson. 1992. Assessing the economic value of traditional medicines from tropical rain forests. Conservation Biology 6:128–130.

Barkalov, V.Y. 1980. New species of vascular plants for the Kuril Islands (in Russian). Botanitcheskyi Zhurnal 65:1802–1808.

Barkalov, V.Y. 1988. High mountain vascular plants of Kuril Island. In: S.S. Kharkevich, ed. Vegetation of High Mountain Ecosystems in the USSR (in Russian), pp. 159–177. Far Eastern Branch of [Russian] Academy of Science Press, Vladivostok.

Bromley, G.F., N.G. Vasiljev, S.S. Kharkevich, and V.A. Nechaev. 1977. Flora and Fauna of Ussuri nature reserve (in Russian). Nauka Press, Moscow.

Chemekov, Y.F. 1975. West Okhotiya. (Special issue of "History of Relief Genesis in Siberia and Far East") (in Russian). Nauka Press, Moscow.

Curtin, C.G. 1993. The evolution of the U.S. National Wildlife Refuge system and the doctrine of compatibility. Conservation Biology 7:29–38.

Daily, G.C. and P.R. Ehrlich. 1992. Population, sustainability, and earth's carrying capacity. Bioscience 42:761–771.

Ehrenfeld, D. 1993. The making of conservation biology. Conservation Biology 7:743–745.

Elyakov, G.B., ed. 1993. Long-Term Program of Nature Conservation and Rational Use of Natural Resources in Primorye Region. (Ecological Program). Parts 1, 2 (in Russian). Dalnauka Press, Vladivostok.

Ginatulina, L.K. 1992. Genetic differentiation among chum salmon, *Oncorhynchus keta* (Walbaum), from Primorye and Sakhalin. Journal of Fish Biology 40:33–38.

Ginatulina, L.K. and A.A. Ginatulin. 1992. Studies of mitochondrial and repetitive nuclear DNA in salmonid fish. Nordic Journal of Freshwater Research 67:92–93.

Gorovoj, P.G., ed. 1972. Flora and Vegetation of "Kedrovaya Pad" Reserve (in Russian). Far Eastern Branch of [Russian] Academy of Science Press, Vladivostok.

Hara, H. 1959. An outline of the phytogeography of Japan. In: H. Hara and H. Kanai, eds. Distribution Maps of Flowering Plants in Japan, Fasc. 2, maps 101–200. Inoue, Morikawacho, Tokyo.

Harris, L. 1984. The Fragmented Forest. University of Chicago Press, Chicago.

Harrison, R.L. 1992. Toward a theory of inter-refuge corridor design. Conservation Biology 6:293–295.

Japan Forest Technical Association (Nihon Ringyo Gijutsu Kyokai). 1964. Illustrated Important Forest Trees of Japan, pp. 22–24. Chikyu Shuppan, Tokyo.

Kharkevich, S.S., ed. 1978. Flora and Vegetation of Ussuri Reserve (in Russian). Nauka Press, Moscow.

Kharkevich, S.S. ed. 1985. Vascular Plants of Soviet Far East. Vol. 1 (in Russian). Nauka Press, Leningrad.

Kharkevich, S.S. and T.G. Buch. 1994. The emerald necklace of the marine biological station "Vostok" (in Russian). Komarov's Lectures, XL. Dalnauka Press, Vladivostok.

Kharkevich, S.S. and S.K. Cherepanov, eds. 1981. Key of Vascular Plants of Kamchatka Region (in Russian). Nauka Press, Moscow.

Kharkevich, S.S. and N.N. Katchura. 1981. Rare Plants of Soviet Far East and Their Protection (in Russian). Nauka Press, Moscow.

Kolesnikov, B.P. 1947. Larch forests of middle-Amur plain (in Russian). Trudy DV Basy Akademii Nauk SSSR 1.

Kolesnikov, B.P. 1956. Korean pine forests in the Far East (in Russian). Trudy Dalnevostochnogo Filiala Sibirskogo Otdelenia Akadimii Nauk 2(4).

Komarov, V.L., ed. 1964. Flora of the SSSR. 1934–1964. Vol. 1–30 (in Russian). Academy of Sciences of the USSR Press, Moscow, Leningrad.

Komarova, T.A. and N.B. Prochorenko. 1993. Change of the age structure of the bush and woody liana cenopopulations during post-fire successions (in Russian). Botanitcheskyi Zhurnal 78:72–80.

Kostenko, V.A., P.A. Ler, V.A. Nechaev, and Y.V. Shibaev, eds. 1989. Rare Vertebrates of the Soviet Far East and Their Protection (in Russian). Nauka Press, Leningrad.

Krassilov, V.A. 1977. Geological history of the Pacific realm and paleontological investigations (in Russian). In: V.A. Krassilov, ed. Organic Evolution in the Circum-Pacific Belt, pp. 5–9. Academy of Sciences of the USSR, Far-Eastern Scientific Centre Press, Vladivostok.

Krishtofovich, A.N. 1958. Origin of Flora of Angarida. (Materialy po Istorii Flory i Rastitelnosty SSSR. III) (in Russian). Academia Nauk SSSR Press, Moscow.

Kurentsov, A.I. 1961. In the Refuges of Ussuri Relicts (in Russian). Primorye Press, Vladivostok.

Kurentsova, G.E. 1968. Vegetation of Primorye Region (in Russian). Far Eastern Press. Vladivostok.

Lindsberg, G.U. 1965. The great deluges (in Russian). Nauka i zhizn 8:124–128.

MacArthur, R.H. and E.O. Wilson. 1967. Theory of Island Biogeography. Princeton University Press, Princeton.

Markov, M.S. and Y.M. Pushcharovsky, eds. 1980. Tectonics of the Continental Margins of the North-West Pacific (in Russian). Nauka Press, Moscow.

Parmusin, Y.P. 1964. Physico-geographical division of the Far East. In: N.A Gwozdetski, ed. Physico-Geographical Division of the USSR (Siberia and Far East), pp. 130–233 (in Russian). Moscow University Press, Moscow.

Petrovsky, V.V. 1988. Vascular plants of Wrangel Island (flora conspect) (in Russian). Preprint, Magadan, Institute of Biological Problems of North. Far Eastern Branch of [Russian] Academy of Sciences Press, Vladivostok.

Pikunov, D.G. and V.G. Korkishko. 1992. Leopard of Far East (in Russian). Nauka Press. Moscow.

Probatova, N.S. 1993. Gramineous Plants of Russian Far East (in Russian). Thesis, Institute of Biology and Soil Sciene. Far Eastern Branch of [Russian] Academy of Sciences,. Vladivostok.

Rabinowitz, D., S. Cairns, and T. Dillon. 1986. Seven forms of rarity and their frequncy in the flora of the British Isles. In: M.E. Soule, ed. Conservation Biology: The Science of Scarcity and Diversity, pp. 182–204. Sinauer, Sunderland, MA.

Redford, K.H. 1992. The empty forest. Bioscience 42:412–422.

Regional Act. 1993. Tentative regulations in the territory of traditional nature use—territory which is populated with Primorye small peoples (in Russian). Regional Soviet of Peoples Deputies, Vladivostok.

Salwasser, H. 1990. Sustainability as a conservation paradigm. Conservation Biology 4:213–216.

Satravkin, M.N. and V.V. Bogatov. 1983. About a mollusc fauna in the middle Bureya River. In: Molluscs. Systematics, Ecology, and Distribution, pp. 145–146 (in Russian). Nauka Press, Leningrad.

Sokolov, V.E. and E.E. Syroechkovsky, eds. 1985. Nature Reserves of the Far East. Reserves of the USSR, Separate issue (in Russian). Mysl Press, Moscow.

Solotov, M.I. 1976. Core- and circular structures of Priamurje. In: L.M. Parfenov and S.M. Tilman, eds. Tectonics of the East Part of the Soviet Asia Academy of Sciences of the USSR, pp. 3–33 (in Russian). Far-Eastern Scientific Centre Press, Vladivostok.

Stevens, G.C. 1992. The elevational gradient in altitudinal range: an extension of Rapoport's latitudinal rule to altitude. American Naturalist 140:893–911.

Stone, D.E. 1993. Biodiversity in the tropics. Biology International (special issue) 29:37–47.

Symbiosphere: Ecological complexity for promoting biodiversity. An international cooperative research project. 1993. Biology International (special issue) 29:5–11.

Templeton, A.R. 1986. Coadaptation and outbreeding depression. In: M.E. Soule, ed. Conservation Biology: The Science of Scarcity and Diversity, pp. 105–116. Sinauer, Sunderland, MA.

Tolmachev, A.I. 1959. About the Sakhalin Island Flora (in Russian). Komarov's Lectures, XII. Academia Nauk Press, Moscow.

Tolmachev, A.I., ed. 1974. Key of Vascular Plants of Sakhalin and Kuril Islands (in Russian). Nauka Press, Leningrad.

Vorobjev, D.P. 1982. Key of Vascular Plants in the Vladivostok Neighbourhood (in Russian). Nauka Press, Leningrad.

Voronov, V.G. 1963. Mammals of Kuril Islands (in Russian). Nauka Press, Leningrad.

Voroshilov, V.N. 1982. Key of Plants of Soviet Far East (in Russian). Nauka Press, Moscow.

Vyshin, I.B. 1990. Vascular Plants of Mountains of Sikhote-Alin (in Russian). Far Eastern Branch of [Russian] Academy of Sciences Press, Vladivostok.

Walker, B.H. 1992. Biodiversity and ecological redundancy. Conservation Biology 6:18–23.

Wilson, E.O. 1992. The Diversity of Life. Harvard University Press, Cambridge.

Younes, T. 1993. Biodiversity research needs and opportunities, the role of international scientific organizations: the Diversitas example. In: International Symposium on Ecological Perspective of Biodiversity. Abstracts, p. 31. Center for Ecological Research, Kyoto University, Kyoto.

Zhuralev, Y.N., O.L. Burundukova, O.M. Koren, Y.A. Zaytseva, and E.V. Kovaleva. 1996a. *Panax ginseng* C.A. Meyer: biodiversity evaluation and conservation. In: W.G. Bailey et al., eds. Proceedings of the International Ginseng Conference, 1994. Vancouver, British Columbia.

Zhuralev, Y.N., M.M. Kozyrenko, E.V. Artyukova, G.D. Rounova, T.I. Muzarok, and G.B. Elyakov. 1996b. Typing Ginseng by Means of RAPD-PCR (in Russian). Doklady RAN. Proceedings of the Russian Academy of Sciences.

14

Singapore: A Case-Study for Tropical Rain Forest Fragmentation and Biodiversity Loss

IAN M. TURNER

Introduction

Forests throughout the tropics are being cleared at a rate that is causing considerable alarm to many people concerned with the fate of the biological diversity of
our planet. The common pattern of human influence on tracts of tropical rain forest is a massive reduction of the forest area and a fragmentation of the habitat into
small remnant patches. Tropical rain forests are the most species-rich ecosystems
on Earth. Therefore, it would seem very likely that tropical rain forest fragmentation will lead to local and ultimately global extinction of species. Alarming predictions of massive levels of extinction from tropical deforestation have been
made (e.g., Myers 1988; Raven 1988). However, a relatively much smaller number of extinctions have actually been documented (Simon 1986; Heywood and
Stuart 1992; Smith et al. 1993), and there remains a need to clarify the influences
of fragmentation on tropical rain forest biodiversity.

Singapore is a small state in the humid tropics that has undergone massive
deforestation and habitat fragmentation and may provide a good model to study
these processes because it has a relatively well-known flora and fauna. In this
chapter I review what is known about the changes in biodiversity in Singapore
during the past 150 years and relate it to findings from other tropical areas.

Singapore: A Highly Fragmented Tropical Landscape

The Republic of Singapore (103°50'E 1°20'N) is a small state in the aseasonal
tropics of South-East Asia. Until about 1820, its 620 km^2 were covered almost
entirely with uninterrupted, highly diverse, lowland tropical rain forest. Soon after
this date, however, large-scale deforestation began, mostly as a result of clearance
for agriculture, and by the end of the nineteenth century primary forest cover
probably occupied no more than 5% of the total land area (Corlett 1991a). Singapore's early agricultural boom did not maintain momentum, and large areas of
cultivation were abandoned; thus, secondary forest has developed on many sites,
currently covering about 5% of the Republic. However, the twentieth century has

seen further declines in the old-growth forest until today barely 200 ha of forest, no more than 0.2% of the land area, could be described as primary (Corlett 1992a). The primary forest is represented by about 50 ha within Bukit Timah Nature Reserve and a number of much smaller fragments, most of which are scattered through the secondary forest of the Central Catchment Area.

Not unexpectedly, this massive level of disturbance and destruction of the natural vegetation has had a major impact on the native flora and fauna of Singapore. Fortunately, from a scientific viewpoint, Singapore has had active resident field biologists since its founding, and their records of the biota, while not perfect, provide rare data on the historical changes in a tropical flora and fauna. Using such information, it has been possible to document a large number of extinctions from Singapore. Singapore is separated from the Malay Peninsula by straits only 600 m wide at their narrowest; therefore, it is not surprising that there is little or no evidence of endemism in the Singapore biota, and thankfully the Singapore extinctions represent local, not global, species losses.

It has been suggested that 594 of 2277 native vascular plants have become extinct in Singapore, 478 of these being forest species (Turner et al. 1994). The epiphytic orchids have been the group in the flora worst hit, losing 132 of the 142 species recorded (Turner et al. 1994). The records are less good for the fauna, and only certain groups of the vertebrates can be accurately assessed for species loss. Lim and Ng (1990) have reported the disappearance of 23 of the 52 native freshwater fish species. The Singapore avifauna has suffered from 106 extinctions in a total of 383 species; 87 of these are forest birds (Hails 1987). The losses include all the hornbills, trogons, all but one barbet, and most of the woodpeckers and babblers (Corlett 1992b).

About one-third of the nonvolant, nonmarine mammals have been exterminated (Yang et al. 1990); these include all the large mammals such as the tiger (*Panthera tigris corbetti*), leopard (*Panthera pardus fusca*), sambur (*Cervus unicolor equinus*), and barking deer (*Muntiacus muntjak peninsulae*). Singapore therefore has probably lost more than a quarter of what could be called its "terrestrial macrobiota." It would seem likely that species of invertebrates have also been lost, but there is no reliable evidence for this. Not only has this substantial proportion of the native biota been eradicated, but many other species are currently severely threatened with extinction (Ng and Wee 1994). Their populations are reduced to very low numbers, and in many cases there is a very great chance of extinction in the near future.

Mechanisms of Species Loss

Evidence is mounting to show that patches of tropical rain forest show considerable losses in biodiversity over time if they become isolated from continuous forest (Willis 1979; Karr 1982; Lovejoy et al. 1986; Newmark 1991; Bierregaard et al. 1992; Leigh et al. 1993). A small fragment will support a smaller population of a given species than a larger fragment. As a fragment becomes very small, populations will drop below viable levels and extinction will ensue. Tiny relict patches may contain "ecologically extinct" populations of species doomed because of their small size. Small populations may also be more liable to size fluctuations, which will inevitably include local extinctions.

Some of the extinctions that have occurred in Singapore can be clearly ascribed to certain causes, others are more a matter for speculation. A number of the large vertebrates were probably hunted to extinction, e.g., the tiger, leopard, deer, and hornbills (Corlett 1992b). Collecting for the horticultural trade may have hastened the decline of *Cycas rumphii* (Turner et al. 1994). However, most of the extinctions were probably the result of habitat loss rather than direct human pressure on population levels. Some species may have been severely affected because all their habitat was cleared or at least heavily disturbed. For example, the mangrove and estuarine trees of Singapore had a rich epiphytic orchid flora (32 species) that has been entirely eradicated because all the old-growth stands have been cleared from Singapore's coastline (Turner et al. 1994).

There is evidence that forest fragments in Singapore have suffered reductions in diversity. The Singapore Botanic Gardens' Jungle is a 4-ha remnant of primary forest located within the Botanic Gardens. Preliminary results of a survey (I.M. Turner and H.T.W. Tan, unpublished data) of the Jungle indicated that about 60% of the native plant species which have been recorded historically (as indicated by the presence of specimens in the Botanic Gardens' Herbarium) are now extinct.

An interesting observation from Singapore is how long small populations manage to hold out and how even tiny fragments deteriorate very slowly. The 4-ha Botanic Gardens' Jungle has lost a lot of plant species, yet it still has a flora of more than 200 native species including many primary forest plants. Much of the 50 ha of the Bukit Timah Nature Reserve is superficially indistinguishable from extensive primary forest, yet both fragments have been excised from continuous forest for well over 100 years. What is clear is that diversity remains very high only in the primary forest fragments; the secondary forests are depauperate even after a century of succession (Turner et al. 1994). The extensive secondary forest stands of the Central Catchment Area are made up of about 40 common tree species (Corlett 1991b), which is species rich in comparison to most temperate forests but still substantially less diverse than the primary forest fragments.

Edge Effects

The edges of primary forest fragments can represent major gradients of environmental change (Murcia 1995). The interior of the rain forest is relatively dark, cool, and humid, but outside the forest the atmosphere may be bright, hot, dry, and quite turbulent. Relative increases in photosynthetically active radiation (Williams-Linera 1990; MacDougall and Kellman 1993), and air temperature and decreases in relative humidity have been observed at the edges of tropical forests compared to their interiors (Kapos 1989). Edge effects on the microclimate have been demonstrated up to 40 m in from the forest boundary (Kapos 1989). These edge effects in the physical environment may have a direct influence on the forest community. Greater tree mortality and a greater rate of disturbance near forest edges have been documented for tropical forest fragments (Lovejoy et al. 1986; Laurance 1991; Leigh et al. 1993).

The edges of primary forest fragments in Singapore appear to be relatively stable, and if there is a greater rate of tree mortality near the periphery of fragments it cannot be much higher than normal as there is no major observable difference in forest structure or composition near fragment boundaries. Surprisingly, the very small true fragment of the Gardens' Jungle has kept its edges intact very well. By chance or design, the policy of planting *Dracaena fragrans* was hit upon and seems to have had a positive effect by producing a green wall around the edge that is maintaining a shaded microclimate close to the forest margin. The absence of strong edge effects in Singapore may reflect the island's particularly equable climate. Strong winds are rare, and rainfall is very reliable with no marked dry season. Drying of the forest's internal microclimate might happen as a result of fragmentation, and it could have an effect on the ecosystem. The reduction in epiphyte diversity in Singapore is a possible consequence, although the shade-tolerant epiphytic pteridophytes have suffered a significantly lower degree of extinction than the often-succulent epiphytic orchids (Turner et al. 1994), possibly indicating that drying of the forest microclimate has not been a factor of great importance in Singapore.

Second-Order Effects

The loss of a certain species within a fragment will inevitable influence those species that interacted with it. Such losses may have major repercussions, producing what have been termed second-order effects. Little, if any, adequate research has been conducted in the field, but a number of anecdotal cases point to fragmentation-related changes in community structure as being of great importance and interest (Terborgh 1992). No systematic observations on second-order effects have been made in Singapore, but it is possible to speculate about their presence on the basis of casual inspection of the various fragments.

There has been considerable emphasis in the tropical forest literature on the close links between animal and plant species in the forests, particularly in pollen and seed dispersal interactions (Bond 1994). This has led a number of authors to postulate that elimination of certain key animal groups from forests will initiate a cascade of further extinctions as plant species no longer have the pollinators and seed dispersers they require to complete their life cycles. Observations in Singapore show that a probable reduction in diversity of pollinators and a definite loss of particular seed dispersers has not had a clear or rapid effect on tree regeneration. A few tree species are suffering from a failure of dispersal. Several of the large-seeded Myristicaceae and Leguminosae drop their entire fruit crop beneath the parent tree (I.M. Turner, personal observation; D.J. Metcalfe, personal communication), probably because of the absence of the usual disperser animal such as hornbills. However, the continued presence of troops of long-tailed macaques

(*Macaca fascicularis fascicularis*), which can act as dispersers for a wide range of plant species at Bukit Timah (Corlett and Lucas 1990), must have helped continue dispersal activities after fragmentation. Bukit Timah does not give the impression of being strongly degraded or in imminent danger of collapse of biodiversity after at least a century of isolation. It seems likely that only a small proportion of the flora is so specialized in its pollinators or seed dispersers that it does not have some animal species still present to do the job after fragmentation.

An increase in seed predator numbers may be having a more profound effect. The elimination of the larger predatory vertebrates in small forest fragments may allow numbers of certain small mammals to increase dramatically. The density of squirrels (*Callosciurus notatus singapurensis* and *Sundasciurus tenuis tenuis*) in the Botanic Gardens' Jungle appears to be high (reaching pestilential proportions in the gardens) and these, with parrots and other small mammals, achieve high seed consumption rates. This is probably the cause of the failure of dipterocarp regeneration in the Botanic Garden's Jungle. A fruiting event in 1990 among the dipterocarps failed to establish any seedlings despite a high viability rate in the seeds produced. Carnivore disappearance has allowed seed predators to become extremely efficient at destroying seed crops in the Jungle. A few species of *Calophyllum* (particularly *Calophyllum ferrugineum*) have been the major exception to this observation, and now considerable areas of the Jungle consist of dense stands of *Calophyllum* poles; possibly *Calophyllum* fruits are distasteful to squirrels. A detailed inventory of a 2-ha forest plot at Bukit Timah also shows a few species to be very common, with a smaller number of rare species than might be expected (J.V. La Frankie, personal communication).

Larger fragments may support populations of medium-sized predators, possibly at densities elevated from those in continuous forest because of the absence of the largest species. This so-called mesopredator release has been implicated in reduced diversity of birds and small mammals in fragments (Loiselle and Hoppes 1983; da Fonseca and Robinson 1990; Sieving 1992; Laurance et al. 1993), particularly via predation from birds' nests. The major loss of forest birds from Singapore might be related to increased predation, although intolerance of frequent human disturbance may also be to blame.

Immigration of Exotics

Tropical rain forest fragments may be more susceptible to invasion by alien species than continuous forest, largely because such species are frequently more tolerant of the conditions of the exposed areas outside the fragments than are the native forest species. Continuous forest areas may be resistant to invasion because conditions may not favor establishment of the aliens, and much of the forest will be outside the range of dispersal of those species. However, small fragments are in range, and the strong tide of alien invaders may overwhelm the native community (Janzen 1983; Simberloff 1992). The greater frequency of disturbance around fragments may also favor the invaders (Laurance 1991).

One of the clear reasons for the loss of plant diversity in the Botanic Gardens' Jungle has been a reduction in the rate of regeneration of many of the native tree species, particularly the large dipterocarps. The most obvious cause of the prevention of regeneration has been the rampant growth of a number of exotic herbs and climbers that have smothered or crowded out tree regeneration (Turner and Tan 1992). These species include *Thaumatococcus daniellii, Costus lucanusianus, Dioscorea sansibarensis, Tanaecium jaroba, Thunbergia grandiflora,* and the native *Smilax setosa.* The alien species were either deliberate or accidental introductions of ornamental or potentially economically valuable species. Alien invasion is not very apparent at Bukit Timah Nature Reserve, the largest remaining fragment of primary forest in Singapore with a core area of about 50 ha. Alien plant invasion is restricted to the occurrence of the melastome herb *Clidemia hirta* and the fern *Adiantum latifolium* along some of the pathsides within the forest. Neither appear to be having a major impact on the forest, which supports Whitmore's (1991) hypothesis that tropical rain forest is generally impervious to exotic invasion unless the forest is very disturbed or fragmented.

Many exotic animals have also become naturalized in Singapore (Corlett 1992b), but as with the plants, they are nearly all absent from undisturbed forest but often dominate the manmade habitats that cover much of Singapore. Their presence may be preventing the native species expanding into sites outside the remnant forest fragments.

Ecosystem Resilience

Observations from fragmented, defaunated lowland rain forest in Singapore show that loss of plant diversity does occur but is slow to happen, presumably because most rain forest plants are long lived. Considerable defaunation does not appear to have had a rapid or direct affect on plant community structure unless the fragments are very small and isolated. Similarly, alien plants are only a threat to small fragments. In an equable climate, edge effects appear to be less severe than those reported from other tropical areas. The rarity of high-intensity disturbance in Singapore may mean that the forest ecosystem is less resilient to disturbance than those of more frequently and severely disturbed areas. Certainly the increase in diversity with time in secondary forest areas, even when contiguous with primary forest fragments, is slow in Singapore, indicating a general inability of the plant species to spread rapidly into uncolonized areas.

Puerto Rico has suffered from very major deforestation of a similar magnitude to Singapore, also mostly occurring in the nineteenth century, yet very little of its native flora has become extinct (Lugo 1988), and less than 12% of its avifauna (Brash 1987). The plant species seem to have rapidly colonized secondary stands. In Singapore, extinction has been more dramatic for a similar degree of deforestation. Could it be that the forest community of Puerto Rico, which is subject to hurricanes crossing the Caribbean, is inherently more resilient to major disturbance than the lowland dipterocarp forest of environmentally equable Singapore? Alternative explanations for the disparity in levels of extinction between Singapore and Puerto Rico, such as differences in degree of human impact on fragments or absolute fragment sizes, might be valid and require more detailed analysis to be confirmed. The comparison, however, highlights the possibility that tropical forest communities may vary in resilience to fragmentation.

Conclusions

Tropical deforestation and its corollary of habitat fragmentation are the major threat to global terrestrial biodiversity, although incontrovertible scientific evidence for fragmentation-related local extinction is surprisingly scarce. For plants, at least, postfragmentation relaxation of species numbers is a slow process even in fragments of just a few hectares (Turner et al. 1994). This indicates that small fragments may be of considerable conservation value as final refuges for species (Magsalay et al. 1995). The continued existence of tiny populations in such sites will allow time for conservation strategies and management options to be formulated, aimed at preventing global extinction and erosion of genetic diversity within the species concerned. Large vertebrates vanish from fragments much sooner because of their requirement for more extensive habitat, their lower population densities, and a higher frequency of direct human exploitation. Singapore provides little evidence to support prophecies of rapidly increasing spirals of extinction in fragments resulting from the elimination of key partners in interspecific mutualisms.

The local natural disturbance regime is likely to influence the inherent resistance and resilience of the endemic forest community. The fragmentation process is the result of major disturbance; a community that has evolved under an intensive natural disturbance regime may be less affected, or recover more quickly from the fragmentation process. On the other hand, where natural disturbance is frequent and intense, fragment edges may be more susceptible to postisolation contraction.

It is realistic to expect that in many areas of the humid tropics the rain forest is, or soon will be, present only as relict fragments. There is an urgent need to know how viable these ecosystem fragments are in the long term, particularly with respect to the levels of biodiversity they will support. Observations from Singapore indicate that major losses of biodiversity are almost inevitable, but even so fragments are more species rich than secondary forest and undoubtedly merit conserving.

Acknowledgments. I am grateful to Daniel Metcalfe for discussions about forest fragments in Singapore and to Richard Corlett for suggesting a number of improvements to earlier drafts of the paper.

Literature Cited

Bierregaard, R.O., T.E. Lovejoy, V. Kapos, A.A. dos Santos, and R.W. Hutchings. 1992. The biological dynamics of tropical rainforest fragments. Bioscience 42:859–866.

Bond, W.J. 1994. Do mutualisms matter? Assessing the impact of pollinator and disperser disruption on plant extinction. Philosophical Transactions of the Royal Society, Series B 344:83–90.

Brash, A.R. 1987. The history of avian extinction and forest conversion on Puerto Rico. Biological Conservation 39:97–111.

Corlett, R.T. 1991a. Singapore. In: N.M., Collins, J.A. Sayer and T.C. Whitmore eds. The Conservation atlas of Tropical Forests: Asia and the Pacific, p. 211–215. Macmillan, London.

Corlett, R.T. 1991b. Plant succession on degraded land in Singapore. Journal of Tropical Forest Science 4:151–161.

Corlett, R.T. 1992a. The angiosperm flora of Singapore 1. Introduction. Gardens' Bulletin, Singapore 44:3–21.

Corlett, R.T. 1992b. The ecological transformation of Singapore, 1819–1990. Journal of Biogeography 19:411–420.

Corlett, R.T. and P.W. Lucas. 1990. Alternative seed-handling strategies in primates: seed-spitting by long-tailed macaques (*Macaca fascicularis*). Oecologia 82:166–171.

da Fonseca, G.A.B. and J.G. Robinson. 1990. Forest size and structure: competitive and predatory effects on small mammal communities. Biological Conservation 53:265–294.

Hails, C. 1987. Birds of Singapore. Times Editions, Singapore.

Heywood, V.H. and S.N. Stuart. 1992. Species extinctions in tropical forests. In: T.C. Whitmore and J.A. Sayer, eds. Tropical Deforestation and Species Extinction, pp. 91–117. Chapman & Hall, London.

Janzen, D.H. 1983. No park is an island: increase in interference from outside as park size decreases. Oikos 41:402–410.

Kapos, V. 1989. Effects of isolation on the water status of forest patches in the Brazilian Amazon. Journal of Tropical Ecology 5:173–185.

Karr, J.R. 1982. Avian extinction on Barro Colorado Island, Panama: a reassessment. American Naturalist 119:220–239.

Laurance, W.F. 1991. Edge effects in tropical forest fragments: application of a model for the design of nature reserves. Biological Conservation 57:205–219.

Laurance, W.F., J. Garesche, and C.W. Payne. 1993. Avian nest predation in modified and natural habitats in tropical Queensland: an experimental study. Wildlife Research 20:711–723.

Leigh, E.G., S.J. Wright, E.A. Herre, and F.E. Putz. 1993. The decline of tree diversity on newly isolated tropical islands: a test of a null hypothesis and some implications. Evolutionary Ecology 7:76–102.

Lim, K.K.P. and P.K.L. Ng. 1990. A Guide to the Freshwater Fishes of Singapore. Singapore Science Centre, Singapore.

Loiselle, B.A. and W.G. Hoppes. 1983. Nest predation in insular and mainland lowland rainforest in Panama. Condor 85:93–95.

Lovejoy, T.E., R.O. Bierregaard, A.B. Rylands, J.R. Malcolm, C.E. Quintela, L.H. Harper, K.S. Brown, A.H. Powell, G.V.N. Powell, H.O.R. Schubart, and M.B. Hays. 1986. Edge and other effects of isolation on Amazon forest fragments. In: M.E. Soulé, ed. Conservation Biology: The Science of Scarcity and Diversity, pp. 257–285. Sinauer, Sunderland, MA.

Lugo, A.E. 1988. Estimating reductions in the diversity of tropical forest species. In: E.O. Wilson, ed. Biodiversity, pp. 58–70. National Academy Press, Washington, DC.

MacDougall, A. and M. Kellman. 1992. The understorey light regime and patterns of tree seedlings in tropical riparian forest patches. Journal of Biogeography 19:667–675.

Magsalay, P., T. Brooks, G. Dutson, and R. Timmins. 1995. Extinction and conservation on Cebu. Nature (London) 373:294.

Murcia, C. 1995. Edge effects in fragmented forests: implications for conservation. Trends in Ecology and Evolution 10:58–62.

Myers, N. 1988. Tropical forests and their species: Going, going…? In: E.O. Wilson ed. Biodiversity, pp. 28–35. National Academy Press, Washington, DC.

Newmark, W.D. 1991. Tropical forest fragmentation and the local extinction of understorey birds in the Eastern Usambara Mountains, Tanzania. Conservation Biology 5:67–78.

Ng, P.K.L. and Y.C. Wee. 1994. The Singapore Red Data Book. The Nature Society (Singapore), Singapore.

Raven, P.H. 1988. Our diminishing tropical forests. In: E.O. Wilson ed. Biodiversity, pp. 119–122. National Academy Press, Washington, DC.

Sieving, K.E. 1992. Nest predation and differential insular extinction among selected forest birds of Central Panama. Ecology 73: 2310–2328.

Simberloff, D. 1992. Species-area relationships, fragmentation, and extinction in tropical forests. Malayan Nature Journal 45:398–413.

Simon, J.L. 1986. Disappearing species, deforestation and data. New Scientist 110:60–63.

Smith, F.D.M., R.M. May, R. Pellew, T.H. Johnson, and K.R. Walter. 1993. How much do we know about the current extinction rate? Trends in Ecology & Evolution 8:375–378.

Terborgh, J. 1992. Maintenance of diversity in tropical forests. Biotropica 24:283–292.

Turner, I.M. and H.T.W. Tan. 1992. Ecological impact of alien plant species in Singapore. Pacific Science 46:389–390.

Turner, I.M., H.T.W. Tan, Y.C. Wee, bin Ibrahim Ali, P.T. Chew, and R.T. Corlett. 1994. A study of plant species extinction in Singapore: lessons for the conservation of tropical biodiversity. Conservation Biology 8:705–712.

Whitmore, T.C. 1991. Invasive woody plants in perhumid tropical climates. In: P.S. Ramakrishnan, ed. Ecology of Biological Invasion in the Tropics, pp. 35–40. International Scientific Publications, New Delhi.

Williams-Linera, G. 1990. Vegetation structure and environmental conditions of forest edges in Panama. Journal of Ecology 78:356–373.

Willis, E.O. 1979. The composition of avian communities in remanescent woodlots in Southern Brazil. Papéis Avulsos de Zoologia (Sao Paulo) 33:1–25.

Yang, C.M., K. Yong, and K.K.P. Lim. 1990. Wild mammals of Singapore. In: L.M. Chou and P.K.L. Ng, eds. Essays in Zoology, pp. 32–41. Zoology Department, National University of Singapore, Singapore.

15

Management of Biodiversity in Aquatic Ecosystems: Dynamic Aspects of Habitat Complexity in Stream Ecosystems

YASUHIRO TAKEMON

Introduction: Spatial Scale and Nonequilibrium Concept in Stream Community Management

Biodiversity in a particular area is influenced by several processes on different spatiotemporal scales, for example, biogeographic processes on a continental scale, regional (climatic and geological) processes, and local processes governed by local biotic and abiotic factors (Schluter and Ricklefs 1993). Species diversity in one locality is an outcome of these processes, which act on different time scales. Although ecologists may study each process separately, understanding of a single process is not sufficient for biodiversity management. To restore faunas and floras in artificially altered environments, for example, geographic and regional characteristics of biodiversity in the area concerned should be taken into account and brought to bear on management plans.

Local processes of communities are a part of mesoscale conditions in both abiotic and biotic aspects (Holt 1993; Fisher 1994). Stream ecosystems represent a convenient model for the concept because they possess a hierarchical structure composed of a drainage basin, floodplain, riffle-pool unit or reach, and microhabitats (Frissell et al. 1986; Church 1992). Local and mesoscale processes of stream communities may correspond to riffle-pool units and to drainage basins, respectively.

Relationships between local and mesoscale processes in stream communities have been considered in the river continuum concept (Vannote et al. 1980; Minshall et al. 1992; Chergui and Pattee 1993). This concept emphasizes functional relationships of communities downstream to those upstream based on the successional changes in functional guilds and resource conditions along the gradient of physical conditions. The gradient of functional guilds is not always found (Statzner and Higler 1985), but longitudinally defined communities generally occur with characteristic patterns of species replacement (Perry and Schaeffer 1987; Tanida and Takemon 1993). This suggests that any biodiversity management cannot ignore mesoscale patterns, even though functional relationships within the system are not fully known.

The controversy surrounding the river continuum concept notwithstanding, maintenance mechanisms of local river communities are too poorly known to be useful for management purposes. The gap between earlier ecological theories of community structure and the knowledge required for environmental management may partly stem from the paradigm of equilibrium theory in which the role of interspecific interactions is emphasized to explain community structure (Reice 1994). Although biological interactions such as competition and predation may be vital processes for structuring communities under equilibrium conditions, their importance will decrease under nonequilibrium conditions. Several empirical data on community structures in patchy aquatic habitats have pointed to the importance of random processes of patch formation by populations rather than biological interactions (Tokeshi and Townsend 1987; Schmid 1992; Tokeshi 1994). Because stream ecosystems are under the impact of frequent disturbances of various kinds and intensity, such as spates and droughts, physical conditions creating environmental patches for settlement by populations may be more important for maintaining high species diversity (Reice 1994).

Previous theories on the roles of disturbance have also assumed equilibrium conditions before disturbance: for example, the habitat templet concept (Southwood 1977), intermediate disturbance hypothesis (Connell 1978), dynamic equilibrium model (Huston 1979), and bimodal model (Menge and Sutherland 1987). Because of the emphasis on biological interactions, these theories seem to miss another role of disturbance, that is, promoting habitat alteration and rearrangement (Sousa 1984; White and Pickett 1985). In stream ecosystems, apart from resetting competitive interactions by removing organisms (McAuliffe 1983; Resh et al. 1988), disturbance often creates spatial and temporal heterogeneity, which is important for maintaining species diversity (Denslow 1985; Doeg et al. 1989; Lake et al. 1989). This role of disturbance may be more important under nonequilibrium conditions, but even in equilibrium conditions disturbance may interact with biological processes in natural communities (Sousa 1985; Lake 1990). Therefore, disturbance as an agent of creating and conditioning environmental patches for settlement by populations should be given more emphasis in community ecology and in biodiversity management.

In this chapter, the importance of local habitat diversity and mechanisms for maintaining diversity are explored using examples of stream geomorphology and longitudinal patterns of animal communities in Japanese streams. I briefly review the relationships of faunas and the structures of riffle-pools, bars, and hyporheic zones in the middle reaches of streams. Following this, 11 types of microhabitat and their spatial arrangement within a riffle-pool unit are described on the basis of preliminary results derived from mapping surveys of organisms and their habitats in a Japanese mountain stream. Consideration is given to functional roles of stream geomorphology in the life histories and community structures of stream organisms.

Relationships between stream geomorphology and longitudinal patterns of stream faunas are then considered, with particular reference to a mayfly species that shifted its longitudinal distribution after dam construction. Mechanisms for maintaining longitudinal zonation are discussed in relationship to habitat structures produced by sedimentation processes.

Habitat Structure Within a Reach

Riffle-Pool Structure As a Habitat Unit

Many studies on stream faunas of the upper and middle reaches have shown that a riffle-pool combination is a structural unit of habitat (Kani 1944, 1981; Kawanabe et al. 1956; Mizuno and Gose 1972; Rabeni and Jacobson 1993), which is important for stream biodiversity on local scales.

Most stream animals need a set of different habitats as their requirements change with stages of their life cycle (Yuma and Hori 1990; Holomuzki and Messier 1993). An extreme case may be seen in migratory salmonids, which require a whole river system and the ocean to complete their life cycles. However, within a specific life stage, they rely on a certain part of a riffle-pool unit (Newbury and Gaboury 1993). For example, the dominant individuals of *Oncorhynchus masou ishikawai* (Jordan et McGregor) hold foraging territories near the pool inlet (Nakano 1995) where the density of drifting invertebrates is very high (Furukawa-Tanaka 1992). In contrast, the spawning redds of *Oncorhynchus masou masou* (Brevoort) and *Salvelinus leucomaenis* (Pallas) f. *japonicus* (Hilgendorf) are located in the transition zone between a pool and a riffle (Maruyama 1981). This part of a pool-riffle unit is also used as a spawning site by the cyprinid fishes *Phoxinus oxycephalus* (Sauvage and Dabry) and *Zacco temmincki* (Temminck et Schlegel) (Mizuno et al. 1958; Katano 1990) and by the ayu fish *Plecoglossus altivelis altivelis* Temminck et Schlegel (Nishida 1978). Their preference for the transition zone as spawning sites may be related first to the softness of the substrate (i.e., inorganic debris), which is easy to dig, and second to the abundant supply of oxygen-rich water permeating the substrate, which is expected to raise the survival rate of eggs and larvae (Mizuno and Gose 1972).

The shore area of the pool–riffle transition is also selected for oviposition sites by the mayflies *Ephemera strigata* Eaton (Ban et al. 1988; Takemon 1989) and *Epeorus ikanonis* Takahashi (Takemon 1990), although the microhabitats used by mature nymphs are located in different parts of the stream unit. The eggs of these mayflies have a smooth surface that facilitates their penetration into the hyporheic zone at the pool–riffle transition. Females of the cranefly *Eriocera longifurcata* Alexander also come to the same area and lay eggs directly in the hyporheic after digging themselves into the sediment (Kani 1952; Y. Takemon, personal observations). These insects may also benefit from the rich oxygen supply in this habitat. The streambed of this area is also important for the larvae of riffle-inhabiting caddisflies that happen to be carried in the drift because the existence of numerous interstices in the substrate make their settlement easier (Nishimura 1984).

All these observations show that the pool–riffle transition zone plays multiple roles in stream faunas. These roles seem to be connected with a well draining the hyporheic waters that derives from the porous substrate of this zone. The transition zone is generally located at the upper part of the bar structure in a riffle; as sedimentation exceeds erosion during spates where the bar structure develops, the porosity of the substrate and the well drainage of the hyporheic waters may be also structured during spates.

Heterogeneity in the Hyporheic Zone

The existence of the hyporheic zone has been known for a long time (Hynes 1968; Godbout and Hynes 1982; Williams 1984), but it has become a major focus of research in stream ecology only recently (Valett et al. 1993). White (1993) defined it hydrologically as a middle zone between the channel waters above and the groundwaters below and presented a conceptual model of the longitudinal structure of the hyporheic zone in a riffle-pool unit that is composed of downwelling zones and upwelling zones. Hendricks (1993) stated that, in terms of chemical properties, the former were more characteristic of surface water, the latter more of groundwater, and that together they formed a retention unit of nutrients and organic material. Although true hyporheic inhabitants are few among stream animals, many species temporarily depend on this habitat at early stages of their life cycles (Godbout and Hynes 1982; Williams 1984). The early instar larvae of such aquatic insects may benefit from oviposition at the downwelling zone for immigration into the hyporheic. The selected oviposition sites of aquatic insects with burrowing larvae mentioned in the previous section suggest that the hyporheic zone is at the pool–riffle transition. This seems to be related to the high density of hyporheos in the upper part of a riffle compared with that in the lower part (Godbout and Hynes 1982). To test this hypothesis, further investigations are necessary on the physicochemical and biological heterogeneity of hyporheic zones connected with a riffle-pool structure.

Arrangement of Microhabitats Within a Riffle-Pool Unit

Classification of microhabitats within a stream reach has been attempted with reference to various characteristics: the geomorphological regularity of stream features (Kani 1944, 1981), combination of geomorphology and flow regimes (Takahashi 1990), morphology of riffles and pools (Kawanabe et al. 1956; Kershner and Snider 1992), substrate types (Holomuzki and Messier 1993), substrate and current combined (Barmuta 1989), and various landscape features (Percival and Whitehead 1929; Kuusela 1979; Jenkins et al. 1984). Other than the hyporheic zone already mentioned, distinctive faunas and patterns of habitat use have been reported for stream margins (Kani 1944; Kuusela 1979; Ormerod 1988), mosscovered bedrock (Huryn and Wallace 1988), macrophytes (Tokeshi and Pinder 1985), woody debris (Gregory and Davies 1992), tree and grass roots (Jenkins et al. 1984), litter-packs (Egglishaw 1964; Brusven and Rose 1981), overhead tree canopies (Hawkins et al. 1982), and backwaters and billabongs (Kuusela 1979; Jenkins et al. 1984; Outridge 1988).

Considering the unique fauna and function of each microhabitat, all of these may affect local species diversity in stream ecosystems. However, these microhabitats are interrelated and cannot exist independently. For example, the distribution of woody debris is regulated by channel morphology, but the debris in turn alters the arrangement of riffles, pools, and substrates (Gregory et al. 1994). When we inquire into the management theory of stream biodiversity, the following three questions remain to be resolved: (i) spatial and temporal patterns of microhabitat arrangement on local scales; (ii) longitudinal patterns of arrangement; and (iii) maintenance mechanisms of habitat structures in stream ecosystems.

Takemon and Tanida (1993) described the microhabitat use of stream fauna in the middle reaches of Takami River, a tributary of the Yoshino River, in central Japan. An environmental map of the observation area, which included a riffle-pool unit 40 m long, was made by recording stream geomorphology, vegetation, water depth, current velocity, and substratum types, using 1 x 1 m grid sections (Fig. 15.1). Distribution of benthic animals was surveyed by sampling with the Surver net at 40–80 sites in the area. Behavioral data on feeding, mating, oviposition, and sheltering of adults of aquatic insects, fishes, frogs, and birds were recorded using the map.

Figure 15.2 shows a preliminary result of the survey conducted in March 1991. Because of the asymmetrical features of the stream bend, the microhabitats of benthic animals tended to be restricted to one side of the stream or to particular areas within the riffle-pool unit. The microhabitats listed in the following discussion were distinctive in terms of species composition or particular behavioral characteristics of inhabitants. The number heading each microhabitat listed here corresponds to that shown in Figure 15.2.

1. *Hyporheic zone over the entire bar:* young nymphs of the mayflies *Paraleptophlebia chocorata* Imanishi and *Drunella yoshinoensis* (Gose) were collected from the entire area of accumulated sediments. The nymphs may occur widely in the hyporheic zone without aggregating in a particular part of the bar structure.

2. *Hyporheic zone along the point bar edge:* the larvae of Elmidae, *Eriocera* sp., *Gibosia* sp., and species of Naididae were collected only from the inner side of a stream bend. Their densities were particularly high along the edge of the point bar that developed near the right shore of the study area. Because these larvae are also inhabitants of the hyporheic zone, their aggregated distribution in the inner side of a stream bend indicates a peculiarity of the hyporheic zone in this part. Relationships to physicochemical conditions of the area is a subject of future study.

3. *Shore areas of accumulated gravel:* mature nymphs of the mayflies *Ameletus kyotoensis* Imanishi, *A. montanus* Imanishi, and *Paraleptophlebia spinosa* Ueno occurred on the shallow shore of gravel substrate. They have probably moved into this area for emergence. Thus, shallow gravel shores are important not only as nymphal habitat but also as emergence sites for some taxa.

4 *Shore areas of rocky substrata:* the nymphs of the stonefly, *Cryptoperla japonica* (Okamoto) were restricted to the shore areas of rocky substrata. Their den-

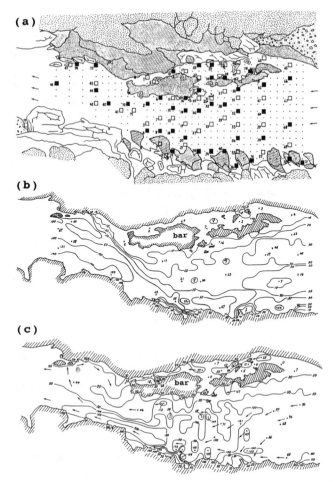

FIGURE 15.1a–c. Maps of the observation area in Takami Stream in the Yoshino River measured on 12 March 1991. (a) Map of the shore area and bankside vegetation. *Dot, closed square,* and *open square* represent the measuring site of environmental factors, sampling site of benthic animals not yet analyzed, and that already analyzed, respectively. *Numerals* beside the *square* show the sample numbers. *Shaded areas with dots, with solid and broken lines, with broken lines,* and the *other open areas* represent sand and gravel ground, area occupied by willow trees, tree canopies, and rocks, respectively. *Arrows* show flow direction. (b) Distribution pattern of water depth (cm) in the area. *Shaded area,* shore land and submerged stones. (c) Distribution pattern of current velocity (cm/s) in the area. *Arrows* show flow direction.

sity was high along the shorelines on the outer side of the stream bend. The nymphs were found frequently on the wet surface of rocks just above the water level mark.

5. *Accumulated stones in a riffle of high flow:* the stony substrata in the high-flow riffle (current speed > 150 cm s⁻¹) harbored a unique set of benthic animals, including *Epeorus uenoi* Matsumura, six species of Blepharocerinae, and

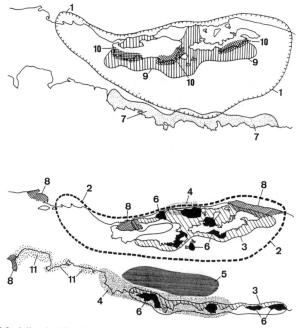

FIGURE 15.2. Microhabitat distribution of stream animals detected by the quantitative sampling surveys and field observations in the study area in Takami Stream. The number of each microhabitat corresponds to its description in the text. The boundary of each microhabitat was drawn by enclosing sites where each set of animals was collected or a particular behavior was observed.

species of Diamesinae. They were restricted to the narrow area in the outer bend of the riffle.

6. *Litter-pack:* the litter-pack habitat, a mass of leaves and twigs trapped between stones or roots of trees, was characterized by the occurrence of a particular species of Elmidae and a high density of the mayfly nymphs of *Cincticostella nigra* (Ueno). The litter-packs on the inner side of the stream bend were found between stones while those on the outer side occurred in the roots of willow trees along the shoreline.

7. *Moss-mat on the surface of bedrock:* the bedrock surface along the outer shoreline of the stream bend was covered by the thick mat of a moss, *Rhynchostegium riparioides* (Hedw), which grew not only along the water's edge but also underwater. Larvae of the caddisflies *Hydropsyche orientalis* Martynov, *Micrasema* sp. MC, and *Micrasema hanasensis* Tsuda inhabited the moss-mat in high density, and nymphs of *Cincticostella nigra* were also abundant. The moss-mat was found only on exposed bedrock washed by strong currents, as reported in a woodland stream in eastern Tennessee (Steinman and Boston 1993).

8. *Side pools:* the side pool habitat is defined as a small stagnant water found on depressed shores or behind rocks. In the study area it occurred on the lee side

of the big rocks in the pool and along the right shore, where litter, detritus, and silt were deposited on sandy substrata. This habitat was characterized by the presence of nymphs of the mayfly *Ephemera strigata,* dragonflies of the family Gomphidae, the sand lamprey *Lampetra (Lethenteron) reissneni* (Dybowski), and the fry and young cyprinid fishes of *Zacco temmincki* and *Phoxinus logowski* f. *oxycephalus.*

9. *Submerged stones:* a total of 24 individuals of the singing-frog *Rhacophorus japonicus* (Hallowell) were found under stones in the study area. They stayed under the submerged stones in the daytime. Their distribution was limited to the stony substrata of the bar in the inner half of the stream bend. This area was also used frequently as a feeding place by the wagtail *Motacilla cinerea* Leach.

10. *Pool–riffle transition:* oviposition sites of the mayfly *Ephemera strigata* were concentrated on the outer shore of the bar. The density of ovipositing females was highest at the upper end of the bar in the pool–riffle transition. Females of the cranefly *Eriocera* sp. were also found to oviposit at the same site. Judging from the contour of the streambed and flow directions in the study area (see Fig. 15.1), this area may correspond to the downwelling zone of the stream water.

11. *Rock caves in a pool:* there were several rock caves along the left side of the pool, which were used for shelter by fishes. One of them held more than 20 individuals of *Leuciscus hakonensis* Gunther and 3 individuals of *Oncorhynchus masou macrostomus* (Gunther). All rock caves suitable as fish refugia were located at the outer bend of the stream where the bank and substrata were strongly eroded.

These microhabitat types were detected by the distributional ranges of distinctive species and particular behavioral characteristics of the inhabitants. Microhabitat classification based on quantitative analyses of species composition will be studied in the future. Even considering this lack, these observations indicate a microhabitat heterogeneity within a single riffle-pool unit. The exclusive use of one microhabitat type by a species may be explained in two different ways. Supposing equilibrium conditions, the microhabitats described here may represent realized niches established through biological interactions, either predator–prey or competitive relationships or both.

Many works on the microdistribution of stream animals have suggested the effects of predator–prey interactions (Peckarsky 1984; Soluk and Collins 1988; Huang and Sih 1991; Holomuzki and Messier 1993; Wiseman et al. 1993). For example, the litter-pack habitat acts as a refuge from fish predation for some invertebrates (Brusven and Rose 1981; Holomuzki and Hoyle 1990). On the other hand, under nonequilibrium conditions, physical factors affecting environmental patchiness of organisms may explain the distribution of microhabitats (Reice 1985, 1994). For example, litter-pack habitats are structured dynamically by physical disturbance regimes: a litter-pack trapped in the watercourse at a debris dam persists at low water levels but will be flushed away and deposited in stagnant pools when the water level rises. Thus, distribution patterns of invertebrates associated with litter-packs can be predicted by the disturbance regimens experienced.

The foregoing discussion suggests that explanations based on both equilibrium and nonequilibrium conditions are feasible and that processes derived from both conditions could affect community structures at the same time or on different time scales. For example, a litter-pack habitat existing under low-disturbance regimes for a long time may encompass biological processes, although the habitat structure and also the spatial pattern of biological processes themselves ultimately reflect the most recent disturbance event. Therefore, historical data on habitat structures and monitoring surveys of microhabitat arrangement on local scales are important for a full understanding of community structures. In this respect, the stochastic patch dynamics model (Tokeshi 1994) incorporating the random processes of patch formation will be a powerful concept for future investigations on the mechanistic roles of disturbance in natural communities.

Although the importance of habitat heterogeneity in community structure has been pointed out by some ecologists (Denslow 1985; Doeg et al. 1989; Lake et al. 1989), the concept of "heterogeneity" is still vague and is difficult to incorporate in theoretical models of community ecology (Tokeshi 1994). Quantification of microhabitat diversity may be an essential process for substantial models of heterogeneity. The mapping method of this study is effective for taking quantitative data on microhabitat diversity.

Longitudinal Variation in Riffle-Pool Structures

Past studies on the longitudinal zonation of stream faunas have identified many physicochemical factors, including water temperature (Illies 1961; Vannote et al. 1980; Ward and Stanford 1982), chemical composition of water (Minshall and Minshall 1978), order of streams (Bronmark et al. 1984), stream hydraulics such as shear stress (Newbury 1984; Statzner and Higler 1986; Statzner and Borchardt 1994), and substratum types and stability (Chutter 1969; Minshall 1984). The importance of each environmental factor, however, must vary among populations and communities, and the processes of longitudinal zonation in community structure cannot necessarily be explained by a single factor because most of these factors are interrelated in a complex manner (Hildrew and Giller 1994). Recent studies using multivariate techniques for analyzing the patterns of community classification and ordination have revealed the importance of chemical components related to acidity (Ormerod and Edwards 1987; Rutt et al. 1990; Hildrew and Giller 1994). These analyses, based on correlation among factors, however, do not necessarily explain causality. From the applied perspectives of conservation and restoration of biodiversity in particular, it is necessary to consider maintaining processes of patterns of zonation rather than correlation among factors. The environmental factors need to be considered as a whole in which spatial and temporal structures are involved.

It is worthwhile to note that most of the factors of stream environments vary more or less concurrently with stream geomorphology. An example of the longitudinal distribution of aquatic insects connecting with stream geomorphology has been found in a mayfly species (Takemon 1989). In the tributaries of the Kamo River in Kyoto, in central Japan, the mayfly *Ephemera strigata* was recorded only from the Kurama Stream and was absent from the Kibune Stream in the 1930s (Imanishi 1941). In the 1980s, however, I found the species inhabiting the upper reaches of the Kibune Stream (Fig. 15.3). During the intervening 50 years many small dams were constructed, resulting in 41 dams in about 8 km of the stream. In the Kurama Stream (Fig. 15.4b), pools and point bars developed at each bend of the stream. In these reaches oviposition by the mayfly was observed along the shore of the pool–riffle transition where the canopy was open. The ovigerous females gathered in a small area where the dead bodies of spent females accumulated.

In contrast, in the upper reaches of the Kibune Stream, pools and riffles formed a stepwise sequence without point bars (Fig. 15.4a), and areas of sediment accumulation were found only close to the upper or lower side of the small dams. Numerous oviposition sites occurred independently of the small dams in the reaches below Station B, while in the upper reaches of the Kibune Stream they were restricted to the neighborhood of the dams (Fig. 15.3) because females of this species fly upstream as far as 1.1 km at most during ~30 min before oviposition and search for oviposition sites within a distance of several hundred meters of stream reaches (Takemon, unpublished data). When I counted the flying females at Station G, located out of the upper border of distributional range of the species, on 8 May 1987, 50 females were flying upstream across the station but the same number of females were observed to fly downstream later in the same day. This indicated that at least 50 females of 81 that oviposited near Station E (Fig. 15.3) decided their oviposition sites after searching in the upper reaches beyond Station G (Takemon, unpublished data).

These observations indicate that the upper limit of the distribution of the mayfly shifted with dam construction, which promoted the sedimentation of streambed materials. Here the distribution pattern of the mayfly is connected with the behavioral process of oviposition. As stated, oviposition sites are located on the shores of pool–riffle transition zones where the canopy is open. On the other hand, these sites may be determined proximately by the level of diffused light reflected on the water surface along the shore, while the water quality of the hyporheic zone, which affects the survival of offspring, operates as the ultimate factor. It should be noted that the diffused light reflection of shore water and the water quality of the hyporheic zone do not exist separately.

Although further investigation is necessary to determine the proximate and ultimate factors for site selectivity in oviposition behavior, it is clear that the topology of the shores of pool–riffle transition zones is highly important for the mayfly population. As discussed earlier, the pool–riffle transition zones reflect sedimentation processes producing bar structures. They are also maintained dynamically by sediment movement. The longitudinal distribution patterns of stream animals seem to have a strong relationship to the stream geomorphology derived from the sedimentation processes. This relationship suggests that a stream management scheme should include control of the magnitude and frequency of sediment transport in a carefully coordinated manner.

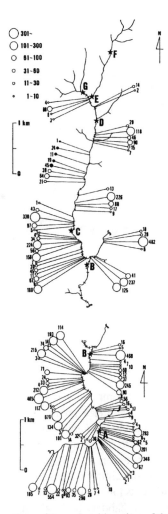

FIGURE 15.3. Distribution pattern of the oviposition sites of the mayfly *Ephemera strigata* along the Kibune Stream (Stations C–F) and the Kurama Stream (below Station B) observed on 8 May 1987. *Size of open circles* corresponds to the number of spent females counted at each oviposition site in the morning on 9 May; although females oviposit in the evening, their total number can be counted the next morning because they die at the oviposition sites. The *crossing bars* along the streams indicate small dams ranging from 4 to 10 m in height.

Summary

A total of 11 microhabitat types was recognized within a riffle-pool structure, based on a detailed mapping survey of the fauna and their habitats in the middle reaches of a Japanese mountain stream. Distribution of these microhabitats was

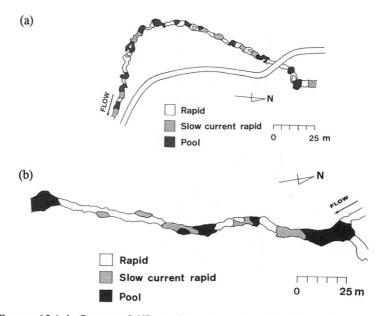

FIGURE 15.4a,b. Patterns of riffle-pool arrangement in (a) the Kibune Stream and (b) the Kurama Stream. *Slow current rapid* (*hatched portions*) means the transitional parts from pool to riffle (after Takemon 1985).

not uniform within the riffle-pool unit and was strongly affected by geomorphological features. Several processes of erosion and deposition of streambed material seemed to be involved in the creation and maintenance of these microhabitats on different time scales and were necessary for producing and maintaining the microhabitats. Relationships between stream geomorphology and longitudinal patterns of distribution of stream animals were discussed with particular reference to a mayfly species that shifted its longitudinal distribution after dam construction. Patterns of distribution of microhabitats both within a reach and along reaches indicate that habitat structures are maintained dynamically by sediment movement. This suggests that a stream management scheme should include the control of the magnitude and frequency of sediment transport in a carefully coordinated manner.

Acknowledgments. I am grateful to Dr. K. Tanida for allowing me to use data and concepts produced through our cooperative research, to Dr. A. Tamaki for information for references, and to an anonymous reviewer and Drs. M. Tokeshi and M. Higashi for thoughtful criticism and useful suggestions. This study was partly supported by the Foundation for Earth Environment, by the Foundation for Management of River Environment, by the Grants-in-Aid of Scientific Research to me (02954019 and 06854046) and to Dr. K. Tanida (63540522), and to Prof. H. Kawanabe (5748006 and 58121004), and by the Grant-in-Aid of Scientific Research on Priority Areas (#319), Project "Symbiotic Biosphere: An Ecological

Interaction Network Promoting the Coexistence of Many Species", and forms a part of guest scientist activities, Center for Ecological Research, Kyoto University.

Literature Cited

Ban, R. and KINKI Aquatic Insects Research Group. 1988. The life cycle and microdistribution of *Ephemera strigata* Eaton (Ephemeroptera: Ephemeridae) in the Kumogahata River, Kyoto Prefecture, Japan. Verhandlungen Internationale Verein Limnologie 23:2126–2134.

Barmuta, L.A. 1989. Habitat patchiness and macrobenthic community structure in an upland stream in temperate Victoria, Australia. Freshwater Biology 21:223–236.

Bronmark, C., J. Herrmann, B. Malmqvist, C. Otto, and P. Sjostrom. 1984. Animal community structure as a function of stream size. Hydrobiologia 112:73–79.

Brusven, A.M. and T.S. Rose. 1981. Influence of substrate composition and suspended sediment on insect predation by the torrent sculpin, *Cottus rhotheus*. Canadian Journal of Fisheries and Aquatic Sciences 38:1444–1448.

Chergui, H. and E. Pattee. 1993. Flow and retention of particulate organic matter in riparian fluvial habitats under different climates. Hydrobiologia 251:137–142.

Church, M. 1992. Channel morphology and typology. In: P. Calow and G. E. Petts, eds. The Rivers Handbook, Vol. 1, pp. 126–143. Blackwell Scientific, Oxford.

Chutter, F.M. 1969. The effects of silt and sand on the invertabrate fauna of streams and rivers. Hydrobiologia 34:57–76.

Connell, J.H. 1978. Diversity in tropical rainforests and coral reefs. Science 199: 1302–1310.

Denslow, J.S. 1985. Disturbance-mediated coexistence of species. In: S.T.A. Pickett and P.S. White, eds. The Ecology of Natural Disturbance and Patch Dynamics, pp. 307–323. Academic Press, San Diego.

Doeg, T.J., P.S. Lake, and R. Marchant. 1989. Colonization of experimentally disturbed patches by stream macroinvertebrates in the Acheron River, Victoria. Australian Journal of Ecology 14:207–220.

Egglishaw, H.J. 1964. The distributional relationship between the bottom fauna and plant detritus in streams. Journal of Animal Ecology 33:463–476.

Fisher, S.G. 1994. Pattern, process and scale in freshwater systems: some unifying thoughts. In: P.S. Giller, A.G. Hildrew, and D.G. Raffaelli, eds. Aquatic Ecology—Scale, Pattern and Process, pp. 575–591. Blackwell Scientific, Oxford.

Frissell, C.A., W.J. Liss, C.E. Warren, and M.D. Hurley. 1986. A hierarchical framework for stream habitat classification: viewing streams in a watershed context. Environmental Management 10:199–214.

Furukawa-Tanaka, T. 1992. Optimal feeding position for stream fishes in relation to invertebrate drift. Humans and Nature 1:63–81.

Godbout, L. and H.B.N. Hynes. 1982. The three-dimensional distribution of the fauna in a single riffle in a stream in Ontario. Hydrobiologia 97:87–96.

Gregory, K.J. and R.J. Davies. 1992. Coarse woody debris in stream channels in relation to river channel management in woodland areas. Regulated Rivers: Research & Management 7:117–136.

Gregory, K.J., A.M. Gurnell, C.T. Hill, and S. Tooth. 1994. Stability of the pool-riffle sequence in changing river channels. Regulated Rivers: Research & Management 9:35–43.

Hawkins, C.P., M.L. Murphy, and N.H. Anderson. 1982. Effects of canopy, substrate composition, and gradient on the structure of macroinvertebrate communities in cascade range streams of Oregon. Ecology 63:1840–1856.

Hendricks, S.P. 1993. Microbial ecology of the hyporheic zone: a perspective integrating hydrology and biology. Journal of the North American Benthological Society 12:70–78.

Hildrew, A.G. and P.S. Giller. 1994. Patchiness, species interactions and disturbance in the stream benthos. In: P.S. Giller, A.G. Hildrew, and D.G. Raffaelli, eds. Aquatic Ecology—Scale, Pattern and Process, pp. 21–62. Blackwell Scientific, Oxford.

Holomuzki, J.R. and J.D. Hoyle. 1990. Effect of predatory fish presence on habitat use and diel movement of the stream amphipod, *Gammarus minus*. Freshwater Biology 24:509–517.

Holomuzki, J.R. and S.H. Messier. 1993. Habitat selection by the stream mayfly *Paraleptophlebia guttata*. Journal of the North American Benthological Society 12:126–135.

Holt, R.D. 1993. Ecology at the mesoscale: the influence of regional processes on local communities. In: R.E. Ricklefs and D. Schluter, eds. Species Diversity in Ecological Communities, pp. 77–88. University of Chicago Press, Chicago.

Huang, C. and A. Sih. 1991. An experimental study on the effects of salamander larvae on isopods in stream pools. Freshwater Biology 25:451–459.

Huryn, A.D. and J.B. Wallace. 1988. Community structure of Trichoptera in a mountain stream: spatial patterns of production and functional organization. Freshwater Biology 20:141–155.

Huston, M. 1979. A general hypothesis of species diversity. American Naturalist 113:81–101.

Hynes, H.B.N. 1968. Further studies on the invertebrate fauna of a Welsh mountain stream. Archiv für Hydrobiologie 65:360–379.

Illies, J. 1961. Versuch einer allgemeinen biozonotischen Gliederung der Fliessgewasser. Internationale Revue der Gesamten Hydrobiologie 46:205–213.

Imanishi, K. 1941. Mayflies from Japanese torrents X. Life forms and life zones of mayfly nymphs. II. Ecological structure illustrated by life zone arrangement. Memoirs of the College of Science, Kyoto Imperial University, Series B 16:1–35.

Jenkins, R.A., K.R. Wade, and E. Pugh. 1984. Macroinvertebrate-habitat relationships in the River Teifi catchment and the significance to conservation. Freshwater Biology 14:23–42.

Kani, T. 1944. Ecology of torrent–inhabiting insects. In: H. Furukawa, ed. Insects, Vol. 1, pp. 171–317 (in Japanese). Kenkyu-sha, Tokyo.

Kani, T. 1952. Biological notes on *Eriocera longifurca* Alexander in late spring. In: M. Mori, ed. Ecological Studies on Stream Insects by T. Kani, pp. 176–182 (in Japanese). Kiso Kyo-iku Kai, Nagano, Japan.

Kani, T. 1981. Stream classification. In: Ecology of Torrent-Inhabiting Insects (1944; an abridged translation). Physiology and Ecology, Japan 18:113–118.

Katano, O. 1990. Dynamic relationships between the dominance of male dark chub, *Zacco temmincki,* and their acquisition of females. Animal Behaviour 40:1018–1034.

Kawanabe, H., D. Miyadi, S. Mori, E. Harada, H. Mizuhara, and R. Ogushi. 1956. Ecology of natural stock of ayu, *Plecoglossus altivelis* (in Japanese). Contributions to Physiology and Ecology of the Faculty of Science, Kyoto University, No. 79.

Kershner, J.L. and W.M. Snider. 1992. Importance of a habitat-level classification system to design instream flow studies. In: P.J. Boon, P. Calow, and G.E. Petts, eds. River Conservation and Management, pp. 179–193. Wiley, Chichester.

Kuusela, K. 1979. Early summer ecology and community structure of the macrozoobenthos on stones. Acta Universitatis Ouluensis Series A, Scientiae Rerum Naturalium 87, Biologica 6.

Lake, P.S. 1990. Disturbing hard and soft bottom communities; a comparison of marine and freshwater environments. Australian Journal of Ecology 15:477–488.

Lake, P.S., T.J. Doeg, and R. Marchant. 1989. Effects of multiple disturbance on macroinvertebrate communities in the Asheron River, Victoria. Australian Journal of Ecology 14:507–514.

Maruyama, T. 1981. Comparative ecology on the fluvial forms of *Salmo (Oncorhynchus) masou masou* (Brevoort) and *Salvelinus leucomaenis* (Pallas) (Pisces, Salmonidae). 1. Structure of spawning redds and spawning sites in Kamidani, River Yura (in Japanese). Japanese Journal of Ecology (Sendai) 31:269–284.

McAuliffe, J.P. 1983. Competition, colonization patterns, and disturbance in stream benthic communities. In: J.R. Barns and G.W. Minshall, eds. Stream Ecology, pp. 137–156. Plenum Press, New York.

Menge, B.A. and J.P. Sutherland. 1987. Community regulation: variation in disturbance, competition, and predation in relation to environmental stress and recruitment. American Naturalist 130:730–757.

Minshall, G.W. 1984. Aquatic insect-substratum relationships. In: V.H. Resh and D.M. Rosenberg, eds. The Ecology of Aquatic Insects, pp. 358–400. Praeger, New York.

Minshall, G. W. and J. N. Minshall. 1978. Further evidence on the role of chemical factors in determining the distribution of benthic invertebrates in the River Duddon. Archiv für Hydrobiologie 83:324–355.

Minshall, G.W., R.C. Petersen, T.L. Bott, C.E. Cushing, K.W. Cummins, R.L. Vannote, and J.R. Sedell. 1992. Stream ecosystem dynamics of the Salmon River, Idaho: an 8th-order system. Journal of the North American Benthological Society 11:111–137.

Mizuno, N. and K. Gose. 1972. Stream Ecology (in Japanese). Tsukiji-shokan, Tokyo.

Mizuno, N., H. Kawanabe, D. Miyadi, S. Mori, H. Kodama, R. Ogushi, and Y. Furuya. 1958. Life history of some stream fishes with special reference to four cyprinid species (in Japanese). Contributions to Physiology and Ecology of the Faculty of Science, Kyoto University No. 81.

Nakano, S. 1995. Individual differences in resource use, growth and emigration under the influence of a dominance hierarchy in fluvial red-spotted masu salmon in a natural habitat. Journal of Animal Ecology 64:75–84.

Newbury, R.W. 1984. Hydrologic determinants of aquatic insect habitats. In: V.H. Resh and D.M. Rosenberg, eds. The Ecology of Aquatic Insects, pp. 323–357. Praeger, New York.

Newbury, R. and M. Gaboury. 1993. Exploration and rehabilitation of hydraulic habitats in streams using principles of fluvial behaviour. Freshwater Biology 29:195–210.

Nishida, M. 1978. Spawning habits of the dwarf ayu-fish in Lake Biwa (in Japanese). Bulletin of the Japanese Society of Scientific Fisheries 44:577–585.

Nishimura, N. 1984. Ecological studies on the net-spinning caddisfly, *Stenopsyche marmorata* Navas. 6. Larval and pupal density in the Maruyama River, Central Japan, with special reference to floods and after flood recovery processes. Physiology and Ecology, Japan 21:1–34.

Ormerod, S.J. 1988. The micro-distribution of aquatic macroinvertebrates in the Wye river system: the result of abiotic or biotic factors? Freshwater Biology 20:241–247.

Ormerod, S.J. and R.V. Edwards. 1987. The ordination and classification of macroinvertebrate assemblages in the catchment of the River Wye in relation to environmental factors. Freshwater Biology 17:533–546.

Outridge, M.P. 1988. Seasonal and spatial variations in benthic macroinvertebrate communities of Magela Creek, Northern Territory. Australian Journal of Marine and Freshwater Research 39:211–223.

Peckarsky, B.L. 1984. Predator-prey interactions among aquatic insects. In: V.H. Resh and D.M. Rosenberg, eds. The Ecology of Aquatic Insects, pp. 196–254. Praeger, New York.

Percival, E. and H. Whitehead. 1929. A quantitative study of the fauna of some types of stream-bed. Journal of Ecology 17:282–314.

Perry, J.A. and D.J. Schaeffer. 1987. The longitudinal distribution of riverine benthos: a river dis-continuum? Hydrobiologia 148:257–268.

Rabeni, C.F. and R.B. Jacobson. 1993. The importance of fluvial hydraulics to fish-habitat restoration in low-gradient alluvial streams. Freshwater Biology 29:211–220.

Reice, S.R. 1985. Experimental disturbance and maintenance of species diversity. Oecologia 67:90–97.

Reice, S.R. 1994. Nonequilibrium determinants of biological community structure. American Scientist 82:424–435.

Resh, V.H., A.V. Brown, A.P. Covich, M.E. Gurtz, H.W. Li, G.W. Minshall, S.R. Reice, A.L. Sheldon, J.B. Wallace, and R.C. Wissmar. 1988. The role of disturbance in stream ecology. Journal of the North American Benthological Society 7:433–455.

Rutt, G.P., N.S. Weatherly, and S.J. Ormerod. 1990. Relationships between the physico-chemistry and macroinvertebrates of British upland streams: the development of modeling and indicator systems for predicting fauna and detecting acidity. Freshwater Biology 24:463–480.

Schluter, D. and R.E. Ricklefs. 1993. Species diversity. An introduction to the problem. In: R.E. Ricklefs and D. Schluter, eds. Species Diversity in Ecological Communities, pp. 1–10. University of Chicago Press, Chicago.

Schmid, P.E. 1992. Population dynamics and resource utilization by larval chironomidae (Diptera) in a backwater area of River Danube. Freshwater Biology 28:111–127.

Soluk, D.A. and N. C. Collins. 1988. A mechanism for interference between stream predators: responses of the stonefly Agnetina capitata to the presence of sculpins. Oecologia 76:630–632.

Sousa, W.P. 1984. The role of disturbance in natural communities. Annual Review of Ecology and Systematics 15:353–392.

Sousa, W.P. 1985. Disturbance and patch dynamics on rocky intertidal shores. In: S.T.A. Pickett and P.S. White, eds. The Ecology of Natural Disturbance and Patch Dynamics, pp. 101–124. Academic Press, San Diego,.

Southwood, T.R.E. 1977. Habitat, the templet for ecological strategies? Journal of Animal Ecology 46:337–365.

Statzner, B. and D. Borchardt. 1994. Longitudinal patterns and processes along streams: modelling ecological responses to physical gradients. In: P.S. Giller, A.G. Hildrew, and D.G. Raffaelli, eds. Aquatic Ecology—Scale, Pattern and Process, pp. 113–140. Blackwell Scientific, Oxford.

Statzner, B. and L.W.G. Higler. 1985. Questions and comments on the river continuum concept. Canadian Journal of Fisheries and Aquatic Sciences 42:1038–1044.

Statzner, B. and L.W.G. Higler. 1986. Stream hydraulics as a major determinant of benthic invertebrate zonation patterns. Freshwater Biology 16:127–139.

Steinman, A.D. and H.L. Boston. 1993. The ecological role of aquatic bryophytes in a woodland stream. Journal of the North American Benthological Society 12:17–26.

Takahashi, G. 1990. A study on the riffle-pool concept. Transactions, Japanese Geomorphological Union 11:319–336.

Takemon, Y. 1985. Emerging behaviour of *Ephemera strigata* and *E. japonica* (Ephemeroptera: Ephemeridae). Physiology and Ecology, Japan 21:17–36.

Takemon, Y. 1989. Emerging behaviour, mating and oviposition behaviour, and longitudinal distribution in ephemeran mayflies. In: A. Shibatani and L. Tanida, eds. Japanese Aquatic Insects (in Japanese), pp. 29–41. Tokai Daigaku Shuppan, Tokyo.

Takemon, Y. 1990. Reproductive ecology of the mayfly *Epeorus ikanonis* (Ephemeroptera, Heptageniidae). Ph.D. thesis, Kyoto University, Japan.

Takemon, Y. and K. Tanida. 1993. Environmental elements for recovery and conservation of riverine nature. In: M. Anpo, ed. The Proceedings of the International Symposium on Global Amenity, pp. 349–356. University of Osaka Prefecture, Osaka.

Tanida, K. and Takemon, Y. 1993. Trichoptera emergence from streams in Kyoto, central Japan. In: C. Otto, ed. Proceedings of the 7th International Symposium on Trichoptera, pp. 239–249. Backhuys, Leiden.

Tokeshi, M. 1994. Community ecology and patchy freshwater habitats. In: P.S. Giller, A.G. Hildrew, and D.G. Raffaelli, eds. Aquatic Ecology—Scale, Pattern and Process, pp. 63–91. Blackwell Scientific, Oxford.

Tokeshi, M. and L.C.V. Pinder. 1985. Microhabitats of stream invertebrates contrasting leaf morphology. Holarctic Ecology 8:313–319.

Tokeshi, M. and C.R. Townsend. 1987. Random patch formation and week competition: coexistence in an epiphytic chironomid community. Journal of Animal Ecology 56:833–845.

Valett, H.M., C.C. Hakenkamp, and A.J. Boulton. 1993. Perspectives of the hyporheic zone: integrating hydrology and biology. Introduction. Journal of the North American Benthological Society 12:40–43.

Vannote, R.L., G.W. Minshall, K.W. Cummins, J.R. Sedell, and C.E. Cushing. 1980. The river continuum concept. Canadian Journal of Fisheries and Aquatic Sciences 37:130–137.

Ward, J.V. and J.A. Stanford. 1982. Thermal responses in the evolutionary ecology of aquatic insects. Annual Review of Entomology 27:97–117.

White, D.S. 1993. Perspectives on defining and delineating hyporheic zones. Journal of the North American Benthological Society 12:61–69.

White, P.S. and S.T.A. Pickett. 1985. Natural disturbance and patch dynamics: an introduction. In: S.T.A. Pickett and P.S. White, eds. The Ecology of Natural Disturbance and Patch Dynamics, pp. 3–13. Academic Press, San Diego.

Williams, D.D. 1984. The hyporheic zone as a habitat for aquatic insects and associated arthropods. In: V.H. Resh and D.M. Rosenberg, eds. The Ecology of Aquatic Insects, pp. 430–455. Praeger, New York.

Wiseman, S.W., S.D. Cooper, and T.L. Dudley. 1993. The effects of trout on epibenthic odonate naiads in stream pools. Freshwater Biology 30:133–145.

Yuma, M. and M. Hori. 1990. Seasonal and age-related changes in the behaviour of the genji firefly, *Luciola cruciata* (Coleoptera, Lampyridae). Japanese Journal of Entomology 58:863–870.

16

Biodiversity: Interfacing Populations and Ecosystems

Simon A. Levin

Introduction

We are in the midst of a global environmental crisis, of a magnitude unknown since humans first populated the planet. Species and genetic diversity are being lost at record paces, although we still do not have the tools to do more than provide crude estimates of the magnitude of the tragedy. Indeed, we barely understand what it is that we are losing.

The argument for preserving biodiversity can be made on a wide variety of grounds. Most straightforward is the aesthetic argument: "Biological diversity should be conserved as a matter of principle, because all species deserve respect regardless of their use to humanity" (IUCN/UNEP/WWF 1991). What we see in nature is the result of a long evolutionary process, and extinction is forever. We are keepers of a rich biological heritage, and bound not to destroy it.

The aesthetic mandate for preservation is the purest, but an equally compelling case rests on the utilitarian aspects of biodiversity. Such a rationale necessarily assumes some system of valuation, based to large extent on an anthropocentric perspective. It leads inevitably to a view that some components of biodiversity are more important than others and that some species are more important than others. The aesthetic argument does not draw such distinctions.

The idea that humans might make decisions regarding which species to preserve and which to sacrifice is an arrogance that does not sit well on our shoulders. Indeed, the logic taken to its extreme has influenced the current debate on the remaining stocks of smallpox virus, the destruction of which would be the first case of humans deliberately causing the extinction of a species. To preserve a virus in the name of biodiversity is, for most people, not a logical imperative; but that the argument even is advanced is evidence of how strongly the basic preservation ethic appeals to people's sensitivities. The World Charter for Nature, adopted by the United Nations General Assembly in 1982, states that "Every form of life is unique, warranting respect regardless of its worth to man, and, to accord other organisms such recognition, man must be guided by a moral code of action." The basic principle (Noss and Cooperrider 1994) is that the most fundamental argument for the preservation of biodiversity is "appreciation of wild creatures and wild places for themselves."

The utilitarian argument rests on the services, direct and indirect, that are provided to humans by biodiversity. "Species...are all components of our life support system. Biological diversity also provides us with economic benefits and adds greatly to the quality of our lives" (IUCN/UNEP/WWF 1991). For example, one-fourth of all prescription drugs contain active ingredients originally derived from wild plants (Noss and Cooperrider 1994; WRI/IUCN/UNEP 1992). The World Wildlife Fund (1991) estimated that only 2% of the quarter-million described species of vascular plants have been screened for potentially useful chemical compounds. How many cures for human diseases lie undiscovered in nature's treasures, and how many will never be identified before they are destroyed? Loss of genetic diversity by the destruction of natural habitats is, in the words of E.O. Wilson, "the folly our descendants are least likely to forgive us."

Even more basically, we exploit living resources for food and fiber. Wild species contribute an estimated 4.5% to the gross domestic product of the United States (Prescott-Allen and Prescott-Allen 1986), and fisheries provided 100 million tons of food worldwide in 1988 (FAO 1988). Our overexploitation of some stocks, especially of marine fisheries, is cause for immediate concern. Many of the most desired fish species, such as salmon, Atlantic bluefin tuna (Safina 1993), and cod and haddock (Hammer et al. 1993; NOAA 1992) have experienced huge declines. The implications are manifold: reduced availability and higher costs, increased energy inputs to extract the declining resources, and serious socioeconomic effects (Regier and Baskerville 1986; Barbier et al. 1994).

That we exploit biodiversity directly for a variety of resources and products is clear, and well understood. More subtle and less appreciated, but of equal importance, are the ecosystem services nature provides to humans, services that rely on biodiversity. Forests and other natural systems mediate climate, and more generally regulate the flows of essential as well as undesirable materials. Benthic communities sequester toxic substances, preventing them from reaching humans; submerged aquatic vegetation performs similar functions in mediating the flow of materials and stabilizing habitat. Coral reefs and algal communities buffer coastlines against the ravages of ocean dynamics. Biological diversity will have a great deal to do with how natural and managed ecosystems respond to global change, although we are only beginning to understand the magnitude of the influence.

Measuring Biodiversity

The measurement of biodiversity may seem simple in concept; it is, however, very complicated in practice. The most straightforward approach, and the most easily communicated, relies simply on species counts (e.g., May 1992). No statistic is more compelling in discussion of the topic than to recount the accelerating loss of species. At least 71 species of vertebrates and more than three times as many species of full plants have gone extinct in the United States and Canada during the past 500 years (Nature Conservancy 1992; Russell and Morse 1992; Noss and Cooperrider 1994). Total extinction rates for all species are estimated to be in the tens to hundreds of thousands of species per year in the tropics alone (Wilson 1988; Diamond 1990). According to Wilson (1985), these extinction rates are about 400 times those recorded through recent geological time and are accelerating. Worldwide, of course, the figures are much higher (Barbier et al. 1994). For purposes of communicating the magnitude of the problem, there is no better and clearer method than simply to list the lost species. And although we cannot know exactly how many species we can afford to lose, it is clear that our tolerance for lost species diversity is limited:

Ecosystems, like well-made airplanes, tend to have redundant subsystems and other "design" features that permit them to continue functioning after absorbing a certain amount of abuse. A dozen rivets, or a dozen species, might never be missed. On the other hand, a thirteenth rivet popped from a wing flap, or the extinction of a key species involved in the cycling of nitrogen, could lead to a serious accident. (Ehrlich and Ehrlich 1981)

There are, however, several limitations to an approach that relies on species counts. Only a small fraction of species have been identified, and this fraction varies dramatically from one taxon to another. We are, at best, in the position of estimating species numbers from catch and effort data. Furthermore, the definition of a species has always been problematic and is virtually meaningless for bacteria and other organisms that reproduce clonally and may exchange much genetic information horizontally across clones. For these, a "species" is simply a level of taxonomic aggregation, largely arbitrary. And even for sexual organisms for which the diploid species may be defined fairly unequivocally, it must be recognized that species differ substantially in terms of how much genetic diversity they embody.

Wilson (1992) has written,

Although the species is generally considered to be the "fundamental unit" for scientific analysis of biodiversity, it is important to recognize that biological diversity is about the variety of living organisms *at all levels*—from genetic variants belonging to the same species, through arrays of species, families and genera, and through population, community, habitat and even ecosystem levels. Biological diversity is, therefore, the "diversity of life" itself.

For plant communities, an ecologist would measure both the within- and among-community components of diversity (Whittaker 1975) and recognize each as contributing to diversity. Similarly, in the measurement of biodiversity, one must recognize the diversity within species as well as the diversity in terms of number of species, or do away with the notion of species entirely in favor of "continuum" measures of the genetic and functional diversity of communities. Such continuum measures are more robust and probably more nearly represent the functional complexity of the system and its ability to respond to perturbations.

Barbier et al. (1994) wrote "Although species extinction is an important manifestation and indicator of biodiversity loss, it is not the crux of the problem...Although species extinction is the most fundamental and irreversible manifestation of biodiversity loss, the more profound implication is for ecological functioning and resilience." Indeed, one view (Steneck and Dethier 1994; Hay 1994) is that details about species composition represent noise, obscuring our perception about what communities really are. The validity of this interpretation, of course, is a matter of scale: Much of the detail about species composition will be irrelevant in terms of influences on ecosystem properties (Levin 1992).

One of the advantages of a continuum description is that it allows easily for weighting of diversity components according to their importance. But importance to what? How to weight a unit depends on its contribution to some performance criterion. That requires both a decision as to what is important and a scientific determination of the relationship between the unit and overall system properties. I return to this in later sections, without purporting to give a complete answer. Nor would I attempt to make such arguments to Aldo Leopold, who in 1953 wrote "The last word in ignorance is the man who says of an animal or plant: 'What good is it?'...To keep every cog and wheel is the first precaution of intelligent tinkering."

Another and equally difficult problem in the measurement of biodiversity is the problem of spatial scale (Levin 1995). The ecosystem is not well defined as a spatial unit, and the measurement of diversity is very much conditioned by the scale of investigation (e.g., MacArthur and Wilson 1967; Levin 1992). Numerous studies (e.g., Williamson 1988; Hubbell and Foster 1986) have characterized the relationship between the area surveyed and the number of species counted; the shape of the species–area curve is a fundamental aspect of the description of diversity, capturing much more than simply the total number of species in the community (even if the boundaries of the community were known). Whatever controls the shape of this curve controls biodiversity, but we are only beginning to understand the relative importance of factors such as fragmentation, isolation, migration, and mutation (e.g., May 1986; Tilman et al. 1994; Durrett and Levin 1996; Hubbell 1997). An extremely promising approach to such matters is the use of individual-based models to help bridge the gap between our understanding of how individuals respond to varying conditions and the patterns observed on broad spatial scales.

The Importance of Species in Ecosystems

The role of species in ecosystems can be addressed only with regard to how one characterizes the ecosystem and its features. No single perspective is the unique, right one for valuation of ecosystem services, and one simply must recognize the bias that an individual perspective imposes on the assessment of the role of biodiversity. Beyond the spatial ambiguity already raised is the dichotomy between the perspectives of a population or community biologist, for whom biotic structure and organization are central, and an ecosystem scientist, for whom flows and exchanges assume precedence.

The strongest functional argument for directing attention to individual species is provided by the notion of keystone species, as introduced by Paine (1966). In his classic experiments in the marine intertidal, removal of a single starfish species, a top predator, resulted in qualitative changes in food web structure, and in other ecosystem properties. In the rocky intertidal system studied by Paine, the predator controlled the subdivision of essential resources among competing species, largely through local exclusion of a dominant competitor. Work by Estes and Palmisano (1974) showed that similar dramatic changes were associated with the removal of the sea otter from nearshore ecosystems on the U.S. West Coast. The absence of the sea otter led to increases in urchin abundance, resultant decline in offshore kelp beds, and the shifting of the local fisheries from finfish to shellfish. Like the starfish, the otter was a keystone predator whose removal led to qualitative shifts in ecosystem properties and determined "the integrity of the community and its unaltered persistence through time, that is, stability" (Paine 1969).

The starfish- and otter-dominated ecosystems are powerful and convincing models for how particular ecological communities are organized and led to more general studies of the role of disturbance in maintaining community structure (Levin and Paine 1974; Paine and Levin 1981). But how general are these examples? How broadly can the notion of a keystone species (or "critical species"; Paine 1994) be extended, beyond the top predators in these examples? The challenge has been one of the most attractive ones for community theorists ever since Paine's classic and seminal paper.

Castilla and Paine (1987) reviewed the keystone concept as it applies to the marine intertidal, and show a wide application of the notion for those systems. More generally, Bond (1994) summarized a quarter-century of research on keystone species—predators, competitors, mutualists, pathogens, among others—demonstrating a diversity of situations in which individual species play critical roles, at least in determining community structure. As regards ecosystem properties, however, the case is less compelling for the role of particular species. One of the most likely candidates for examination would seem to be nitrogen fixers, such as the leguminous shrub *Myrica faya,* which invaded Hawaiian lava fields (Vitousek and Walker 1989). *Myrica* was able to invade in large part because it is a nitrogen fixer (through a mutualism with an endosymbiont), and thus it might be expected that its introduction would have had major effects at least on nitrogen cycling. Alterations in local nitrogen dynamics, however, do not seem to have had any major effect on other species abundances; direct competition by *Myrica* was a much more important factor (Mueller-Dombois and Whitteaker 1990). Biodiversity clearly is important in mediating ecosystem properties, but it is difficult to cite examples where individual species are documented to have made a difference.

Where keystone species exist and can be identified, they provide a handle on understanding community structure. More generally, however, the notion of the keystone species must be broadened. Functional groups, in which individual species may be to some extent functionally interchangeable, control critical system properties. The argument for the maintenance of biological diversity is equally strong in either case. Where keystone species exist, their protection obviously is vital to the maintenance of system integrity. But even when functional groups play the same role, the reduction of diversity within those groups reduces redundancy and the ability of the functional group to withstand stress.

The notion of a functional group is a very attractive one. From the point of view of systems theory, it is to be expected that large ensembles of interacting components will self-organize into clusters that interact more strongly among themselves than with other such clusters, and that the within-group dynamics will occur on much shorter scales than dynamics among groups (Simon and Ando 1961; Iwasa et al. 1987, 1989). Such hierarchical organization is characteristic of ecosystems (O'Neill et al. 1986; Paine 1980, 1994; Holling 1992). Soil microorganisms clearly form a keystone or critical group, in the sense of Paine's definition. But there is a great deal of functional redundancy within groups (Hay 1994; Lawton and Brown 1994; Steneck and Dethier 1994); indeed, the great majority of soil microorganisms cannot be identified to species. An important exception to the general view that large groups of species perform similar functions, and can substitute for one another, might be provided by the genus *Rhizobium,* which is of major importance in nitrogen fixation, along with bacteria of the family Azobacteraceae; these form a functional group. There is functional redundancy within this group, but *Rhizobium* plays a vital and nonsubstitutable role. Perhaps only our ignorance of microbial taxonomy limits the list of such examples, but it remains that the notion of functional groups is a valuable and appropriate concept for the understanding of the control of many critical ecosystem processes. Meyer (1994), in an excellent review of functional groups of microorganisms, categorizes microorganisms according to function, and concludes that "it is likely that we would not realize the extinction of a specific microbial species in nature unless a complete functional group was affected."

Schulze (1982) developed a functional group approach to vegetation, building on properties such as life form and phytosociological association. Solbrig (1994), in his approach to functional groups for plants, relies on the interplay between evolution and plant adaptive strategies as an organizing principle. In many other cases, functional groupings may make much more sense than taxonomic ones for understanding community organization, even if that implies a clustering that divides species; for example, in marine ecosystems, one might lump juveniles of diverse benthic-feeding fish into one cluster and adults into another.

A fundamental point made by Solbrig, and which is broadly applicable to the notion of functional groupings in any ecosystem, is that even when the roles of species within groupings cannot be distinguished, diversity and redundancy within groupings are critical features of the system's ability to respond to change and disturbance. Thus is provided one of the most compelling arguments for the maintenance of biodiversity: in the short term, elimination of redundancy within groups may result in no noticeable change in system dynamics. Over time, however, systems with reduced within-group diversity will be less able to respond to change and more likely to exhibit collapse. This is not to say that ecosystems have evolved these traits as adaptive strategies, but rather that diversity and associated stability are emergent properties of these self-organizing systems and of the evolution of the component species.

Steneck and Dethier (1994) have provided one of the most compelling studies to date of the utility of the notion of functional groups. Drawing from experiments and experiences in subtidal algal communities in Maine, Washington state, and the Caribbean, they developed a view of the organization of such communities which argues that functional groupings of taxonomically distinct species share morphological attributes, and that convergent biogeographic patterns in ecosystem organization may be discerned clearly when the approach of functional groupings is taken. Environments are characterized by productivity and herbivore pressure, and these prove sufficient for the prediction of algal community composition. Moreover, communities viewed in terms of functional groupings prove much more stable and predictable than when viewed in terms of species composition. The regularity seen in these communities is reminiscent of the regularity seen in the organization of trophic webs (Pimm 1982; Sugihara 1984; Cohen et al. 1990) when attention is on the macroscopic properties of those webs rather than on the identities of individual species.

Biodiversity and Ecosystem Function

Although it may be difficult to demonstrate the importance of individual species to ecosystem properties, biodiversity more generally conceived certainly is critical to ecosystem properties such as net primary productivity. The points made in the last section direct attention to a problem of fundamental importance: How much detail, in terms of the species composition or even the genetic composition of a community, is necessary and sufficient to understand and maintain the dynamics as observed at the ecosystem level? The answer depends on the scale of interest (Levin 1992). Short-term dynamics are more likely to be governed by gross properties of community structure, while longer term dynamics, involving the response of systems to disturbance and feedbacks, will involve much more of the detail of community structure.

One of the challenges facing scientists today is an understanding of the responses of ecosystems to global change and the associated feedbacks to climate systems. General circulation models, tools to predict changes in climate, have reached very sophisticated stages of development but do not in themselves incorporate biological detail. In coupling such models to the dynamics of regional vegetation, simplifying assumptions must be made (Melillo et al. 1993; Potter et al. 1993; Schimel et al. 1994). The usual approach is to ignore spatial and compositional detail, representing the vegetation in grid cells hundreds of kilometers on a side by a homogeneous superspecies or "big leaf," parameterized by averaging the properties of extant species or, worse, by assuming that the vegetation evolutionarily will achieve a cooperative optimum performance in some physiological function.

A cooperative optimum is a poor approximation to reality because, by ignoring competition within a group, one overestimates the collective performance of the group; the competitive equilibrium may be far below what a coalition would produce in terms of physiological performance. Communities do not evolve as superorganisms, but rather represent in their properties the evolution of their components. On the other hand, while assumption of an evolutionary optimum will overstate performance, the static approach based on current composition may understate at least the short-term response. The biotic heterogeneity within a group will lead naturally to selection and consequent shifts in system properties, usually resulting in an increase in short-term mean performance; nonlinearities would be expected to modify this on longer time scales. Indeed, by introducing diversity within the vegetational functional group and allowing for shifts and feedbacks from natural selection, Bolker et al. (1995) found that the classical static assumption led to a significant underestimation of the response of the biota in terms of increase in basal area, on time scales of 50–100 years.

How much biodiversity is important, in theory as well as in practice? There is no simple answer. On the very shortest time scales, average properties of systems will dominate, and biotic detail is probably irrelevant. But as the example just given demonstrates, short may mean very short: 50–100 years is not a long time in the lifetimes of forests, or in the horizons of interest to humans. On the slightly longer scales, biotic diversity and consequent feedbacks may fundamentally alter the responses of systems to stress. The elaboration of the functional group approach is central to the problem of scaling from the species to the ecosystem (Koerner 1994), but no formula for relevant detail is possible without reference to the problems and scales of interest. For net primary productivity, as is likely to be the case for any system property, biodiversity matters only up to a point; above a certain level, increasing biodiversity is likely to make little difference (Baskin 1994).

Understanding what is appropriate biotic detail requires understanding the relationship between the structure and function of ecosystems. The health of an ecosystem is measured in terms of both its biotic composition and the flow of elements among its compartments. Yet an understanding of the interconnections among these is woefully lacking. Among all studies that have been carried out to examine the effects of diversity on ecosystem properties, even the best nonagricultural examples are equivocal in their conclusions (Vitousek and Hooper 1994). Despite this paucity of information, there is hope for substantial progress in the next decade. "The background knowledge and techniques are now in place to begin a rigorous examination of the effects of species diversity on ecosystem biogeochemistry" (Vitousek and Hooper 1994). The problem clearly is one of fundamental importance in our evaluation of the role of biodiversity.

Conclusion

Where do we go from here? Much information, both from the empirical and from the theoretical literature, suggests strong links between biological diversity and the structure and function of ecosystems. In some systems, particular "critical" or "keystone" species can be clearly identified as linchpins whose removal can lead to cascading effects in system properties. In other systems, the notion of the keystone species must be replaced by that of the critical functional group, but many of the lessons are the same. When species are at issue, genetic diversity within them governs their capability for resilience in the face of perturbations; when functional groups are involved, resilience comes from diversity within those groups. Efforts to measure diversity solely in terms of numbers of species, therefore, while a logical place to start, miss much essential detail.

On the other hand, although we know that diversity is important, we cannot quantify how important...yet. As Schulze and Mooney (1994) argued, "There is evidence that biotic diversity at levels ranging from genetic diversity among populations to landscape diversity is critical to the maintenance of natural and agricultural ecosystems. We still know little, however, about the critical thresholds of diversity and the conditions or time scales over which diversity is particularly important. Given the rapid declines in biodiversity, research programs must be planned promptly in a manner that allows results to be effectively incorporated into policies if we are to maintain the biodiversity of the globe."

How should diversity be measured, and how do system attributes respond to changes in diversity? We are at the threshold of being able to provide answers to these questions, through combined empirical and theoretical studies; Lawton (Chapter 12, this volume) reports on some of the most exciting recent progress in this direction. We are not yet at a point, however, where definitive statements can be made. The population–ecosystem interface has been one of the most challenging areas for conceptual development ever since the notion of the ecosystem was introduced by Tansley (1935), but sociological and other barriers have limited progress. The times are changing, as Vitousek and Hooper (1994) pointed out, and we are now equipped and of a mind to develop the essential theoretical linkages. In part, this must come from a proper theory of ecosystem development and evolution, in which system properties are seen to emerge from the self-organizing development of ecosystems and landscapes (Holling 1992), within the context of the evolution of individual species (Levin et al. 1990; Holt 1994).

Acknowledgments. I am pleased to acknowledge valuable comments by Ann Kinzig and Eric Klopfer, and the U.S. Department of Energy under DOE grant DE-FG02-94ER 61815, Princeton University.

Literature Cited

Barbier, E.B., J.C. Burgess, and C. Folke. 1994. Paradise lost? The Ecological Economics of Biodiversity. Earthscan, London.

Baskin, Y. 1994. Ecosystem function of biodiversity. Bioscience 44:657–660.

Bolker, B.M., S.W. Pacala, C. Canham, F. Bazzaz, and S.A. Levin. 1995. Species diversity and ecosystem response to carbon dioxide fertilization: conclusions from a temperate forest model. Global Change Biology 1:373–381.

Bond, W.J. 1994. Keystone species. In: E.D. Schulze and H.A. Mooney, eds. Biodiversity and Ecosystem Function, pp. 237–254. Springer-Verlag, Berlin.

Castilla, J.C. and R.T. Paine. 1987. Predation and community organization on Eastern Pacific temperate zone, rocky intertidal shores. Revista Chilena de Historia Natural 60:131–151.

Cohen, J.E., F. Briand, and C.M. Newman. 1990. Community Food Webs: Data and Theory. Springer-Verlag, Berlin.

Diamond, J.M. 1990. Playing dice with megadeath. Discover 11:55–59.

Durrett, R.T. and S.A. Levin. 1996. Spatial models for species area curves. Journal of Theoretical Biology 179:119–127.

Ehrlich, P.R. and A.H. Ehrlich. 1981. Extinction. The Causes and Consequences of the Disappearance of Species. Random House, New York.

Estes, J.A. and J.F. Palmisano. 1974. Sea otters: their role in structuring nearshore communities. Science 185:1058–1060.

FAO (Food and Agriculture Organization of the United Nations). 1988. Current Fisheries Statistics. FAO, Rome.

Hammer, M., A.M. Jansson, and B.-O. Jansson. 1993. Diversity, change and sustainability: implications for fisheries. AMBIO 22:97–105.

Hay, M. 1994. Species as 'noise' in community ecology: do seaweeds block our view of the kelp forest? Trends in Ecology & Evolution 9:414–416.

Holling, C.S. 1992. Cross-scale morphology, geometry and dynamics of ecosystems. Ecological Monographs 62:447–502.

Holt, R.D. 1994. Linking species and ecosystems. In: C.G. Jones and J. H. Lawton, eds. Linking Species and Ecosystems, pp. 273–279. Chapman & Hall, New York.

Hubbell, S. 1997. A Theory of Biogeography and Relative Species Abundance. Princeton University Press, Princeton. (To appear.)

Hubbell, S.P. and R.B. Foster. 1986. Biology, chance, and history, and the structure of tropical rain forest tree communities. In: J. Diamond and T.J. Case, eds. Community Ecology, pp. 314–330. Harper & Row, New York.

IUCN/UNEP/WWF (The World Conservation Union), UNEP (United Nations Environment Programme), WWF (World Wide Fund for Nature). 1991. Caring for the Earth: A Strategy for Sustainable Living. IUCN/UNEP/WWF, Gland, Switzerland.

Iwasa, Y., S.A. Levin, and V. Andreasen. 1987. Aggregation in model ecosystems. I. Perfect aggregation. Ecological Modelling 37:287–302.

Iwasa, Y., S.A. Levin, and V. Andreasen. 1989. Aggregation in model ecosystems. II. Approximate aggregation. IMA Journal of Mathematics Applied in Medicine and Biology 6:1–23.

Koerner, C. 1994. Scaling from species to vegetation: the usefulness of functional groups. In: E.D. Schulze and H.A. Mooney, eds. Biodiversity and Ecosystem Function, pp. 117–140. Springer-Verlag, Berlin.

Lawton, J.H. and V.K. Brown. 1994. Redundancy in ecosystems. In: E.D. Schulze and H.A. Mooney, eds. Biodiversity and Ecosystem Function, pp. 255–270. Springer-Verlag, Berlin.

Leopold, A. 1953. Round River. Oxford University Press, New York.

Levin, S.A. 1992. The problem of pattern and scale in ecology. Ecology 73:1943–1967.

Levin, S.A. 1995. Scale and sustainability: a population and community perspective. In: M. Munasinghe and W. Shearer, eds. Defining and Measuring Sustainability: The Biogeophysical Foundations, pp. 103-114. The United Nations University, New York; The World Bank, Washington, DC.

Levin, S.A. and R.T. Paine. 1974. Disturbance, patch formation and community structure. Proceedings of the National Academy of Sciences of the United States of America 71:2744–2747.

Levin, S.A., L.A. Segel, and F. Adler. 1990. Diffuse coevolution in plant-herbivore communities. Theoretical Population Biology 37:171–191.

MacArthur, R.H. and E.O. Wilson. 1967. The Theory of Island Biogeography. Princeton University Press, Princeton.

May, R.M. 1986. The search for patterns in the balance of nature: advances and retreats. Ecology 67:1115–1126.

May, R.M. 1992. How many species inhabit the earth? Scientific American 267:42–48.

Melillo, J.M., A.D. McGuire, D.W. Kicklighter, B. Moore, C.J. Vorosmarty, and A.L. Schloss. 1993. Global climate-change and terrestrial net primary production. Nature (London) 363:234–240.

Meyer, O. 1994. Functional groups of microorganisms. In: E.D. Schulze and H.A. Mooney, eds. Biodiversity and Ecosystem Function, pp. 67–96. Springer-Verlag, Berlin.

Mueller-Dombois, D. and L.D. Whitteaker. 1990. Plants associated with *Myrica faya* and two other pioneer trees on a recent volcanic surface in Hawaii Volcanoes National Park. Phytocoenologia 19:29–41.

Nature Conservancy. 1992. Extinct vertebrate species in North America. Unpublished draft list, March 4, 1992. The Nature Conservancy, Arlington, VA.

NOAA (National Oceanic and Atmospheric Administration). 1992. Our living oceans: report on the status of U.S. living marine resources. NOAA Technical Memorandum NMFS-F/SPO-2, NOAA, Washington, DC.

Noss, R.F. and A.C. Cooperrider. 1994. Saving Nature's Legacy: Protecting and Restoring Biodiversity. Island Press, Washington, DC.

O'Neill, R.V., D.L. DeAngelis, J.B. Waide, and T.F.H. Allen. 1986. A Hierarchical Concept of Ecosystems. Princeton University Press, Princeton.

Paine, R.T. 1966. Food web complexity and species diversity. American Naturalist 100:65–75.

Paine, R.T. 1969. A note on trophic complexity and community stability. American Naturalist 103:91–93.

Paine, R.T. 1980. Food webs: linkage, interaction strength and community infrastructure. Journal of Animal Ecology 49:667–685.

Paine, R.T. 1994. Marine Rocky Shores and Community Ecology: An Experimentalist's Perspective. Ecology Institute, Oldendorf/Luhe, Germany.

Paine, R.T. and S.A. Levin. 1981. Intertidal landscapes: disturbance and the dynamics of pattern. Ecological Monographs 51:145–178.

Pimm, S.L. 1982. Food Webs. Chapman & Hall, New York.

Potter, C.S., J.T. Randerson, C.B. Field, P.A. Matson, P.M. Vitousek, H.A. Mooney, and S.A. Klooster. 1993. Terrestrial ecosystem production: a process model based on global satellite and surface data. Global Biogeochemical Cycles 7:811–841.

Prescott-Allen, C. and R. Prescott-Allen. 1986. The First Resource. Yale University Press, New Haven.

Regier, H.A. and G.L. Baskerville. 1986. Sustainable redevelopment of regional ecosystems degraded by exploitive development. In: W.C. Clark and R.E. Munn, eds. Sustainable Development of the Biosphere, pp. 75–101. Cambridge University Press, Cambridge.

Russell, C. and L. Morse. 1992. Extinct and possibly extinct plant species of the United States and Canada. Unpublished report; review draft, 13 March 1992. The Nature Conservancy, Arlington, VA.

Safina, C. 1993. Bluefin tuna in the West Atlantic: negligent management and the making of an endangered species. Conservation Biology 7:229–234.

Schimel, D.S., B.H. Braswell Jr., E.A. Holland, R. McKeown, D.S. Ojima, T.H. Painter, W.J. Parton, and A.R. Townsend. 1994. Climatic, edaphic and biotic controls over storage and turnover of carbon in soils. Global Biogeochemical Cycles 8:279–293.

Schulze, E.D. 1982. Plant life forms and their carbon, water and nutrient relations. In: O.L. Lange, P.S. Nobel, C.B. Osmond, and H. Ziegler, eds. Physiological Plant Ecology. II. Water Relations and Carbon Assimilation, pp. 616–676. Springer-Verlag, Berlin.

Schulze, E.D. and H.A. Mooney. 1994. Ecosystem function of biodiversity: a summary. In: E.D.Schulze and H.A. Mooney, eds. Biodiversity and Ecosystem Function, pp. 497–509. Springer-Verlag, Berlin.

Simon, H. and A. Ando. 1961. Aggregation of variables in dynamic systems. Econometrica 29:111–138.

Solbrig, O.T. 1994. Plant traits and adaptive strategies: their role in ecosystem function. In: E.D. Schulze and H.A. Mooney, eds. Biodiversity and Ecosystem Function, pp. 97–116. Springer-Verlag, Berlin.

Steneck, R.S. and M.N. Dethier. 1994. A functional group approach to the structure of algal-dominated communities. Oikos 69:476–498.

Sugihara, G. 1984. Graph theory, homology and food webs. In: S.A. Levin, ed. Mathematical Population Biology, pp. 83–101. American Mathematical Society, Providence.

Tansley, A.G. 1935. The use and abuse of vegetational concepts and terms. Ecology 16:284–307.

Tilman, D., R.M. May, C.L. Lehman, and M.A. Nowak. 1994. Habitat destruction and the extinction debt. Nature 371:65–66.

Vitousek, P.M. and D.U. Hooper. 1994 . Biological diversity and terrestrial ecosystem biogeochemistry. In: E.D. Schulze and H.A. Mooney, eds. Biodiversity and Ecosystem Function, pp. 3–14. Springer-Verlag, Berlin.

Vitousek, P.M. and L.R. Walker. 1989. Biological invasions by *Myrica faya* in Hawaii: plant demography, nitrogen fixation, ecosystem effects. Ecological Monographs 59:247–265.

Whittaker, R.H. 1975. Communities and Ecosystems, 2nd Ed. Macmillan, New York.

Williamson, M. 1988. Relationship of species number to area, distance and other variables. In: A.A. Myers and P.S. Giller, eds. Analytical Biogeography, Chapter 4. Chapman & Hall, London.

Wilson, E.O. 1985. The biological diversity crisis. Bioscience 35:700–706.

Wilson, E.O. 1988. The current state of biological diversity. In: E.O. Wilson and F.M. Peter, eds. Biodiversity, pp. 3–18. National Academy of Sciences Press, Washington, DC.

Wilson, E.O. 1992. The Diversity of Life. Penguin Press, London.

WRI/IUCN/UNEO (World Resources Institute/The World Conservation Union/United Nations Environmental Program). 1992. Global Biodiversity Strategy: Guidelines for Action to Save, Study, and Use Earth's Biotic Wealth Sustainably and Equitably. WRI/IUCN/UNEP, Washington, DC.

World Wide Fund for Nature (WWF). 1991. The Importance of Biological Diversity. Yale University Press, New Haven.

Index